Wireless Technologies and the National Information Infrastructure

Office of
Technology
Assessment

Congress
of the
United States

Recommended Citation: U.S. Congress, Office of Technology Assessment, *Wireless Technologies and the National Information Infrastructure*, OTA-ITC-622 (Washington, DC: U.S. Government Printing Office, July 1995).

For sale by the U.S. Government Printing Office
Superintendent of Documents, Mail Stop: SSOP, Washington, DC 20402-9328
ISBN 0-16-048180-5

Foreword

The United States is in the midst of a fundamental restructuring of its communications and information technology infrastructure. Congress, the executive branch, and the states all are attempting to determine how to combine the country's many different networks—telephone, computer, cable television, cellular telephone, satellite, and broadcasting—into a broader National Information Infrastructure (NII). The private sector is spending billions of dollars developing systems to bring a wide variety of improved services to businesses and consumers. Wireless technologies, including radio and television broadcasting, satellites, cellular and other mobile telephones, and a variety of data communication systems, make up one of the most vibrant elements of this new telecommunications order, and will offer the American people new and more flexible ways to communicate with each other, access information resources, and receive entertainment. However, realizing the benefits of wireless technologies while avoiding potential obstacles and adverse consequences will require a long-term commitment to overseeing the changes now being set in motion. Government and private sector representatives must cooperate to ensure that wireless and NII goals and policies work together.

This report examines the role wireless technologies will play in the emerging NII and identifies the challenges that policymakers, regulators, and wireless service providers will face as they begin to more closely integrate wireless systems with existing wireline networks. The report provides Congress with a broad overview of the wireless technologies being developed and discusses the technical, economic, and public policy issues associated with deploying them. Potential policy options are presented to help ensure a smooth transition to an integrated wireline/wireless NII. The report also discusses some of the technical and social implications of the widespread use of wireless technologies—paying particular attention to the profound changes that wireless systems may cause in patterns of mobility.

OTA appreciates the assistance of the project advisory panelists, workshop participants, and contractors who contributed to the study. OTA also thanks the many representatives of industry; federal, state, and local government officials; and members of the public who were so generous with their attention and advice. OTA values their perspectives and comments; the report is, however, solely the responsibility of OTA.

ROGER C. HERDMAN
Director

Advisory Panel

Alfred F. Boschulte
President and Chairman
NYNEX Mobile
    Communications Co.

Timothy J. Brennan
Resources for the Future

Steven D. Dorfman
President
Hughes Telecommunications
    and Space Co.

Francis J. Erbrick
Senior Vice President
United Parcel Service

Susan Hadden[1]
Professor
LBJ School of Public Affairs
The University of Texas at Austin

Ellwood R. Kerkeslager
Vice President
AT&T

Rob Kling *(Chairman)*
Professor
University of California, Irvine

Jim Lovette
Principal Scientist
Apple Computer, Inc.

John Major
Senior Vice President
Motorola, Inc.

Howard Miller
Senior Vice President, Broadcast
Public Broadcasting Service

Alex Netchvolodoff
Vice President, Public Policy
Cox Enterprises, Inc.

Stewart D. Personick
Assistant Vice President
Bellcore

William W. Redman, Jr.
Commissioner
North Carolina Utilities
    Commission

W. Scott Schelle
Chief Executive Officer
American Personal
    Communications

Jim Strand
President
Lincoln Telecommunications

William F. Sullivan
General Manager
KPAX-TV

Laurel L. Thomas
Telecommunications Consultant

Marilyn B. Ward
Division Commander
Orlando Police Department

Daniel Weitzner
Deputy Director
Center for Democracy and
    Technology

---

[1]Deceased.

Project Staff

Peter D. Blair
Assistant Director, OTA
*Industry, Commerce, and
    International Security Division*

James W. Curlin
Program Director*
*Industry, Telecommunications,
    and Commerce Program*

Andrew W. Wyckoff
Program Director
*Industry, Telecommunications,
    and Commerce Program*

PRINCIPAL STAFF

David Wye
Project Director

Todd La Porte
Analyst

Alan Buzacott
Analyst

Greg Wallace
Research Analyst

Jean Smith
Editor

ADMINISTRATIVE STAFF

Liz Emanuel
Office Administrator

Karolyn St. Clair
PC Specialist

Diane Jackson
Administrative Secretary

Karry Fornshill
Secretary

CONTRACTORS

Bruce Egan
Consultant

Glenn Woroch
University of California, Berkeley

Philip Aspden and James Katz
Bellcore

George Morgan
Center for Wireless
    Telecommunications
Virginia Polytechnic Institute and
    State University

*Until September 1994

Contents

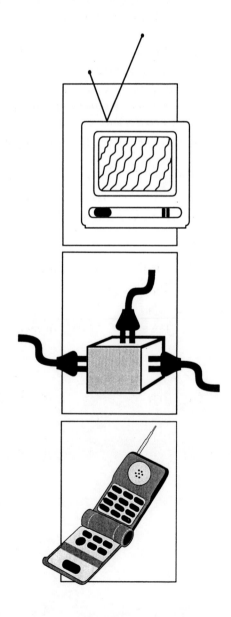

Executive Summary

Over the next five to 10 years, wireless technologies will dramatically reshape the communications and information infrastructure of the United States. New radio–based systems now being developed will use advanced digital technologies to bring a wide array of services to both residential and business users, including ubiquitous mobile telephone and data services and many new forms of video programming. Existing wireless systems, including radio and television broadcasting, cellular telephony, and various satellite and data networks, will also convert to digital technology. This will allow them to improve the quality of their services, expand the number of users they can serve, and offer new information and entertainment applications. Before the benefits of these wireless systems can be realized, however, technical, regulatory, and economic uncertainties must be resolved. This report examines the role wireless communication technologies will play in the evolving National Information Infrastructure (NII), examines the challenges facing policymakers and regulators as wireless becomes a more integral part of the telecommunications and information infrastructure, and identifies some of the longer term implications of the widespread use of wireless systems and services.

BACKGROUND

The public's imagination has been captured by notions of an "information superhighway." Newspaper articles, television advertisements, and technical journals are filled with visions of communication services that allow people to transmit and receive phone calls, computer files, images, and even movies; people working anywhere—at the beach, in their homes, or in their cars; and hundreds of channels of entertainment programming, includ-

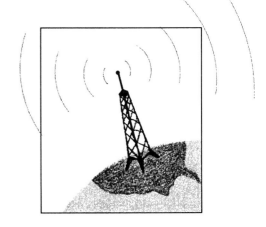

ing movies on demand. The foundation for these visions, the technologies that will make them possible, is formally known as the National Information Infrastructure. The NII is conceived as a ubiquitous, interconnected series of telecommunications networks and computer–based services that will allow every home and business in the nation to access a never–before–seen array of advanced communication, information, and entertainment services and applications. Some also see the NII as part of a larger Global Information Infrastructure (GII) that would link the countries of the world in an even wider network.

Wireless technologies will play an important role in realizing these visions. In the past several years, wireless technologies and services have become one of the fastest growing segments of the telecommunications industry. U.S. cellular phone companies add 28,000 customers each day, and in recent years have achieved annual growth rates that, in some cases, surpass 40 percent. The Federal Communications Commission (FCC) recently raised almost $10 billion for the U.S. Treasury from auctions of radio frequencies that will be used to deliver next–generation mobile telephone and data services. Sales of small dishes to receive television programming directly from satellites have been brisk, and large businesses have been installing computer networks connected by satellites to keep track of sales and to deliver interactive employee training.

ADVANTAGES OF WIRELESS TECHNOLOGY

Consumers and businesses have found that wireless technologies have unique capabilities that allow them to do things they either cannot do with wire–based systems, or cannot do efficiently and/or cost–effectively. First and foremost, wireless technologies make mobile communication possible. People can use cellular telephones to make and receive phone calls while they are walking, driving, flying, in a boat, or on a train. They can use computers equipped with radio modems to send and receive data and electronic mail. They can stay connected wherever they are.

Second, radio–based systems can offer more flexible and affordable access to the nation's information/communication resources—not only for mobile users, but also for those who may be tied to a particular location. Wireless technology can extend wire–based (telephone, cable television, and computer network) systems and provide services to people who could not receive them before. Satellites already deliver video—television programs, movies, and special events—directly to homes, and many new systems and services are being developed that could make portable phones and computers as ubiquitous as today's wired phones. Broadcasting technology will continue to be one of the easiest methods for delivering information to large numbers of people over a wide area.

WIRELESS TECHNOLOGIES AND THE NII

In order to realize the benefits these technologies offer, many technical, economic, regulatory, and social issues will have to be addressed as new wireless technologies and systems are integrated into the nation's communication and information infrastructure. For residential and business users, the influx of wireless service alternatives will magnify and intensify the changes brought about by the breakup of the Bell Telephone System in 1984. New entrants will challenge the historical monopolies in local phone and cable television service, offering comparable packages of services at similar or lower prices using satellites or land–based radio towers. In the near future, users will have a dizzying range of services to choose from, but not all systems will be compatible, and moving information between networks may be difficult. Standards that will allow this diverse mix of networks to interoperate are still several years away. As a result, however, confusion may be common as users are confronted with choices unknown in the past.

Rapid technological change, uncertainty about what customers really want and will pay for, and an outdated regulatory structure that is in the process of being overhauled all contribute to a dynamic, but chaotic, marketplace. Despite the "hype" surrounding new services, many of the systems that get the most attention—personal communications services (PCS) and low–Earth orbiting satellites (LEOs), for example—are not yet operational, although some experiments have been conducted. Technical challenges still exist in deploying new wireless systems, and, as one executive put it: "Much of the Buck Rogers stuff is going to take awhile."[1] Customer demand continues to be unfocused. Service providers believe there is a great untapped demand for wireless—and especially mobile—services, but no one knows exactly what customers want and will be willing to pay for. From a regulatory and economic standpoint, the role of wireless in the NII is similarly unclear. Companies will compete in some markets while cooperating in others, and the structure of the industry and its relation to the wireline network and companies will likely change in response to market forces.

The evolution to a competitive environment will be challenging for the different segments of the wireless industry. Competition for customers in the future NII is expected to be fierce—profit margins are likely to be low, many service providers may not survive, and industry consolidation is likely. Some analysts have questioned whether existing populations can support a plethora of mobile service providers that includes cellular, specialized mobile radio (SMR), paging, PCS, and satellite–delivered communications. The economics (cost structure and demand) of some of these services are poorly understood, and researchers are just beginning to explore systems and services that have not even begun operating.

In part because of these underlying uncertainties, the structure of the industry is likely to be remarkably fluid. Over the next several years, many new companies will enter the various wireless markets—some will fail, some will succeed, and some will be bought up by larger concerns or merge with competitors. This dynamism will be fueled by changing economic conditions, changes

in regulation, and technical opportunities for integrating systems and services. Administration policies, legislation now being debated in Congress, and evolving state and local regulations will have an important impact on how competition will develop and the role of wireless technologies in the NII.

IMPLICATIONS AND POLICY ISSUES

As the United States becomes a more mobile and information–intensive society, policymakers and regulators will face a number of challenges in bringing the benefits of wireless and the NII to all potential users. This report identifies a number of issue areas that policymakers should be aware of as the NII develops and wireless technologies become a more integral part of it. In general, given the recent successes of some wireless industry segments, and the nascent state of development in other segments, government action currently is indicated in only a limited number of areas.

- **Universal service.** Wireless technologies can extend service to those who do not have it. Through competition with established providers, they may lower prices for many different applications, making a wider range of services affordable to many people. The evolving definition of universal service, however, is critical for wireless providers. If universal service comes to be defined as ubiquitous two–way broadband access, as some groups propose, almost all wireless systems will be disadvantaged because they currently are unable to provide this level of service.

- **Interconnection and standards.** To realize the vision of the NII, the various systems and networks that comprise today's communications infrastructure—telephone, cable television, satellites, broadcasters, and cellular telephony—will have to connect with each other and with the new wireless communications systems now being developed. While this may

---

[1]Comments of John Clendenin, *Radio Communications Report*, vol. 13, No. 7, Apr. 11, 1994, p. 27.

not be technically difficult, the process of developing the standards that will govern these interconnections and allow different networks to work together is becoming increasingly difficult. The result may be a patchwork of systems that do not work together or that makes it difficult for users to exchange information across multiple networks.

- **Wireless and NII policymaking.** Early plans for the NII were dominated by visions of fiber-optic networks crossing the nation and linking every home and business; as a result, the unique contributions of wireless technologies went largely unnoticed. On the other hand, policies guiding the development of wireless systems were actively formulated, but generally were not placed in an NII context. Today, policymakers in Congress, the FCC, and the executive branch are more actively promoting wireless as an integral part of the NII.

- **Spectrum policymaking.** The rush to wireless technologies and systems has created congestion in many popular bands of radio frequencies. Given the potential demand for wireless —especially mobile—services, it is likely that spectrum will continue to be in short supply for some applications. Policymakers are having difficulty balancing the needs of existing services with emerging applications. New ways of managing the spectrum may be needed.

- **Research.** The uncertainties surrounding almost all aspects of wireless development and use are exacerbated by the lack of research on fundamental issues, including the characteristics of mobility, the economics of the wireless industries, and the possible health effects of wireless devices and systems. The longer term implications of the use of wireless services on personal lives and business productivity and organization are unknown. Especially important are questions relating to scale—as more people and businesses use wireless services, new and unexpected effects are likely to emerge. For example, interference between different wireless devices could become a more serious problem.

- **Federal/state/local jurisdiction.** As wireless systems have become more common, questions relating to regulation and jurisdiction have grown more controversial. The federal government has a long-standing responsibility to promote nationwide communications systems that benefit the public. This goal, however, is increasingly coming into conflict with the historic rights of states to govern communications services within their borders and the efforts of local governments to maintain control over how local lands are used. Finding locations for the antennas required to provide cellular or future personal communication services, for example, is becoming more difficult as communities seek to exert control over where the towers can be built.

Introduction and Policy Issues 1

Wireless communications technologies are poised to bring dramatic changes to the nation's telecommunications and information infrastructure, reshaping how people communicate, access information, and are entertained. These technologies, which use radio waves instead of wires to transmit information, already play an important part in the daily lives of almost all Americans. For more than 70 years, radio and television broadcasters have entertained and informed millions of people each day. Satellites connect the countries of the world, allowing people to converse, share information, and transact business. Most recently, cellular telephones have extended the reach of the public telephone system to people who are on the move or beyond the reach of traditional telephones.

Over the next several years, use of wireless technologies is expected to grow dramatically as a wide range of new radio-based communication, information, and entertainment services and applications is introduced, and the prices of both equipment and services fall. Some of the wireless systems now being developed include: 1) terrestrial and satellite-based telephone systems that will allow people to make and receive calls from almost any point on Earth, 2) digital television that promises clearer images and better sound, 3) digital radio broadcasting that will offer crystal clear sound as well as a range of information services, and 4) a wide range of data communications systems that expand the reach of computer and online services. These emerging wireless technologies, along with existing wireless services, will become an integral part of the nation's evolving telecommunications and information infrastructure—more formally known as the National Information Infrastructure (NII).

Wireless systems offer many benefits for individuals and businesses, but a number of challenges must be overcome before wireless technologies can be effectively integrated into the NII. Residential and business users, for example, will have a wider range of communication, information, and entertainment services to choose from, but systems may not work together and switching between service providers could be difficult. Wireless companies will offer many new technologies and services, but competition is likely to be intense in many markets and the long-term outcome of current policy initiatives—on interconnection of networks, universal service, and industry structure—remains uncertain. Some wireless technologies will complement existing services and networks, but many will also compete with the traditional communications and information providers—telephone companies, computer networks, broadcasters, and cable television companies. The economics of wireless systems are not yet well understood. In this uncertain and rapidly changing environment, policymakers and regulators will have to be vigilant in monitoring the effects of policies and rules already put in place.

Finally, the deeper implications of the widespread use of wireless technologies and services are not well understood. With the exception of television and radio broadcasting (and perhaps cellular telephony), radio-based systems have not yet penetrated deeply into the social and organizational fabric of American society and business. This is expected to change rapidly as technologies come into more widespread use as true mass-market products. Once large-scale use begins, the hidden impacts—both positive and negative—of wireless access and mobility will become clearer. While the benefits of ubiquitous communications and a wider range of services are important, potential problems remain regarding security, privacy, health effects, and social/organizational upheaval—including widening the gap between informa-

tion and communication "haves" and "have-nots." Technical, regulatory, and economic policy decisions will be required to ensure that the benefits of wireless are realized to the fullest extent possible, while minimizing the potential disadvantages for individuals, business, and society as a whole.

REQUEST FOR THE STUDY

The initial focus of NII initiatives was primarily on wireline technologies. Some visions of the NII seemed to ignore wireless technologies completely, failing to recognize the unique benefits that wireless systems offer. Other views of the NII—declared "technology neutral"—addressed wireless technologies as just another delivery method, but generally failed to take into account the special challenges that wireless solutions will pose for a national communications infrastructure. Most NII plans concentrated on developing the necessary infrastructure primarily through the expansion of the existing telephone network, cable television systems, and national computer networks (such as the Internet and the National Research and Education Network). Even today, most observers and telecommunications analysts believe that the backbone of the NII—the high-capacity links that will bind together the disparate networks that will make up the NII—will be primarily based on fiberoptic technology.[1]

The role of wireless technologies in the NII, however, has never been fully developed by either the Administration or Congress. Wireless proponents, especially in the broadcasting and satellite communities, have attempted to have their systems more directly included in NII discussions, and their efforts have been somewhat successful. Wireless technologies are generally recognized by most policymakers as an important way to access the NII, but the general bias toward wire-based NII systems remains. To broaden understanding of these issues, the Committee on Science, Space, and Technology (now the Committee on Science)

---

[1] Joseph N. Pelton, "CEO Survey on the National Information Infrastructure," *Telecommunications*, vol. 28, No. 11, November 1994.

> **BOX 1-1: Scenario—Wireless Technologies in Health Care**
>
> Scenarios scattered throughout this chapter sketch some possible visions of what wireless technologies and systems can and cannot do, discuss some of the implications of their widespread use, and provide some of the potential downsides. These scenarios are set in the not-too-distant future, and, in fact, many of the applications described below are already being tested or deployed.
>
> Ellen, a nurse in a big city hospital, does her rounds with an electronic clipboard. After checking her patient's temperature, pulse, glucose levels, and breathing, she enters the data directly on her clipboard. The information is immediately transmitted to the hospital's patient data network via a wireless link between her clipboard and the hospital's computer network.
>
> A doctor wanting to talk to Ellen about dosages for a patient undergoing chemotherapy reaches her on her handheld phone. She is reminded how much easier the phone makes it to stay in touch. Only last year she had to listen for pages on the building loudspeaker, and often had to wait to get to a phone to call back. She calculate once that she spent two hours per week, on average, just waiting to be called back or trying to get in contact with the doctors on duty.
>
> As she is checking on another patient, Ellen's pager signals that a staff meeting is beginning. Work schedules and patient loads are going to be reorganized and Ellen is opposed to one of the changes being proposed. She wants to canvas her colleagues and mobilize the opposition, but prefers to do this face to face, because it is a delicate matter. She calls up the personnel locator program on her electronic clipboard, which indicates that three of the 14 day shift staff are in the nurses' lounge. One of them is new—she can't recall the face, so she asks the hospital's computer for a photo.
>
> Midway through the meeting, Ellen's pager signals that she is wanted in the emergency room receiving area :a gunshot victim and multiple automobile accident victims are being brought in simultaneously. Preliminary information on the patients is being sent in from the ambulance, so Ellen calls the emergency room receiving program. As she is running to the receiving area, she is informed that the gunshot victim is a white male, 23 to 26 years old, his blood pressure is dropping rapidly, his blood type is B negative, he is likely to be a diabetic, and he has been taking antidepressant medication. Quickly, she grabs the appropriate IV units on the way down the hall, and is not surprised to see the other medical staff who will attend to this patient already there. In the emergency room, instant communication is crucial—a quick response and good information saves lives.
>
> SOURCE: Office of Technology Assessment, 1995.

of the House of Representatives and Representative Michael Oxley asked the Office of Technology Assessment (OTA) to study the role of wireless technologies in the emerging NII.

SCOPE OF THE STUDY
This report considers how wireless systems and services can contribute to the development of a national information infrastructure and what specific impacts the NII, as presently conceived, may have on the development and deployment of new and existing radio services. Because of the breadth of the subject, not all technologies and issues can be analyzed in detail. Rather, this report is de-

signed to serve as a general introduction to wireless technologies and services and the opportunities and problems they may give rise to in the context of the NII. It surveys most of the major wireless applications now being developed and identifies the most important issues arising from their implementation and use. Issues needing further study are identified. Some policy options for Congress are identified, but are limited primarily to broad issues that could affect the evolution or impacts of wireless technologies.

The study does not discuss generic NII issues—copyright, investment, or information content, for example—nor does it address several aspects of wireless communication that, while critical, are outside the scope of present work. First, the report does not address the special needs and contributions of private radio systems—including those systems used for public safety. Only those systems available for use by the general public or businesses are included. Private radio systems, while often used to meet important public safety and emergency preparedness needs, cannot be used by the public. However, during the course of this study, OTA has noted the challenges facing the public safety community in the use of radio communications to fulfill its various missions, including severe shortages of capacity, incompatible radio systems that hamper cooperation in emergency relief efforts, and rising communication needs in a period of budget cutbacks. These problems deserve much greater attention than they could be given in this report, and should be the focus of a separate inquiry.

The report also does not directly address the global aspects of wireless technologies or the NII. Prior OTA reports on international wireline and wireless communications found that domestic and international telecommunications policy need to be more closely coordinated.[2] OTA continues to believe in the importance of viewing domestic telecommunications policy in an international context, but chose to limit the scope of the present report to domestic issues for purposes of clarity and length. The report uses examples from other countries to illustrate technology advances or policy choices where appropriate. Likewise, OTA recognizes the importance of foreign markets for U.S. wireless equipment manufacturers and service providers. Promoting the competitiveness of U.S. firms in international wireless products and services should be an integral part of domestic policymaking.[3]

BACKGROUND: THE NII

The U.S. telecommunications infrastructure is already among the best in the world, providing high-quality communication, information, and entertainment services to over 90 percent of the population:[4]

- telephone service is available to 93.8 percent of American households;
- cable television service is available to almost 95 percent of U.S. households, 63 percent of whom subscribe;
- 94 percent of U.S. households can receive at least five broadcast television stations;
- radio broadcasting is ubiquitous, with 99 percent of American homes having an average of five radios;
- cellular telephone service is available to about 95 percent of the population, covering 50 percent of the geographic area of the United States (including Alaska, which has large unserved areas);

---

[2] U.S. Congress, Office of Technology Assessment, *The 1992 World Administrative Radio Conference: Issues for U.S. International Spectrum Policy,* OTA-BP-TCT-76 (Washington, DC: U.S. Government Printing Office, November 1991); U.S. Congress, Office of Technology Assessment, *The 1992 World Administrative Radio Conference: Technology and Policy Implications*, OTA-TCT-549 (Washington, DC: U.S. Government Printing Office, May 1993); U.S. Congress, Office of Technology Assessment, *U.S. Telecommunications Services in European Markets*, OTA-TCT-548 (Washington, DC: U.S. Government Printing Office, August 1993); U.S. Congress, Office of Technology Assessment, *Global Communications* (in progress).

[3] Office of Technology Assessment, *Global Communications*, ibid.

[4] Telephone statistics are from A. Bellinfonte, Federal Communications Commission, *Telephone Subscribership in the United States*, April 1995; cable figures from National Cable Television Association, *Cable Television Developments*, spring 1995, and U.S. Bureau of the Census, *Current Population Survey*, March 1994; television broadcast figure from Federal Communications Commission, *Broadcast Television in a Multichannel Marketplace*, June 1991; radio broadcasting figures from Radio Advertising Bureau, *Radio Marketing Guide and Fact Book for Advertisers*, 1994; cellular figures from Tim Rich, Cellular Telecommunications Industry Association, personal communication, May 4, 1995; and computer figures from Times Mirror Center for the People and the Press, *Technology in the American Household: The Role of Technology in American Life*, May 1994.

- 31 percent of American households have a personal computer, 12 percent have a computer with a modem, and about 50 percent of all workers use computers on the job.

It is this base of technology—the existing communications infrastructure—from which the NII will evolve. Technology advances are already improving these systems, especially in terms of capacity, quality, and flexibility. New wireless technologies will extend and expand the use of existing networks, and will create new links to information, allow more flexible communication, and provide connections to new sources of entertainment.

■ History and Purpose of the NII

The concept of a national information infrastructure originally focused on the development of a national computer network, the NREN, that the federal government played a key role in financing and developing.[5] The idea of the information infrastructure broadened, however, as telephone and cable companies—driven by advances in fiberoptics, digital signal processing, and data compression—began to promote their ability to provide a more diverse range of services using their networks.

To make the most of the existing information and telecommunication infrastructure, and to bring the benefits of advanced telecommunications, information, and entertainment services to all U.S. consumers and businesses, government policymakers formally advanced the idea of the NII. In September 1993, the Clinton Administra-

Satellites carry voice, data, and video communications all around the world, linking far-flung business locations, allowing researchers to keep in touch, and bringing television images of far off events to millions of American living rooms.

tion released its *Agenda for Action.*[6] That report established, in broad outline, goals for the development of telecommunications and information resources in the United States, and identified a concept of how the U.S. communications and information infrastructure should evolve. The purpose of the NII, as described by the Administration, is to enable all Americans to access the information they need; when they want it, where they want it—at an affordable price.[7]

To serve this purpose, the Administration has stated that many different technologies and systems will be used where appropriate.[8] In fact,

---

[5] High-Performance Computing Act of 1991 (HPCA), Public Law 102-194.

[6] Department of Commerce, Information Infrastructure Task Force, *The National Information Infrastructure: Agenda for Action*, Sept. 15, 1993.

[7] See, for example, comments of Mike Nelson, Office of Science and Technology Policy, at the Workshop on Advanced Digital Video in the National Information Infrastructure, Georgetown University, Washington, DC, May 10-11, 1994.

[8] As explained in the *Agenda for Action*, the NII is really more than just an interconnected series of telecommunications or computer networks. It encompasses: 1) a wide and ever-expanding range of equipment; 2) the information itself, which may be in the form of video programming, scientific or business databases, images, sound recordings, library archives, and other media; 3) applications and software that allow users to access, manipulate, organize, and digest [information]; 4) the network standards and transmission codes that facilitate interconnection and interoperation between networks; and 5) the people—largely in the private sector—who create the information, develop applications and services, construct facilities, and train others to tap its potential. Department of Commerce, op. cit., footnote 6, pp. 5-6.

BOX 1-2: OTA's Definition of the National Information Infrastructure

In discussing the integration of wireless technologies into the emerging communications infrastructure, OTA adopts a broad definition of the National Information Infrastructure (NII). It includes all the systems and applications necessary for the public to communicate with whomever they want and access the information they desire. The NII will be one-way and two-way, point-to-point and broadcast, and narrowband and broadband. It will be an amalgam of existing systems and services and completely new technologies and applications. Different parts of the NII will serve different functions depending on technology and need, and some systems may serve a multitude of needs. The NII will include satellite systems, fiberoptic cable, terrestrial radio systems, broadcasting, and the telephone and cable television networks, among others.

What will the NII not be? Despite the singular way in which the term is used—the "NII" is not, and will not be, one "thing." Rather, it will be more accurate to think of the NII as a unifying concept or overarching idea that brings together all the different systems, technologies, and applications that are necessary for people to communicate, access information, and be entertained. Just as the transportation infrastructure of this country is more than just the interstate highway system—it consists of roads, railroads, aircraft, passenger cars, trucks, and ships—so, too, will the NII consist of more than just an "information superhighway." It will also include all the different, lower speed "on and off ramps"—the many local connections that provide access to the network.

Nor will the NII be, as some have suggested, a huge collection of completely interconnected networks capable of transmitting interactive voice, data, and video among all businesses and citizens. Rather, the NII will be a collection of many different kinds of systems. Some general-purpose systems may indeed be capable of carrying two-way, high-bandwidth, multimedia communications, but many other systems will carry only certain kinds of information (voice/data, but not video) or will carry it only one-way (broadcasters).

In addition, not all of these different subsystems will be completely or directly interconnected. Rather, the interconnections will be based on practical and/or economic considerations. It may not make sense, for example, to connect a phone system to a television broadcast station. The existing public switched telephone network may serve as a "core" network that serves as a common point of interconnection for many smaller networks. Finally, the NII will not evolve out of the Internet—the name given to a worldwide network of interconnected computers. The Internet will be only one of the many parts comprising the larger concept that is the NII.

SOURCE: Office of Technology Assessment, 1995.

---

most analysts today think of the NII not as a single system, but as a "system of systems" or "network of networks" that will carry voice, data, and video communications to homes, businesses, schools— to people wherever they are. It is unclear, however, just what the public thinks the NII is. In the popular press, it is often referred to as the "information superhighway." This may connote, incorrectly, a *separate* system that is to be built in addition to existing cable, telephone, and computer networks. For purposes of this report, OTA defines the NII quite broadly (box 1-2).

To bring the NII into being, the Administration has identified five overarching policy guidelines that will serve as the framework for developing not only wireline NII services, but wireless systems and applications as well.[9]

1. *Competition* is seen as the engine that will drive private sector investment in the NII, allowing companies to compete on fair and equal terms, while stimulating efficiency and innovation. Competition is also believed to lower costs for consumers, increase choices and diversity in information sources and entertainment, and protect quality and reliability.

2. A commitment to *universal service* seeks to ensure that NII services will be available to all who want them, regardless of income, location, or ethnicity. This commitment has been the foundation of the telephone system for more than 90 years; as a result, almost everyone in the country is able to have a telephone.

3. *Private investment* will be the source of almost all funding for the NII; the government will not build or operate the systems that comprise the NII. Government agencies, however, will operate publicly accessible databases and their own telecommunications and information networks.

4. *Open access* means the networks that will carry the information and entertainment will be open to all users—distributors of programming as well as residential and business consumers.

5. *Flexible government regulation* is recognized as vital to promoting the goals outlined above. Regulations must seek to promote fair competition and private investment in rapidly changing technology and market conditions; they must also protect consumers' interests by ensuring

low-cost services, high reliability, and personal privacy and security.

■ Information Infrastructure Task Force

To guide its development of policies for the NII, the Administration formed the Information Infrastructure Task Force (IITF) in 1993. It is composed of high-level representatives of the federal agencies that play a major role in the development and application of communication and information technologies and those that rely on communication and information technologies to deliver their services. The IITF operates under the aegis of the Office of Science and Technology Policy and the National Economic Council, but is chaired by the Secretary of Commerce. Much of the staff work and administrative support for the IITF is done through the National Telecommunications and Information Administration (NTIA) of the Department of Commerce. To gather private sector input and assist the IITF, President Clinton established an Advisory Council on the NII.[10]

Functionally, IITF's work is divided among three main committees: telecommunications policy, information policy, and applications. These committees have delegated specific tasks or responsibilities for certain issue areas, such as privacy, reliability, universal service, health, etc., to individual working groups.[11] Although several of the working groups may cover wireless technologies in the context of their broader work, none deals specifically with wireless as a separate area. Given this lack of focus, it is unclear to what extent wireless technologies play a role in the com-

---

[9] The *Agenda for Action* originally identified nine principles that would guide the NII initiative: 1) promote private sector investment; 2) extend universal service—ensure that information is available to all at affordable prices; 3) promote technological innovation and new applications; 4) promote seamless, interactive, user-driven operation; 5) ensure information security and network reliability; 6) improve management of spectrum; 7) protect intellectual property rights; 8) coordinate with other levels of government and other nations; and 9) provide access to government information and improve procurement. These nine principles were collapsed into five over time. See, for example, remarks by Vice President Gore at the Federal-State-Local Telecomm Summit, Washington, DC, Jan. 9, 1995.

[10] Clinton, W. J., President, United States, "Executive Order 12864—United States Advisory Council on the National Information Infrastructure," *Weekly Compilation of Presidential Documents* vol. 29, No. 37, Sept. 20, 1993, p. 1771.

[11] For more indepth information on the structure and accomplishments of the Information Infrastructure Task Force, see U.S. Department of Commerce, *National Information Infrastructure: Progress Report September 1993-1994*, September 1994, especially appendix B.

mittees' deliberations, and how well the specific benefits and problems associated with wireless are being considered. Another IITF working group, the Technology Policy Working Group has addressed wireless technology in some of its discussions as part of its mandate to examine cross-cutting technology issues. Government activities and policy initiatives relating to the NII and wireless systems are discussed in more detail in appendix B.

Industry Initiatives

Industry, as the primary builder and operator of the evolving NII, has been an active participant in the policy development process since before the moniker "NII" was attached to the effort. Innumerable industry groups and consortia have produced vision statements and proposals, lobbied Congress, and testified at federal, state, and local hearings on all aspects of the NII. At the same time, all segments of the telecommunications industry, wireline and wireless, have been moving ahead to build their systems. A complete overview of industry activities regarding the NII would be impossible, given the scope and depth of their work and the fact that almost everything industries do could be considered NII-related in one way or another. Such a review is beyond the scope of this report.

The NII Today

The main challenge in building the NII will not be technical—the basic technologies that will form its foundation are already in place or being developed, and standards are being written that will permit different devices and networks to interoperate. The biggest obstacle to moving the NII forward is the lack of consensus on what it should encompass and, as a result, what policies, administrative procedures, and regulations are needed to deploy it. Beyond the broad concepts outlined by the Administration, the vision of the NII has remained vague and somewhat ill-defined. Different interest groups, government leaders, and industry ob-

servers have offered their own visions of what it should be and what needs it should serve. However, no real agreement has been reached, and, in many cases, it has even been difficult to agree on common terms of reference. Some have pointed out that the NII is all things to all people—that definitions are as varied as those who create them.

In the past eight months, the concept of "the NII" has become even more amorphous, eclipsed by broader efforts to overhaul regulation of the nation's telecommunications industries. Some even call the NII "quaint." A subtle shift has occurred that places competition at the center of the telecommunications policy framework rather than the NII. As a result, the NII now seems to be defined as whatever a competitive marketplace creates as a result of deregulated telecommunications and media competition—it has been reduced to a byproduct rather than the result of a specific vision or plan. Policy efforts seem directed more toward meeting NII *goals*—access, diversity, low prices, and interconnection—through the engine of competition as opposed to creating "an NII." In this new environment, wireless and wireline policies are rarely linked explicitly, but they are being developed under the same set of unifying principles—a dedication to competition. Despite this coalescence, however, no long-term vision of how wireless systems will fit into the NII exists or is being developed, and the marketplace is being relied on to sort out the details.

Despite the continued vagueness of the overall NII concept, however, intensive research, experimentation, and other development work is being done on its various parts. Technology vendors and service providers continue to develop and refine technologies and applications they believe will become part of the NII. The federal government has sponsored or organized many discussions—with both public and private sector input—on the issues of universal service, interconnection, and privacy, among others. Many people—in both government and the private sector—have invested considerable time and effort to advance the ideas of the NII, but questions still remain about what it is, what it will do, how much it will cost to develop, and when its benefits will be available.[12]

Some analysts and citizens question the wisdom of pushing ahead with such a massive undertaking while fundamental questions—about the real need for the NII, what its functions will be, and what negative effects it might have—remain unanswered.[13] Many of these same questions also apply to the deployment of wireless in the NII. OTA has argued that deploying technology solutions before assessing the needs of the users is not likely to lead to the best solutions.[14] While such questions are important and valid—and should be carefully considered—events appear to have overtaken this type of carefully planned approach. The NII is already being built, and it would be virtually impossible to stop it. Further, even if one could start over, the rapid pace of technology development has made the concept of "needs" highly individualistic and subject to rapid change—making them difficult to rationally identify and plan for on a broad system level. This report examines some of the important issues surrounding the deployment of wireless systems in the NII, while acknowledging that some of the most fundamental questions about the NII have become moot.

The NII concept has served to focus more attention on telecommunications in general. It has also given added impetus to wireless development efforts, but industry analysts and stakeholders believe that wireless would be just about where it is today even without the NII. Perhaps more importantly, there is a widespread belief that development and use of radio-based systems and technologies will continue to expand dramatically—with or without the NII—as users become more familiar with them and as applications that meet real needs are developed.

WHY WIRELESS?

While estimates of demand and future subscriber rates vary considerably, most analysts believe that wireless telecommunications will become widely available over the next decade. Demand for mobile access to telecommunications networks and services is growing, and many companies—old and new—are rushing to get into the wireless business. But what is driving the trend toward wireless technologies?

Portable computing devices allow users to send and receive electronic mail, access online services and exchange files with other users. The combination of portability and connectivity is driving many new applications of wireless technology.

---

[12] Pelton, op. cit., footnote 1, pp. 27-34. Despite a wealth of conferences, papers, and public hearings, for example, the debate over universal service continues. Different segments of the service provider community remain split over how best to deliver an evolving "universal service." "Universal Service Consensus Eludes NTIA..." *Telecommunications Reports*, vol. 60, No. 52, Dec. 26, 1994.

[13] In comments on this report, one reviewer noted: "In essence, what we are doing is that we are building a system's structure without knowing what its function is or ought to be. When one would design most other systems or for that matter, e.g., a building, one typically would first start with function from which structure follows. With the NII, and with wireless infrastructures as well, I believe we ignore this thinking and we start first with structure....should we not raise the basic question as it will probably be inevitable that many dysfunctions are the result of building a structure, i.e., happily paving the NII?" Rolf T. Wigand, personal communication, Apr. 28, 1995.

[14] U.S. Congress, Office of Technology Assessment, *Linking for Learning: A New Course for Education,* OTA-SET-430 (Washington, DC: U.S. Government Printing Office, November 1989).

TABLE 1-1: U.S. Wireless Telecommunications Service Subscribers and Growth Rates, 1984-94				
	Cellular telephone subscribers (millions)	Cellular telephone growth rate (percent)	Pagers in use (millions)	Pager growth rate (percent)
1984	0.09	–	–	–
1985	0.34	278	4.5	–
1986	0.68	100	5.4	20
1987	1.23	81	6.5	20
1988	2.07	68	7.8	20
1989	3.51	70	9.4	21
1990	5.28	50	11.2	19
1991	7.56	43	13.4	20
1992	11.03	46	15.3	14
1993	16.01	45	19.3	26
1994	19.28*	20	–	–

*Through June 1994. All others at year end.

SOURCES: Cellular Telecommunications Industry Association, Personal Communications Industry Association, Telephone Industry Association, and National Cable Television Association, 1995.

To understand the role radio-based technologies will play in the NII, it is necessary to understand the factors driving the demand for wireless services, as well as the technological capabilities and advances that are making new applications possible. Each of these factors—technology *push* and demand *pull*—is working independently to fuel the rush to wireless, but they also sustain and reinforce each other. This section describes the technical and sociological context in which wireless technologies and services are evolving and that simultaneously underlies the transition to the NII.

∎ Wireless Growth Estimates

Much of the excitement that surrounds wireless communications is based on assumptions analysts and companies make about what people and businesses want, but there is little agreement on how big the potential market for wireless might be. Most analysts base their estimates of future wireless growth on the diffusion of cellular telephone service and, to a lesser extent, on sales of portable computers. The growth rate of cellular telephone

service is high, running about 45 percent per year in the United States until 1994, with comparable rates in other developed countries.[15] Paging, another widely used service, has experienced growth rates of about 20 percent per year for nearly a decade (table 1-1). In another measure of potential demand, NTIA recently completed a study of future spectrum requirements that indicated that more than 400 MHz of additional spectrum was needed to support a growing range of wireless services.[16]

As a result of such findings, there is growing consensus that the demand for some kinds of wireless services is likely to be very high. Some analysts believe that as many as 100 million people will use some type of wireless telecommunications device by the year 2010. The following table of projected demand demonstrates both the trends and the variations in demand, but do not necessarily reflect OTA's assessment of the extent of the market (table 1-2).

All data or forecasts relating to future demand for wireless services must be regarded cautiously. Projections vary widely, reflecting different in-

---

[15] Compound annual growth rates (1990-92) in other countries range from 16 percent in the United Kingdom to 54 percent in Australia to 115 percent in Taiwan. Statistics cited in "ITU Deems Cellular Telephone Growth 'Truly Explosive,'" *Mobile Phone News*, June 20, 1994.

[16] U.S. Department of Commerce, National Telecommunications and Information Administration, *U.S. National Spectrum Requirements: Projections and Trends,* Special Publication 94-31 (Washington, DC: U.S. Government Printing Office, March 1995).

Service	1994 Subscribers (millions)	1994 Penetration (percent)	2000 Subscribers (millions)	2000 Penetration (percent)	2005 Subscribers (millions)	2005 Penetration (percent)
New PCS	–	–	14.8	5.4	39.4	13.1
Satellite	0.1	0.0	1.3	0.5	4.1	1.4
Narrowband/Paging	24.5	9.6	56.2	20.4	92.2	30.7
Dedicated data	0.5	0.2	3.4	1.2	5.7	1.9
Cellular	23.0	9.0	46.9	17.0	65.4	21.8
SMR/ESMR*	1.5	0.6	5.2	1.9	9.0	3.0
Total	34.1	13.4	79.7	28.9	136.3	45.4
Total voice services	14.6	5.7	48.2	17.5	96.5	32.1

TABLE 1–2: Wireless Technologies Subscription Forecast, 1993–2003

Note: The following U.S. population figures were used: 1994—255 million, 2000—275.8 million, 2005—300.3 million.
Note: Total subscriptions include individuals with multiple subscriptions across services (i.e., there are more subscriptions than subscribers).
*SMR/ESMR = Specialized Mobile Radio/ Enhanced Specialized Mobile Radio

SOURCES: Personal Communications Industry Association, "1994 PCS Market Demand Forecast," (Washington, DC: Personal Communications Industry Association, January, 1995); Personal Communications Industry Association, "PCIA 1995 PCS Technologies Market Demand Forecast Update, 1994-2005," (Washington, DC: Personal Communications Industry Association, January, 1995).

dustry definitions, assumptions, and biases. The data are highly uncertain and projection methods themselves crude and imprecise. Great uncertainty underlies all these numbers.

■ Technology Trends and Drivers

Rapid advances in technology are the most visible, and one of the most important, drivers in the development of the NII and wireless services. Over the past five years, advances in information and communications technology have greatly expanded the capabilities and flexibility of existing services, while also making possible a whole range of new services, including wireless. Cellular, PCS, and Enhanced Specialized Mobile Radio (ESMR) services, for example, are the result of improvements in computer processing, battery technology, miniaturization, and new digital signal processing and transmission techniques (box 1-3). New satellite services are the result of advances in digital compression technologies and improved computer processing—in both the provider's network and in consumer equipment. Cur-

rent development efforts promise to bring users even more features and advantages in the future.

Technology advances have a two-fold, somewhat paradoxical, impact on the development of wireless technologies. First, as noted above, advances make new applications and services possible. As new services are introduced and existing services are improved, however, more people use them, sometimes resulting in congestion and "crowding" of the most popular frequency bands. Cellular telephones are now so popular that, in some areas, it can be difficult to place a call during rush hour because the cellular system is full. Technology advances, however, can also help solve these capacity and congestion problems. New technologies enable more efficient use of the spectrum by squeezing more users into existing bands, and by allowing radio frequencies to be shared more easily among different kinds of services.[17] Cellular service providers are now installing digital technology to add capacity to their systems and provide clearer calls.

[17] For a discussion of the range of solutions to spectrum crowding, see U.S. Department of Commerce, National Telecommunications and Information Administration, *U.S. Spectrum Management Policy: Agenda for the Future*, NTIA Special Publication 91-23 (Washington, DC: U.S. Government Printing Office, February 1991), p. 13; and Richard Gould, "Allocation of the Radio Frequency Spectrum," contractor report prepared for the Office of Technology Assessment, Aug. 10, 1990.

BOX 1-3: Wireless Technology Trends

Miniaturization

A key technical factor pushing the development of wireless technologies is the rapidly shrinking size of radio components. Advanced technology has enabled increasing numbers of functions to be performed by a single chip and at higher speeds. This allows manufacturers to produce telephones, pagers, and computers that are smaller, lighter, and consume less power. The limiting factor to the size of some of today's products is no longer the chips needed to make them operate, but the physical characteristics of the people who use them—keys that are too small to easily type on or dial are not very useful.

Battery Technology

The problems associated with powering today's portable devices continue to frustrate and annoy many wireless users. The batteries required to run portable cellular phones and computers are usually heavy and/or provide limited hours of operation, and they can be expensive. A number of developments in battery technology may remedy this situation. Some involve new technologies, such as nickel metal hydride (NiMH) and lithium ion (Lion) batteries. Another solution being developed is a zinc-air battery that draws oxygen from the atmosphere to extend its life to 15 continuous hours. Power-saving solutions that make smarter use of battery power by the devices themselves hold promise for extending battery life further. More power-efficient displays and more efficient sleep modes are examples of ways in which small improvements could yield significant benefits in battery life.[1]

Frequency Reuse

Capacity is a major problem with many mobile communication systems. In any given area, when a specific frequency is in use it cannot be used for other purposes or by other users.[2] Radio waves, however, travel limited distances (see appendix A) before they fade out; beyond that point, a specific frequency can be reused without interfering with the other signal. This is the principle that underlies cellular telephony. Within a geographic area encompassing many cells, the same frequencies might be used up to six times. Shrinking cell sizes and lower transmitter powers, however, are not a permanent solution for increasing capacity. There are limits on how small a cell can be and how low power can go while still maintaining adequate quality.

Use of Higher Frequencies

As the lower frequency bands have become increasingly crowded, engineers have begun to develop technologies that would use higher, less crowded frequencies.[3] As was the case in extending terrestrial frontiers, developing higher frequencies is difficult and expensive. In addition to the cost of developing new devices that will operate at the higher frequencies, transmission problems typically worsen at higher frequencies. Some of those problems, such as increased attenuation due to rain, appear to be surmountable only by brute force—by increasing transmitter power. In satellite systems, power must be increased at both the original transmission (uplink) site on Earth and on the satellite itself. Increased satellite power greatly increases costs.

(continued)

---

[1]Clive Cookson, "Battery Technology: Still an Achilles Heel," *Financial Times Review, Information Technology,* May 3, 1995, p.7.
[2]This is, of course, an oversimplification. Different radio services can be designed in many ways to share spectrum.
[3]For a recent discussion of the upward expansion of usable radio frequencies, see Edmund L. Andrews, "Seeking To Use More of the Radio Spectrum," *New York Times*, Sept. 11, 1991, p. D7.

BOX 1-3: Wireless Technology Trends (Cont'd.)

Satellite Antennas

Advanced satellite antennas permit the use of smaller, less expensive Earth stations by making more efficient use of available satellite power. Such antennas direct the signal toward, and concentrate it in, areas where the intended users are located. Systems with such antennas, called spot beams, also make more efficient use of spectrum than those with large, circular beams which waste satellite power by transmitting beyond the limits of the desired service area. The reduction of signal levels outside the service area permits the same frequencies to be reused by other systems serving nearby areas, in the same way that cellular technology operates. The National Aeronautics and Space Administration's (NASA) Advanced Communications Technology Satellite (ACTS) system uses spot beam techniques, and Motorola's Iridium and the Teledesic system also plan to use them.

Spread Spectrum

Spread spectrum is a modulation technique first developed to hide military communications amid natural noise and other signals. More recently, spread spectrum has been used to permit low-power signals to share spectrum with other services. As the name implies, the original modulating signal is spread over a wide range of frequencies (bandwidth) for transmission. Interference from conventional signals or other spread spectrum signals appear as noise to the system, and can be eliminated.

There are several types of spread spectrum systems. One type, known as direct-sequence spread spectrum, divides a radio signal's energy over a wide range of frequencies so that a little part of the signal appears on each frequency in the band. Frequency-hopping spread spectrum techniques spread a signal out over many frequencies by hopping from frequency to frequency in a sequence synchronized with the receiver. One frequency is not dedicated to one user, and all frequencies can be used more efficiently. As more user/signals are added, however, the noise may eventually become too great for good communications. New adaptations of spread spectrum techniques, including advanced forms of CDMA may help solve some of these problems.[4]

Advanced radio transmission technologies that spread radio signals over extremely wide bandwidths may also provide solutions to transmission and capacity problems. Several companies are working on radios that send and receive over an extremely wide range (up to several GHz) of frequencies, providing greater capacity than today's channel-oriented approach. However, little is known about the operational aspects of these devices, especially the potential interference they could cause to other systems—and spectrum managers believe that implementing such radios, especially in already-assigned bands, could be extremely difficult.

---

[4]Synchronous CDMA, e.g., is being developed for use in future personal communications systems. Jack Taylor, Cylink, personal communication, Mar. 14, 1991.

SOURCE: Office of Technology Assessment, 1995.

Digital Technology

Many recent communications technology improvements are the result of the rapid diffusion and deployment of digital technologies in all aspects of communications and information processing. Digital information is easier to compress, transmit, manipulate, and store; and digitized voice, data, and video are much easier to combine into a wide range of multimedia applications. These advances are fundamentally altering the relationships between previously separate systems and services.

For wireless communication systems, digitally encoded and transmitted information offers sev-

CELLULAR TELECOMMUNICATIONS INDUSTRY ASSOCIATION

With just a personal computer, a radio scanner, and some software, criminals can reprogram cellular phones to steal service from unsuspecting consumers.

eral advantages over analog systems. The greatest benefit of digitizing radio communications is the ability to compress and combine multiple signals. This allows more information to be transmitted in a given time and more users to share a given amount of spectrum, thereby increasing speed and capacity.[18] Applications using digital compression techniques are spreading rapidly in many radiocommunication services. In cellular telephony, for example, digital signal processing and transmission techniques promise capacity up to 10 times that of existing analog cellular systems. Satellite companies are reportedly working on technologies that will combine up to 16 video signals on a single transponder.

Combined with compression, digital transmission techniques allow wireless system operators to exploit the spectrum more efficiently and deploy a wider range of applications serving more users. Digital transmission technologies, including spread spectrum, are a crucial piece of the solution to the spectrum congestion some radio frequency bands are now experiencing. Time Division Multiple Access (TDMA) and Code Division Multiple Access (CDMA), for example, are digital transmission schemes that allow more telephone conversations to be transmitted over a given bandwidth than analog technology allows (see chapter 3). Such schemes will allow Commercial Mobile Radio Service (CMRS) providers to dramatically increase the capacity of their systems and offer a wider range of services. Broadcasters believe that digital compression and transmission technologies will allow them to use their existing spectrum—which currently can carry only one analog channel—to transmit six or more channels of digital television programming (at today's quality), one high-definition television (HDTV) channel, or new information services.[19] Digitized information can also be more easily and effectively encrypted, making conversations and other communications more private, and preventing unauthorized *pirating* of pay services.

Uncertainty

Despite the benefits that new technologies bring, rapid technology advances also cause a great deal of uncertainty among users, manufacturers, service providers, and policymakers. Which technology is best? What is coming next? With technology life spans now measured in months not years, it has become harder for consumers and businesses to decide what services and equipment to buy and when.[20] For policymakers and regulators, rapid change makes policymaking and standards-setting more complex. Several factors underlie the uncertainty that now characterizes wireless technology development.

Much of the uncertainty can be traced to the fact that, despite significant research and development, and a great deal of industry "hype," few of

---

[18] Digital compression works by removing redundant or unnecessary information from the signal. In video transmission, for example, individual elements of the picture that do not change from frame to frame (when the background of a scene remains the same, for example) are not resent for each frame—just a code that tells the receiver/decoder that no change has taken place. This allows less information to be sent, requires less bandwidth, and allows more channels to be transmitted.

[19] Advanced Television (ATV) or Digital Television (DTV) increasingly are being used in place of high-definition television (HDTV).

[20] The recent delay in Bell Atlantic's planned system upgrade is evidence of the uncertainties facing today's service providers.

the new wireless systems are widely available. PCS frequencies have just been auctioned, and services are not expected to be available until late 1995 or 1996; ESMR technology has been plagued by technical problems; only one of the LEOs systems has launched satellites[21] (although experimental satellites have been used for some little LEO-like services), and many of the data services being developed are hampered by slow transmission speeds, incompatible systems and protocols, and a limited selection of applications. As a result, potential users do not know what the new systems will really offer, and technical details remain to be finalized. Lack of real-world operational experience also makes it hard to realistically determine the most efficient wireless access system—and thus to identify potential winners and losers.

In addition, the pace of technology deployment is also uncertain. Although, strictly speaking, the development of core radio-based technologies will not be a barrier to the development of new wireless systems and services, the pace of development and implementation is likely to be slower than most analysts predict, and, in combination with slow standards-setting processes and regulatory change, could slow the deployment of many new systems by at least months and possibly years. Finally, in some newly reallocated bands— the PCS bands, for example—new users are being required to pay to move incumbent users to other frequencies. This process will also be time-consuming and slow the deployment of wireless services.

■ Demand: Why Do Users Want Wireless?

In addition to the *push* provided by rapidly advancing technologies, users—consumers, businesses, and government—have an expanding set of needs and demands that are *pulling* the development of wireless applications. Although each user group has its own specific needs, there are also general factors that are increasing demand. These include the need for mobility and/or portability, easy access, ubiquity, and low cost.

Advantages of Wireless Technology

Wireless technologies have several unique characteristics that make them valuable to both individual users and companies wishing to distribute information. First, radio-based systems can be used to *broadcast* voice, data, and video programming and information to large groups of people over wide areas at relatively low cost. Broadcasting is point-to-multipoint and generally one-way. Radio and television broadcasters have served the American people for decades with news, entertainment, sports, public interest, and emergency programming. Satellite broadcasting promises to extend the reach of local audio and video programming to national, or even international audiences.

Second, wireless systems can serve needs that are not practically or efficiently served by wire-based networks. Both satellite and terrestrial technology, for example, can be used to create a *wireless local loop* to serve extremely remote telephone customers (see chapter 3).[22] Radio technologies can also be used to deliver communications services faster and less expensively than building or extending a wire-based network. Cellular technologies, for example, are being used in many developed and developing countries to bring telephone service to areas that have been unserved. Wireless Local Area Networks (LANs) connect computers where it is too expensive or impractical to install a wire—for example, in a building where asbestos creates construction hazards or an historic site. Many of the nation's schools reportedly have this problem. Wireless systems also allow flexible deployment of people and devices quickly and easily—e.g., to reconfigure a computer net-

---

[21] Orbcomm launched two satellites in March 1995. Both developed technical problems, but were later reported to be operational.

[22] Basic Exchange Telecommunications Radio Service, or BETRS, has been in use for many years to provide telephone service to remote rural residents. US West has also been testing the use of satellites to provide telephone service to remote areas of Wyoming.

BOX 1-4: Scenario—Wireless Technologies in Small Business

Sandra has operated her own plumbing business for the past three years in sprawling Phoenix, Arizona. It is a demanding business—lots of competition, small margins, and customers who can't wait long for service.

Sandra decided early on to minimize her overhead and run her business entirely out of her van, so she bought a portable telephone, a pager, and a laptop computer with a wireless modem. Sandra figures she can be on the job and be able to respond to calls for service, thus keeping business flowing in. Her response time is often very rapid, which customers appreciate. She handles all the estimates, ordering, invoices, and accounts on her laptop, including ordering parts for delivery either to her house, or directly to the jobsite. This means she doesn't have to hire a secretary or maintain an office, keeping her costs down.

Setting all this up was quite a chore for Sandra. She tried to do it on her own, but assembling the right hardware, software, service providers, and actual services proved too difficult. She ended up using a systems integrator, a national franchise operation that could get better deals on components than she could, and even handles the various telecommunications service billings for her. Even though she pays a premium for the service, she figures she will come out ahead because the technologies are just changing too fast for her to keep up.

Because the city is so big and growing so rapidly, Sandra also decided to invest in satellite navigation and route-planning equipment. Traffic can be difficult and time spent on the road is time lost on the job, so the payoff is obvious. She also hopes to expand her business to two vans, and hire her friend Wayne. The navigation gear she has will allow her to keep tabs on him, and coordinate their responses to emergency calls.

Sandra is also the mother of two young school-age children. Because she needs to spend so much time on the road, she stays in touch with them via pocket telephones and pagers. She likes the sense of security it gives them all because she can locate them whenever she needs to, and they can call her (and have twice) or 911 if they feel in any way in danger. But she also worries that they will never know the feeling of really being free and independent, like she was at their age, when the whole neighborhood was her playground.

SOURCE: Office of Technology Assessment, 1995.

---

work without having to move wires, or deploy emergency personnel in times of natural or man-made disasters.

Finally, wireless can serve quite well when communication needs are unpredictable or transitory. Radio-based technologies are ideally suited to providing ubiquitous access in a specific geographic area where a user will be traveling.[23] A mobile repair person, for example, may not know in advance where his or her services will be needed, and will likely need to stay on-site for only a short period of time. This capability allows people to be connected wherever they are, and serves the need to get information or communicate *immediately*. Different types of systems will serve different areas—a building or mall, an office park or

---

[23] In reality, many of the benefits of radio technologies for access and mobility are based on the concept of broadcasting. Broadcasting, in fact, is the mode of communication that allows mobility to take place—no matter where one travels within the range of the signal, the signal is always present. Cellular telephony, although not a broadcasting service like radio or television, uses a broadcast signal to contact the desired person. Similarly, in cases where many users in an area need access or where users will be at different locations—some known, some not—the broadcast radio signal, because it blankets a given area, is what makes ubiquitous access possible.

BOX 1-5: Scenario—Wireless Technologies in the Migrant Community

Jose is a migrant farmworker in West Virginia with strong ties to Miami where his family lives. His job is tough—he spends many hours in the orchards battling poison ivy, insects, and the residue of pesticides sprayed on the trees. He worries about his wife and children—particularly how his eldest daughter is doing in school—and his sister, who has had a series of medical difficulties that have left her unable to work. Jose has always been the responsible family member. Because he speaks English, he often negotiates appointments, visits to the health clinic, and so on for family members. Being able to contact and be contacted by them is essential for his family's survival. Jose gets little time off during his workday; even when he does, he is unable to find a payphone because he is often miles from the nearest town.

But recently, Jose bought a pocket telephone that he takes into the fields with him. Several years ago, this would have been too expensive, but a price war among the mobile telephone companies has put both telephone handset and service charges within his grasp. He uses the phone to call his family in Florida nearly every day. Occasionally he even contacts his widowed mother back home in Mexico, despite the very high international telephone charges (he typically pays about $40 for a four- to five-minute call). Jose also finds it convenient to make appointments for himself and his coworkers at the local clinic in rural West Virginia, to contact the school his daughter attends in Miami, and to call the hospital where his sister's doctors work. It used to be difficult to get a return call because he was not near a telephone, the payphone was busy, it was after business hours, or he had followed the migrant work stream to a different community. Now he feels much more connected to the people and services he needs to live a better life.

SOURCE: Office of Technology Assessment, 1995.

---

downtown area, a metropolitan area, a region of the country, the whole country, and even the whole world. Users will be able to pick and choose the technologies that best meet their needs.

Mobility and Access

The unique advantage that wireless technologies bring to the NII is mobility. Increasingly, users want to communicate wherever they are—while walking, driving, or traveling on a plane, train, or ship.

> [E]very human, even the most committed landlubber, is a sailor of sorts, or else a driver, or a flyer, or at least a pedestrian. After almost a full century of development, the telephone still had a very fundamental shortcoming: telephone wires don't move. People do.[24]

All wire-based services are inherently limited in one important way: they can go only as far as the wire extends. For applications that require mobility, wireless is the only way communications services can be provided, and thus mobility is the most important characteristic and benefit that wireless technologies bring to the NII. Most radio-based services in the NII will function as tetherless, mobile, portable extensions of the wire-based network.

Clearly, mobility is an integral aspect of human activity, but telecommunications services that enable or accommodate mobility are still in their infancy. Strong demand for such services has existed in the past, and business interest in new wireless technologies suggest that future demand is anticipated by many others. However, little is

---

[24] Peter W. Huber, Michael K. Kellogg, and John Thorne, *The Geodesic Network II: 1993 Report on Competition in the Telephone Industry,* (Washington, DC: The Geodesic Co., 1992), p. 1.5.

DRAWING BY GAHAN WILSON; COPYRIGHT 1994, THE NEW YORKER MAGAZINE, INC.

"Who can we call?"

known about the scope or scale of that demand. Few data are available to predict how people will actually use mobile systems, and thus which wireless services are most needed. Better data about mobility and its characteristics would help anticipate the future direction of these technologies as they are brought into the marketplace and the society in greater numbers. Chapter 2 discusses mobility in greater detail.

Wireless technologies can provide more than just mobile services, however. Radio-based systems can provide information, entertainment, and communication services to homes and businesses as well. In this context, wireless technologies are expected to make their greatest contribution as an access point to the resources of the NII, either extending services where wires cannot go—to remote customers, for example—or competing with wireline networks in the provision of traditional communications and entertainment services such as telephone, data communications, and video programming. Satellite systems, for example, can provide end-to-end voice, data, and video services that bypass, and could compete with, the wireline infrastructure entirely.[25] Cellular, PCS, and direct broadcast satellite (DBS) services will compete with wireline alternatives such as the telephone and cable television networks to provide the *last mile* connection to the resources of the NII. *Unlicensed* wireless technologies (see chapter 4)

---

[25] Satellite systems are technically capable of two-way broadband services, but the limited capacity of satellites has meant that such services were largely restricted to large business or government users who could pay for the equipment and satellite time. Such services have not been intended for general commercial (public) use. This may change with the advent of such satellite-based systems as Spaceway and Teledesic (see ch. 5). In combination with the existing telephone network, satellite systems also can deliver interactive services to the home, but with only limited return channel capabilities.

BOX 1-6: Scenario—Wireless Technologies in the Drug Trade

Mike is a major drug distributor in southeast Los Angeles. He moves thousands of kilograms of cocaine worth millions of dollars through his neighborhoods every year. He knows the Drug Enforcement Administration and the police are investigating him. Though Mike is 24 years old, he has never been caught, mostly, he thinks, because he is smart, protects himself, and stays ahead of the cops. He keeps his records (encrypted) on a laptop. He keeps in touch with his information and distribution networks through pagers and stolen and cloned cellular phones. Associates tell him what the cops and other dealers are up to through frequent calls, and he makes each call short so that even if they find him they won't be able to tap him as he moves from cell to cell.

Mike is always on the move in his car because he believes that this makes it harder for the cops to find him and listen in. He has his car searched daily for vehicle location devices, which he thinks might be planted by the police to keep tabs on his movements. He hears that new phone encryption devices cannot be broken even by the government, but he still needs to think about getting one. So far, it's easier just to clone a phone and change the number.

SOURCE: Office of Technology Assessment, 1995.

---

will allow users to create spontaneous, direct connections between their computers or PDAs—allowing them to share data or communicate in limited areas such as a classroom or office—all without connecting to a wired network.

Productivity and Efficiency

For businesses, the bottom line on wireless technology is its contribution to productivity. Although research on the productivity impacts of wireless communication technologies is limited and largely anecdotal, some analyses attribute large productivity gains to the use of wireless, mobile technologies. One analysis estimates the economic contribution of mobile services at five to eight times the cost of ownership.[26]

Another study assessed employees' ability to "recapture" time spent away from the office by using cellular telephones.[27] Table 1-3 shows the annual productivity gains for broad job categories, and assumes conservatively that at least 14 minutes or 10 percent of time away from the office per day is recaptured using cellular telephones.[28] If a sales representative recaptured 20 percent of time away from the office, the productivity gain

TABLE 1–3: Annual Productivity Gains Using Cellular Phones

Occupation	Annual productivity gains
President or chief executive officer	$2,200
Sales or other revenue-generating employee	1,200
Middle management/director/ supervisor	780
Field service person/technician	680

SOURCE: Pactel Cellular, "Cellular Use and Cost Management in Business," study prepared for Pactel Cellular by Yankelovich Partners, Newport Beach, CA, 1993.

---

[26] "Mobile Communications: Europe Lags Behind America," *Intug News*, October 1994, p. 20.

[27] PacTel Cellular, "Cellular Use and Cost Management in Business," study prepared for PacTel Cellular by Yankelovich Partners, Newport Beach, CA, 1993.

[28] Senior executives in the survey reported they were away from their offices 149 minutes per day, and that they used cellular telephones about 10 percent of this time. The study then calculated the annual productivity gain by multiplying time recaptured by the average wage rates for various job classifications.

would be substantially greater, about $3,540. The study also notes that increased accessibility and faster customer response time play an important role in decisions about providing cellular telephone service to employees.

These figures should be viewed with caution. Employees may do many productive things during the time they spend away from the office while not on the telephone. On the other hand, time spent talking on the phone is not necessarily productive. Alternatives to having a cellular telephone, such as using pay phones, are not addressed in the study. Calling the home office too often may reduce an employee's autonomy and incur increased coordination costs for the whole firm, and could reduce productivity overall. Variations in job structure and performance may occur as well; as a result, measuring recaptured time away from the office may not accurately describe the benefits and disadvantages of cellular phones. For example, in addition to improvements in productivity, the increased sense of company control over employees' activities is often a significant element in decisions to adopt wireless technologies.

Evaluating the contributions of wireless technologies to national productivity is even more difficult. Extrapolating from figures like those presented above to make estimates of national productivity enhancements is problematic because the job classifications given are too aggregated to know what they actually contain and how different groups actually use wireless telecommunications. As a result, attributing and separating direct and indirect contributions of wireless and mobile systems and services to gross domestic product are difficult.[29]

Likewise, the contribution of wireless telecommunications to employment levels is likely to be positive, but its magnitude is unclear. Rough estimates by the Cellular Telecommunications Industry Association based on the U.S. experience with cellular telephony suggest that the introduction of PCS and the extension of cellular telephony, SMR, and paging will result in the creation of 280,000 new jobs in these industries over the next decade and approximately 700,000 in related industries, such as manufacturing, retailing, and ancillary services.[30] Estimates of the contribution of wireless telecommunications to economic growth have not been made.

Uncertainty

Beyond the basic characteristics of demand, the fundamental question surrounding the evolution of the NII and new wireless services is: What do users really want? What will they be willing to pay for? Many companies have done marketing studies and some have conducted field trials to determine the answers to these questions. So far, no "killer applications" have emerged. In communications services, quality, reliability, coverage, and low price seems to be most important. In entertainment and data services, there is little consensus about consumer and business demands beyond, possibly, electronic mail. Interactive services have continued to disappoint both users and providers. Nevertheless, proponents point to the success of cellular telephony as evidence of widespread demand for wireless, especially mobile, products and continue to develop services and equipment based on the belief that eventually they will discover what customers really want.

The uncertainties of demand are some of the most important considerations underlying many of the NII policy debates now taking place. Specifying NII services, setting minimum service

---

[29] Estimates range from 2 to 3 percent of GDP to 33 percent, according to a study conducted by MITI (Japan) and reported in "Mobile Communications," op. cit., footnote 26.

[30] These are rough estimates based on proprietary information from firms in the industry, projections of wireless service subscriber rates, extrapolations from the growth and penetration rates of cellular telephony, and estimates of the ratio of direct jobs in cellular service provider companies to indirect jobs in manufacturing, retailing, etc. No effort was made to determine the number of jobs lost, if any, due to substitution of wireless communications for other communications services. These estimates should be considered very tentative; further research is needed.

"None of this seems to be doing me any good at all!"

standards, and defining universal service all hinge on an implicit understanding of what people want and need. Without this understanding, setting specific, long-term policies for NII services is likely to be premature. Because most policymakers and industry representatives believe it would be inappropriate for the federal government to pick technology "winners and losers," regulators also must avoid enacting policies that inadvertently have the same effect. At this early and uncertain stage of wireless development, putting constraints on the industry could stifle valuable development efforts. Open entry and competition—subject to some safeguards for basic consumer protection—may be the best solution, at least in the near term. As the market matures, new regulations and safeguards may be needed based on the experiences of the industry and the users.

POLICY ENVIRONMENT

Over the past several years, technology advances have fundamentally changed the nature of com-munication, information, and entertainment services and the industries associated with them. These changes have put increasing pressure on lawmakers to reform telecommunications regulation, a process in which they are now deeply engaged. The ideological concepts dominating the current public policy debate about telecommunications reform will significantly affect how wireless systems will fit into the NII, determining how and where they can compete with wireline carriers and what rights and responsibilities they will have. Considered together, these two elements, technology and ideology, constitute the policy context for wireless technologies in the NII.

■ Concepts Guiding Policy

Competition

The many networks and systems that make up the current U.S. communications and information infrastructure are widely deployed and access to services is usually physically available even if the services are not taken. Thus, as the development and deployment of the NII moves forward, the challenge for policymakers is to ensure that the benefits of the new services and applications will be available to all those who need or want them, including those who cannot access them now. To accomplish these goals, most policymakers have come to view competition in an open and deregulated (as far as possible) market as the most socially and economically efficient solution for promoting diversity in information sources, keeping quality high and innovation moving, and controlling prices.

However, the form (or level) of competition is being bitterly disputed—what is a "level playing field?" And perhaps more importantly, there seems to be little public policy consideration of the long-term effects of competition and market reliance.[31] There is little doubt that private companies and their consultants have done such analyses, but the information is closely guarded and is not generally available to policymakers and analysts. As a consequence, it is impossible to judge the extent to which these analyses consider long-term (10 years and longer) effects. Policies that promote competition now may ultimately lead to a market structure that consists of a small number of large corporations controlling end-to-end communications of all kinds. Again, little or no research has been done that bears directly on wireless economics and long-term industry structure.[32]

It is also unclear whether a one-shot approach to changing regulatory structures will work. The history of cable television regulation reflects the need to adapt rules to the different stages of industry growth and external (competitive) conditions. What is clear is that industries and technologies

are changing rapidly, suggesting that any new laws/regulations will need to be similarly flexible and allowed to evolve over time. Expectations that a new "Communications Act" can be written that will last another 60 years—as the current one has—may be unrealistic, given the pace of technological, social, and economic change.

Finally, many analysts and public interest groups are concerned that social goals and needs may get lost in a competition-driven policy framework. What safeguards might be needed to promote continued diversity of services and protect consumers from high prices or poor quality? Does the imposition of universal service requirements on wireless businesses threaten their ability to operate? Some analysts believe that complete reliance on the market and competition—where economic and business decisions are paramount—could lead to a situation in which services will not be rolled out evenly, users will not be protected from poor service or confusing service plans, and that service will be available only to those who can afford it. On the other hand, overly aggressive requirements by the federal and state governments could threaten the vitality and even the existence of new competitive services. The private sector's research, development, and investment activities could be stifled if the federal, state, or local governments adopt rules and regulations that inhibit the flexibility to develop new products and services.

Competition, Diversity, and Interconnectivity

Diversity and competition are closely related. Competition is premised on many diverse companies producing goods and/or services. In the current technological climate, the wide range of new services being developed is largely due to the

---

[31] See, for example, Eli Noam, "From the Network of Networks to the System of Systems," *Regulation*, No. 2, 1993. Some commenters argue this statement is too strong. In its policies regarding PCS licensing, the FCC did set limits on cellular participation to address concerns about competition and industry concentration.

[32] Some initial research has been completed. See, for example, Bruce L. Egan, "Economics of Wireless Communications Systems in the National Information Infrastructure (NII)," unpublished contractor report prepared for the Office of Technology Assessment, U.S. Congress, Washington, DC, November 1994; Glenn A. Woroch, "The Evolving Structure of the U.S. Wireless Communications Industry," unpublished contractor report prepared for the Office of Technology Assessment, U.S. Congress, Washington, DC, December 1994.

introduction of new digital technologies. These technologies make it easier to produce, transmit, and store information and data, allowing businesses to combine voice, data, and video on networks that were previously dedicated to only one type of service. This digital convergence has made it easier for companies to invade each other's turf. So, for example, telephone companies want to use their telephone lines to deliver information and video services, and cable companies want to provide telephone service. The rapid development of new technologies has also led new companies to enter the field—utility companies, for example, want to enter the telecommunications business—further increasing the diversity of companies involved. Wireless companies are an increasingly important part of this competitive mix.

The diversity of service providers and users reinforces the importance of connectivity and interconnection. The more different sources of information and entertainment there are, and the more users follow their individual tastes, the greater the need for interconnectivity. In the history of telephony, this is referred to as the externality of networks—the value of the network (to an individual or business) rises as more and more users are connected to it. Today, the value of interconnectivity is higher than ever. Computers, for example, started out primarily as stand-alone devices, but increasingly they are part of networks that allow them to access almost any type of information around the world. Allowing users to access rich and diverse forms of information from a variety of suppliers is at the heart of most views of the NII, and moves issues surrounding interconnection to the forefront of many current policy debates.

■ Technology's Impact on the Policy Environment

Technology and "Convergence"

As noted above, the U.S. communications infrastructure consists of many different technologies and systems. Over the past 100 years, each of these developed independently from the others,

and different regulatory structures were developed to manage the distinct characteristics of each industry. For example, the telephone system, which was designed to provide two-way voice communications, has operated as a virtual monopoly for almost a century based on the principles of common carriage and universal service. The broadcasting industry, both radio and television, has concentrated on delivering one-way information and entertainment to a wide local audience, and has been regulated as a user of a scarce resource, the radiofrequency spectrum. Satellites have been providing national and international connections for voice, data, and television signals for many years. And finally, the cellular industry uses radio waves to extend the reach of the telephone network to mobile users. Each of these industries provided a different service based on different technologies, and consequently was subjected to different rules and regulations.

As a result of digitization and the increasing use of computer processing power in more and more telecommunications applications, however, systems and services that once were separate have now begun to overlap. This *convergence* is not merely the result of combining computers with communications, but of combining many services and applications that historically had been separate. Thus, convergence can be separated into three distinct phenomena:

1. *convergence of technology*, where computer power and communications technologies are integrated to improve functionality and offer new applications. For example, the marriage of computer power to radio technology was crucial in enabling cellular radio to be developed. Computers route calls to the correct cells and handle hand-offs as mobile users move from one cell to another. The networks that allow cellular users to *roam* are actually interconnected computer databases.

2. *convergence of applications*, where voice, data, and video services can be offered over the same network. Today, networks of all kinds—whether originally developed to transmit voice, data, or video—are being improved in order to

carry all kinds of information in many different combinations. By far the most common convergence is between voice and data services. The most obvious example is the use of the telephone network to send data by fax, using a modem, or via new digital transmission technologies such as integrated services digital network (ISDN). Cellular service providers have begun to offer a wider range of data services (see chapter 4), and some of the new LEO systems are designed to carry both voice and data. Because of its high bandwidth requirements, video is less often combined with other services; however, some cable companies offer data services, and satellites are capable of transmitting voice, data, and/or video signals.

3. *convergence of networks and companies* through mergers, acquisitions, and joint ventures. The most obvious example is the recent acquisition of McCaw Cellular by AT&T. This type of convergence is not between similar technologies and providers— a merger of cellular companies, for example—but a combination of systems: AT&T's long-distance (mostly fiberoptic) network with McCaw's local (cellular) systems. In addition to the economic rationale behind the merger—AT&T's desire to avoid paying access charges to local telephone companies—this type of merger indicates that there may be technical and economic efficiencies that make previously distinct communication systems interdependent.

Wireless technologies and companies are playing a central role in much of the convergence activity. In the past several years, wireline service providers of all types—cable, local telephone, and long-distance companies—have shown increasing interest in using wireless technologies to provide new services. The big winners in the recent PCS auctions, for example, were various groups of telephone and cable companies. Wireline companies are also investing in many different kinds of wireless companies and technologies. The Primestar DBS system is owned by a consortium of cable companies, and several telephone companies recently announced large investments in wireless cable companies. Most analysts expect such mergers to continue as the benefits of wireless become more apparent.

Convergence and Policy

The convergence of technologies and services has serious implications for U.S. policymakers at all levels of government who are already engaged in efforts to redefine how telecommunications is treated in this country. As the technological differences that have characterized different modes of communication disappear, new regulations and policies will be needed that are focused more on services and industry/market structure than on technology. This idea was explicitly recognized by Congress when it created CMRS based on the principle of "treating like services alike." Federal and state governments continue to struggle with how to update regulations in order to bring the benefits of new technologies to the widest range of people, while simultaneously promoting fair and open competition among the many different companies that want to provide services. Economic concerns are becoming more important as various segments of the wireless industry mature. Mergers and acquisitions have been going on for several years in the cellular, SMR, and paging industries, and horizontal concentration has already become a concern to regulators.[33] Over the longer term, the effects of market concentration and vertical integration of the sort promised by AT&T/McCaw are uncertain; economists have just begun to sort out the economics of wireless services and how they may interact with wireline services. Trying to anticipate the long-term competitive effects of current deregulatory policies will be difficult.

---

[33] Nextel, for example, was required by the Department of Justice to divest some of its radio licenses in specific cities before it completed acquisitions of its major competitors. This was due to a concern that Nextel would control too much of the SMR market.

FINDINGS AND POLICY OPTIONS

Overall, OTA found that—apart from current regulatory reform efforts—federal government action or assistance is currently needed in a relatively few, but important, areas regarding wireless technologies and their effective integration into the NII. Several factors led to this conclusion. First, the outcomes of current policy initiatives are unclear. The FCC is in the middle of a number of critical proceedings regarding wireless technologies, and Congress is in the midst of completely reshaping the nation's telecommunication industry. Before proceeding with even more far-reaching changes, it may be wise to evaluate the effectiveness of changes already put in motion.

Policy analysis is complicated by the dynamic nature of the industry itself. The structure of most segments of the overall wireless industry is about to change in fundamental and radical ways. Some services are only at a nascent stage. Services such as DBS, for example, have only just begun operating. For others, such as PCS or LEOs, initial regulations have been set, but the systems are still being built and are not yet operational. A final group of services, including Local Multipoint Distribution Service (LMDS) and some satellite services, does not even have final spectrum allocations or operating rules; widespread commercial service is years away. In addition, even the wireless services that have been in existence for many years—radio and television broadcasting, cellular telephone, and satellite services—are facing radically new environments as digital technology and new competitors reshape their traditional ways of doing business. This fact—along with the uncertainties associated with technology development, the regulatory climate, and, most importantly, customer demand for wireless services—puts policymakers and analysts into the difficult position of waiting to see how consumers and markets will react to what has been done so far. Policies designed and implemented based on past assumptions and models of industry structure—monopoly-based or limited competition—are likely to be inadequate to address future

models in which the structure will be quite fluid and unpredictable.

The second factor indicating a limited government role is the large amount of innovation and development now occurring in the wireless industry without benefit of direct government support. Over the last several years, hundreds of companies have begun developing wireless products and services, and most large telecommunications firms have initiated wireless projects as well. Few areas appear to need government financial assistance to develop new technologies or services—with some important exceptions noted below. This represents a change from several years ago, when financial markets were not eager to invest in wireless companies because of their often speculative nature and regulatory uncertainties. Money is now flowing to most segments of the industry, and, in fact, a number of analysts have commented that "wireless is hot" on Wall Street.

Finally, a political commitment to competition is the foundation of current economic and regulatory policy. Many policymakers view competition as a more effective "regulator" of industry than the government rules of the past, and are reluctant to put additional regulatory burdens, however well motivated, on industry. This approach, however, limits government involvement, and the development of the wireless industry needs to be closely monitored to ensure that the public interest is served.

Given these circumstances, determining the appropriate role and level of involvement of the federal government in the wireless industry is difficult. A strong government role could help promote industry growth, encourage diversity and innovation, and protect consumers. Low prices, quality, and security all are important concerns that may or may not be ensured by the market. Forward-looking policy also could anticipate and help diminish any potential future problems. However, a government approach that is too strong could overburden industry and reduce investment.

On the other hand, an approach that is too "hands off," relying too much on private sector initiatives, could actually contribute to uncertainties (in this case primarily regulatory) that slow innovation and development. In the short term, companies may not invest if uncertainties are too great or development resources could be wasted on efforts that are later superseded by new technologies, regulations, or economic conditions. Benefits may take longer to appear. Given that the market has not even begun to operate in significant portions of the wireless industry, it is premature to identify market failures that could indicate policy problems. **In the future, government intervention—through changes in regulation or other incentives—may be needed if market failures develop.**

Despite the uncertainties, it is possible to indicate some specific social and public interest needs that competition and the market are not likely to address effectively, and for which some form of government intervention will be needed. Spectrum management, for example, is one important area requiring government action. Because public uses of the spectrum—public safety, national defense, amateur radio, and education, for example—are not subject to auction provisions (and do not operate as commercial or fee-producing services), there will continue to be an important federal government role in managing the spectrum to accommodate the largest number of services and users while avoiding interference and congestion.

Wireless technologies can also contribute to the achievement of other social and public policy goals where the market may not provide adequate incentives. Two specific examples are: 1) education, which may not have the resources to take advantage of wireless technologies where appropriate, and 2) underserved users and areas—the so-called "information poor," people whose economic status or remote locations may cause them to be underserved by profit-maximizing firms. Proposed legislation now under consider-

ation by Congress addresses the need for connectivity through universal service requirements and provisions for educational institutions.[34] These issues are discussed throughout the report.

Given this environment, the federal government can perform three important functions over the next several years:

- monitor the growth of the industry and competition, and identify any potential market failures or social concerns that arise;
- continue to pursue policies that promote open access to all networks, including goal-setting and encouragement of industry standardization efforts; and
- promote development of new technologies, including ensuring the availability of adequate spectrum for existing and emerging wireless technologies.

The following sections discuss OTA's specific findings and identify several areas of interest and concern for policymakers. Although not every issue requires a policy response, the discussion will provide policymakers with a context for their deliberations and identify possible options for consideration as NII development advances.

Schools can use wireless technologies such as satellites to connect students and teachers to educational resources and peers around the world.

---

[34] See, for example, U.S. Congress, Senate 652, "The Telecommunications Competition and Deregulation Act of 1995," (Washington, DC: U.S. Government Printing Office, June 15, 1995) sections 103, 104, and 310.

❚ Uncertainty Pervades Wireless Technology Diffusion

Rapid technology advances, unfocused user needs, regulatory reorganization, and the nascent state of the wireless industry all combine to make predicting the future of wireless technologies and services highly speculative.[35] These same uncertainties make long-term social and economic implications even more difficult to forecast. **In the case of wireless and the NII, the level of uncertainty is much higher and more pervasive than usual; all aspects of the wireless industry—technologies, markets, and rules—are changing almost constantly.** Defining social and public policy goals in such an environment becomes quite a challenge as the current telecommunication debate in Congress attests. Consumer advocates believe that legislation currently proposed—S. 652 and H.R. 1555—will lead to concentration in various communications and media industries that will reduce diversity and raise prices. Industry proponents and many lawmakers, however, believe that allowing companies to compete and merge will produce lower prices and a wider range of programming. At this time, there is no way to determine conclusively what will happen. The issues will only become clearer once final legislation is passed and companies and consumers begin to react. Many issues are actively being addressed, but many more—some of the most difficult ones involving social and public policy—remain to be identified and resolved.

The uncertainty of demand is particularly important for legislators and policymakers charged with the task of defining rules to regulate various competing services. Overestimating demand for new services, and making such a judgment part of a definition of universal service, could subject companies to higher costs for upgrades or system construction that may not be recoverable through revenues. In addition, the technological limita-

tions of some wireless systems may mean that they cannot—using today's technology—deliver some of the most advanced services, potentially disqualifying them from receiving universal service funding. Conversely, underestimating demand and matching policies to lower expectations may lead to inequities as companies roll out advanced services only to certain users—based on where they live and what they can pay. This could widen the gap between information "haves" and "have nots."

Uncertainty is not unusual in the development and deployment of new technologies, nor is it necessarily a bad thing. Some uncertainty is always involved in developing and marketing new products as manufacturers and service providers struggle to discover what works, what customers will buy, and what they will not. Uncertainty is characteristic of the early stages of innovation as different approaches are tried to solve problems and meet ill-defined demand.

❚ Wireless Technologies Extend and Compete in the NII

Wireless technologies will serve two critical functions as the NII develops: radio-based technologies will *extend* the reach of the NII to places that wire-based technologies do not reach, and wireless systems will provide valuable *competition* to emerging NII service providers. These two functions are not mutually exclusive; in many cases, the same system will provide both. DBS systems, for example, compete locally with cable television suppliers, but they also provide services almost anywhere in the country to those who cannot get cable. Mobile (cellular and PCS) telephone systems extend the NII by providing communications services to people on the move, but are also expected to compete in the provision of telephone service to homes and businesses in the future.

---

[35] This uncertainty is not limited just to wireless. Many aspects of the NII, such as the future of interactive and multimedia services, are similarly unclear. "Demand for Interactive, Multimedia Services Is Unclear..." *Telecommunications Reports*, vol. 60, No. 48, Nov. 28, 1994.

Wireless technologies can *extend* the NII in two important ways. First, they allow users to tap into communication and information networks as they move about. Mobility is a key driver for wireless (see chapter 2). Second, as noted earlier, wireless technologies can extend NII services to places where wire is too costly or difficult to install. This may prove to be especially important as links need upgrading. In this role, wireless systems will help ensure that future universal service goals are met (see chapter 9).

Wireless technologies and systems will also *compete* in the delivery of NII-related services, both among themselves and against wire-based services. Competition is a key principle underlying the NII, and different wireless services have advantages that will allow them to compete effectively in a number of markets. For example, Broadcast, DBS, and Multichannel Multipoint Distribution Service (MMDS), already compete with cable television systems (and each other) across the country, and competition is expected to increase as companies convert to digital and new competitors enter the market for video services. Wireless technologies are also expected to make a substantial impact in the market for voice and data communications, especially where mobility is desired. A good deal of spectrum has recently been allocated for wireless voice and data services and companies have been working on systems for a number of years. Many analysts believe that wireless could become the voice communications technology of choice for many people—eventually becoming a substitute for existing telephone service—because it offers the added advantage of mobility.[36] Over the next five to 10 years, wireless technologies will emerge as significant competi-

tors in most communication, information, and entertainment markets.

The ultimate outcome of a more wide-open competitive marketplace—which technologies and companies will "win" and which will "lose" and what the structure of the various industries will be—cannot be determined at this time. The uncertainties that pervade the development and implementation of wireless technologies, including rapid changes in technology, unfocused consumer/business demand, and regulatory upheaval, all combine to make analysis exceedingly difficult.

The one area in which wireless is not expected to become a significant competitor in the near future is in the provision *to the general public* of two-way, broadband, multimedia communications, including integrated voice, data, and video services.[37] These are the types of high-end applications often discussed as the ultimate objective of NII policymaking and technology development. Wireless technologies are technically capable of providing such services and there is nothing that inherently prevents it, but most existing systems are limited based on past technical and regulatory choices. Two-way voice and data systems, for example, operate with a limited amount of spectrum that was originally allocated before high-bandwidth applications were widely accepted. As a result, most of them cannot be economically upgraded to provide two-way broadband services including multimedia, video telephony, or any other applications requiring high-speed connections.[38] Broadcast and satellites services potentially have enough spectrum, but generally only work one-way—to the consumer. Some of these systems have limited inter-

---

[36] Egan, op. cit., footnote 32.

[37] This discussion is based on Egan, ibid.

[38] Some current and planned systems provide integrated broadband services, but their limited capacities will limit them primarily to business or high-end users in the near term. For example, a few systems currently provide such services, notably satellite systems based on very small aperture terminal (VSAT) technology. However, these systems are not designed for the mass market, and current system capacities could not support a consumer/mass market type of service that would accommodate millions of individual users.

active capabilities—provided either with a small return radio channel or telephone lines—that may make them competitive with wire-based systems and could serve important market demand.

In general, however, to upgrade existing systems for interactive high-bandwidth services, either new spectrum will have to be allocated or new compression techniques developed, or both. New wireless systems that could provide these "bandwidth-on-demand" services on a mass-market level are now being conceived, but are not expected to be available in the near term (see chapter 5). As a result:

> [u]nless there is radical, and, as yet, unanticipated, advances in both wireless access technology and the FCC's spectrum allocations, the future vision of integrated broadband access offering end-user bandwidth-on-demand type service will likely be reserved to the province of wireline technology.[39]

■ Universal Service Definitions Could Disadvantage Wireless Systems

The debate over the future of universal service—what it should include, how much it should cost, how it will be paid for—exposes some of the most difficult questions facing NII policymakers, private sector developers, and public interest groups. Many analysts and consumer advocates strongly believe that interactive, broadband services should be a key component of any future definition of universal service. They maintain that such communications capabilities will be necessary if Americans of all socioeconomic levels are to participate in the social, economic, and political life of the country. However, if such a definition were immediately adopted, there is a potential for over-building the NII based on projected needs (broadband and interactive) that the majority of users currently do not have, and likely will not have for many years.

Depending on how universal service and the NII are defined in the short term—what functions and conditions policymakers impose— and how new requirements are implemented, wireless technologies could become an integral part of the NII or be seriously disadvantaged. The outcome of current universal service debates will affect the role wireless technologies and services can play in the NII in several important ways. According to one researcher:

> The important message for public policy is that, until the service requirements of the universal NII have been specified, the question as to which is preferred, wireline or wireless access service, cannot be answered. If, as many believe, the NII only contemplates socially efficient access to narrowband digital voice and data services, then digital wireless technology is preferred for dedicated subscriber connections to the wireline intercity PSTN. The fact that wireless access costs are lower notwithstanding, the real bonus for the consuming public from this scenario is portability.
>
> If, however, access to broadband service, especially bandwidth-on-demand type access service, must be added to the narrowband service mix for the NII, then wireline access technology is likely to be the winner in the race for preeminence in the NII.[40]

Wireless technologies offer several advantages over wire-based telecommunications systems, but wireline systems also have advantages in delivering some services. On an economic basis, the ability of wireless systems to deliver narrowband voice and data and one-way video (broadcast) services puts them at least on par with wire-based systems, and, in fact, will likely allow wireless to compete directly with wireline in the future.[41] In the delivery of two-way broadband data, video, or multimedia applications, however, wire-based media are still the most cost-effective. In large part, this is a result of the amount of spectrum that

---

[39] Egan, op. cit., footnote 32, pp. 11-12.

[40] Ibid.

[41] Ibid.

has been allocated to radio services historically, the technical limitations of wireless systems, and the phenomenal advances in fiberoptic and digital technologies. Current technical and regulatory constraints simply do not allow two-way broadband wireless services to compete with wire-based systems in the general consumer market. Thus, *at this time,* **a minimum definition of universal service in the NII as interactive, two-way, and broadband could disqualify wireless systems where they would otherwise be most appropriate or efficient.**

In addition, if universal service expectations and definitions are set too high, simpler, lower cost solutions that might profitably stand by themselves may be lost. As a result, those businesses and consumers who have more basic needs could be forced to pay for more than they really want. It is not clear that all information and all communications need to be broadband, interactive, and/or multimedia—particularly in the presence of "cheaper non-integrated alternatives."[42] Some users may not want or need these advanced features. From an economic standpoint, mandating such a high level of service begs economic efficiency questions. Why should companies be forced to build to such a standard? Will customers have to pay for a level of service they do not need and may not use? Do the potential benefits justify the expense?·

In the long term, there can be little doubt that advanced interactive broadband services will play a critical role in the NII, and probably will eventually be included as elements of a future definition of universal service. In the near- and medium-term (five to seven years), however, OTA believes that interactive, broadband capabilities are not likely to be needed by the majority of citizens and should be allowed to evolve as demand warrants. A flex-

ible approach to NII universal service policy would, allow the different parts of the NII—interconnected or not—to grow to meet varying levels of need, while simultaneously ensuring a smoother upgrade path. In fact, many policymakers favor defining universal service in an evolutionary fashion, updating it as services become more ubiquitous and necessary.[43]

Aside from these broader issues, the definition and implementation of new universal service requirements could have a substantial impact on wireless systems and services. The potentially lower cost structures of both terrestrial and satellite-based (and combinations of the two) wireless systems make them an efficient alternative to wire-based media for reaching unserved users in both rural and urban settings (see chapter 3).[44] The current move to deregulate pricing may encourage wireless alternatives because of the increasing emphasis on least-cost technology options, which allow a company to cut its costs through use of more efficient technologies and lower its prices to compete more effectively. However, current subsidy flows and rate-of-return regulations may actually serve as a disincentive to wireless technologies. In addition, "essential telecommunications" (carrier of last resort) obligations, which have been proposed to bring service to areas where no carrier is operating, could harm wireless start-ups that are unable to meet the requirements and, therefore, could not qualify for universal service funds.[45] A much closer examination of these issues is necessary.

Options

Congress has proposed legislation directing the FCC, in consultation with the states, to develop a new (evolving) definition of universal service. NTIA has held several hearings on universal ser-

---

[42] Ibid.

[43] See, for example, S. 652, op. cit., footnote 34, sec. 103.

[44] Ibid.

[45] S. 652 would designate carriers as "essential telecommunications carriers" in specific service areas for purposes of providing universal service. Wireless companies are eligible for this designation. Ibid.

vice. Nothing has yet been decided. In order to protect business and enhance access to NII services for *all* Americans, Congress could:

- Enact proposed legislation directing the FCC and the states to work out a definition of universal service and enforce deadlines for this effort. Before such definition(s) are put in place, Congress may first wish to consider the business impacts and prospects for providing service to the unserved.

- review proposed legislation to ensure that it is fair and competitively neutral. The structure, funding levels, and participation in a new universal service fund will need to be carefully considered to ensure that startup and/or wireless carriers are not unfairly disadvantaged.

- develop its own policies or guidelines for NII development based on hearings held to determine: 1) what services should be available, and 2) what technical capabilities are needed to enable these services to develop. Alternatively, Congress could establish a working group or outside commission to develop recommendations.

■ Interconnection and Standards are Increasingly Important

As a consequence of the boom in wireless technologies and systems, the importance and complexity of interconnection arrangements, standards, and interoperability are about to grow dramatically.

As the National Information Infrastructure develops, policymakers must recognize the importance of wireless access to information and communications services because wireless may become "the first mile on and the last mile off" the information superhighway. Interconnectivity and interoperability are important determi-

nants of whether a product or service can be offered in such an environment. The adoption of standards that make it difficult for wireless technologies to connect with the superhighway will be detrimental to the now well-documented consumer demand for mobile, wireless service.[46]

In thinking about the NII and wireless technologies' role in it, it is important to carefully define some of the major assumptions that underlie the vision. It will be necessary to specify exactly what the "network of networks" means and what implications it has for policies regarding interconnection. For example, the notion of the NII as a seamlessly integrated network of networks is at best simplistic and at worst misleading. The NII initiative does not call for all networks to be directly connected to each other, which would be virtually impossible. Some companies and networks will connect directly, based on business needs. In many cases, however, different networks will interconnect *indirectly* through separate links to existing core networks—the public switched telephone network (PSTN), cable networks, and computer networks—and direct interconnection will not be necessary to enable different systems to interoperate (figure 1-1). The interconnection policies now being debated in Congress and at the FCC are vital in allowing all service providers to connect to other networks (see chapters 6 and 7). Determining which companies should be required to open their networks to interconnection by other carriers is already a hotly contested issue. As new wireless networks and services are deployed and usage increases, more direct interconnections may occur.[47]

Until recently, very few systems, services, or companies connected at all. Cellular telephony is the most visible exception. Over the next several years, however, as a multitude of PCS, ESMR,

---

[46] Center for Wireless Telecommunications, Virginia Polytechnic Institute and State University, "A Survey of Emerging Applications of Wireless Technology," unpublished contractor report prepared for the Office of Technology Assessment, Sept. 15, 1994, p. 4.

[47] For more discussion of changing interconnection arrangements and their implications, see Rob Frieden, "Universal Personal Communications in the New Telecommunications World Order," *Telecommunications Policy,* vol. 19, No. 1, January/February 1995, pp. 43-49.

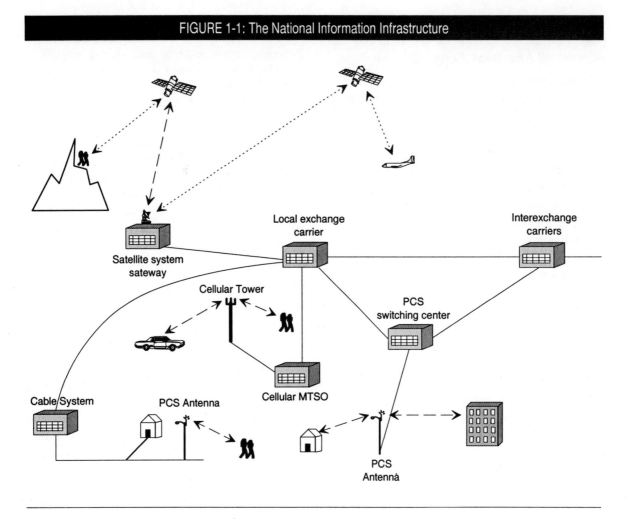

FIGURE 1-1: The National Information Infrastructure

SOURCE: Office of Technology Assessment, 1995.

and satellite communication providers begin offering services, interconnection issues will become critical. In the past, wireless systems have been conceived primarily as adjuncts to the PSTN, and wireless technologies were employed only in special (mobile) circumstances.

In the future, wireless systems and technologies will become an integral part of the overall communications infrastructure, providing not only mobile communications and broadcasting, but a wide range of mobile and fixed services for both businesses and consumers. Interconnection and interoperability arrangements premised on older, asymmetrical relations—where cellular companies pay access charges to local telephone

companies, but not vice versa—will give way to technical and contractual arrangements based on treating wireless carriers as equals. A number of factors will impact the ability of wireless companies to interconnect with the PSTN (and other wireline systems, such as cable television or computer networks), including the different cost structures of radio-based services, rising consumer demand for wireless services, increasing business demands for more integrated communications solutions, and technical advances that may help or hinder greater interconnection and interoperability.

Interconnection and interoperability are widely viewed as the keys to realizing the vision of the

NII—allowing users to easily send voice/data/video across many different types of networks. Today, system interconnection is usually accomplished through the PSTN for voice, and increasingly through the Internet for data. New systems and services are already putting a strain on this arrangement. Many wireline systems, especially data communication systems, operate according to protocols that often do not work well for wireless communication—which is affected by a number of factors not present in wireline systems, including interference from other radio services and propagation losses from rain or even trees. In addition, as new companies have entered the field, the number of proprietary applications and standards has grown. For individual users, sending information across different networks can be difficult, and using software on different systems can be almost impossible. It is unclear what will happen when additional services are developed and different kinds of companies begin to link up. **Developing new standards that accommodate the needs of wireless technologies and that operate across multiple systems will be critical to ensure that the benefits of an interconnected NII are realized. Most analysts expect that technical solutions will be developed, but OTA believes it will take longer than expected to work out many interoperability issues.**

Standards are the critical link that will allow different parts of the NII to work together. One kind of standard describes the connection between consumer devices—radios, televisions, and cellular telephones—and the networks that provide services. These standards benefit consumers by ensuring that their devices will work across different companies' networks. They also enable manufacturers to build one device rather than many different types of equipment for many incompatible systems. Standards also make it easier for the industry to plan and deploy upgrades, allowing consumers and businesses to revise, customize, and improve their systems as their needs dictate.

Other standards govern the connections between networks. While general rules are now well known, a whole range of new companies and in-

terconnection agreements will have to be addressed in the near future, putting pressure on existing interconnection arrangements. For example, the rules that govern the transmission and reception of digital video services are only beginning to be considered. Multiple standards are being developed for transmission of video services in the broadcast, cable, and satellite industries, and there are a number of complex issues, and a range of vested interests, that will have to be addressed before such services are widely available and interoperable. The economic consequences of these decisions are enormous, and will have a vital affect on the broadcast and consumer electronics industries.

A lack of standards, or the proliferation of multiple standards, may undermine the NII goal of interconnectivity. For example, the current analog cellular telephone standard specifies how a cellular phone can "talk" to the cellular network. The fact that the United States settled on one standard for analog cellular telephones many years ago ensures that any phone will work with any cellular network. Today, however, two digital cellular standards are being deployed and up to seven standards are being considered for PCS systems. As a result, it is likely that all phones will not work with all networks (see chapter 6).

The current situation is different from the past because the process of setting standards has become very difficult. Historical standards-setting processes have undergone tremendous change since the breakup of AT&T in 1984. The FCC has largely backed away from aggressive standards-setting, preferring to let industry and/or the marketplace set standards; however, the intense competition that is expected to characterize NII services puts the process of cooperative standards-setting in question. The FCC approach to HDTV is an exception to current practice (see chapter 5). The federal government could play a stronger role in setting standards for interconnection and interoperability, but it is unclear what that role should be. Individual circumstances call for different government responses—there is no well-defined set of procedures that will work in all cases. Some companies prefer a "hands-off" approach by government, while others would like the government to at least set goals or even deadlines for standards. This idiosyncratic, flexible approach to standards-setting is likely to continue.

Options

The tension between fair competition and NII goals for a widely interconnected series of networks is felt most acutely in relation to interconnection, standards, and deregulation issues. The FCC has established a number of different wireless license areas, which do not necessarily coincide, and that do not match the boundaries and regulations set up to govern local and long-distance communication services. To ensure the benefits of NII interconnections, while preserving competitive incentives, Congress could:

- review the regulatory and structural underpinnings of the long-distance industry. Possible congressional actions include: 1) eliminating the Local Access and Transport Area (LATA) boundaries that currently define long-distance service, and/or 2) harmonizing CMRS license areas. These options are not mutually exclusive, and could be pursued as part of a larger redefinition of local/long-distance communications.
- establish guidelines to direct the FCC's standards-setting activities or mandate the FCC to do so. Guidelines could help the FCC determine when to get involved in standards-setting and what its actions should be. In this way, the benefits of early standards-setting could be combined with the flexibility of industry or market-based solutions.
- explicitly allow the FCC greater latitude in preempting state regulations that may slow wireless startup interconnection to the public network and each other. Potential areas for congressional action include: 1) establishment of co-carrier status, rights and obligations; 2) mutual compensation for competing local communications companies; and 3) consistent interconnection arrangements ensured through tariffs or publicly-filed contracts.

■ Integration of Wireless and NII Policymaking is Improving, But...

Integrating NII and wireless communications has been and will continue to be a challenge. Early thinking and policy development regarding the NII focused primarily on wire-based technologies, especially fiberoptic networks. Policies for wireless technologies and systems, meanwhile, developed largely independently of NII initiatives. There has been little formal coordination between government NII efforts and wireless efforts—the two have proceeded along parallel, but seemingly separate, tracks. As a result, many of the issues surrounding wireless technologies, especially broadcasting and satellites, were delayed until long after NII planning efforts got under way, and **no comprehensive vision exists for integrating the wide range of wireless technologies into the NII.**

Wireless technologies were only lightly treated in early legislative and executive branch NII planning. The Administration's *Agenda for Action*, for example, mentions wireless technologies in its nine principles; however, the treatment of wireless is limited, concentrating on spectrum reallocation, use of market principles in assigning spectrum (auctions), and ensuring that small, rural, minority- and women-owned businesses can participate in the auctions—all concepts proposed or required by Congress in previous legislation. One specific effort to combine NII policy development with a wireless focus, the Untethered Networking Group, met with no success (see appendix B).

Several factors contributed to this situation. First, no common vision exists for the development and implementation of radio services in the United States. Wireless policy development is divided between the FCC, which manages private sector and state/local government spectrum use, and NTIA, which manages the spectrum needs of the federal government. This division of responsibilities historically has hampered the development of a clearly defined, comprehensive framework to guide U.S. radiocommunication policy development (see next section). The lack of a unified vision for wireless makes it correspondingly difficult to develop a more comprehensive strategy for the integrating wireless systems into the NII. At a practical level, wireless policy development has been more successful. See appendix B for a discussion of the efforts of the Federal Wireless User's Forum and The Federal Wireless Policy Committee.

The second factor making the integration of NII and wireless policies difficult is that policymaking regarding radio technologies and services has *historically* been separate from wireline policymaking. Radio and television broadcasting networks, amateur radio, and even early satellite systems were developed and operated largely as stand-alone systems, capable of communicating information separately from the wireline networks—there was little need to coordinate wireless and wireline policies.

In addition, the philosophy underlying radiocommunications policy was substantially different from the models applied to wireline services. Unlike the tightly controlled, monopoly-based regulation that characterized the telephone system, wireless systems of all kinds have been much less closely regulated on an economic basis. Companies have been able to set rates, merge, and compete much more freely than most wireline companies. Today, the federal government continues—as part of this long-standing practice—to let market forces play the primary role in deciding how radio frequencies should be used. As wireless technologies become a more integral part of the NII, however, a purely market-based approach to wireless policymaking may prove inadequate. As wireless and wireline systems increasingly connect and the services they offer overlap, the need for integrated policymaking will correspondingly increase.

Finally, the separation of wireless and wireline policymaking is a matter of timing and historical accident. The issues of cable/telephone competition have occupied center stage of the telecommunications debate in this country for almost a decade. It is, therefore, no surprise that the NII has

centered around these industries. Additionally, some wireless supporters charge, policymakers were slow to recognize the potential of wireless systems. Others concede that some wireless industries entered the policy development process late, and their potential contributions were not recognized by government officials.

In the latter part of 1994, however, wireless technologies began to receive more attention as an important, even integral, part of the NII. Officials at the FCC, for example, now refer to wireless as "one lane on the information superhighway." NTIA is in the process of reallocating at least 200 MHz of spectrum as mandated by the Omnibus Budget Reconciliation Act of 1993, and has recently completed a study of the nation's radio spectrum needs for the future. The FCC has proceedings under way in many areas of wireless communication, many of which overlap. It is still unclear, however, how all these initiatives will contribute to the establishment of an interconnected NII.

Options
To maximize the benefits of the NII and minimize inefficiencies and potential adverse effects, wireline and NII policymaking must explicitly recognize and address the unique capabilities and limitations of wireless technologies. Wireless and wireline policymaking need to be more closely coordinated by establishing goals for wireless technologies in the context of the NII, and needs must be prioritized. This does not imply that all NII and telecommunications-related planning—each individual decision—should be centralized and bound together in one master plan. It only suggests that a focused vision of the future could help guide private sector development and implementation efforts. To bring wireless technologies and policy development more directly into the mainstream of NII policymaking, Congress could:

- direct the FCC and NTIA to develop policies and plans—or justify/amend existing plans—for integrating the wide range of wireless systems into the NII. Specific plans could be developed for specific industry segments.

- hold hearings to determine if NII policies or FCC rules currently discourage wireless systems from playing a larger role in NII development. Hearings could also help determine how wireless technologies could more directly contribute to the goals of the NII—universal service, for example.
- mandate more direct coordination of NII and wireless policy development, both within the executive branch and with the FCC. Reporting requirements could be established.

■ Spectrum Policymaking Faces Significant Challenges

Government policymakers and regulators will face an increasingly difficult task in meeting expanding spectrum needs while accommodating existing users. No coordinated framework for making spectrum policy exists, although some long-range planning is taking place.

Technology advances and increasing demand for mobile services have led to the development of a wide range of new and improved wireless services. As a result, however, many portions of the radio frequency spectrum are becoming increasingly congested, leading to what one analyst has called spectrum "pollution."[48] To alleviate overcrowding, and expand the number and variety of wireless applications even further, there has been a sharp increase in demand for radio frequencies.[49] The most valuable frequency bands, however, have already been allocated, and many are heavily used.

Several factors are pushing the increasing demand for spectrum: 1) existing wireless service providers—including broadcasters, satellite companies, and data communication companies—want additional spectrum to expand capacity and

services; 2) new applications now being developed—including digital radio and television broadcasting, terrestrial- and satellite-based communications systems, and data and information messaging systems for mobile and fixed users—will need new frequencies; and 3) communication and entertainment applications will increasingly combine voice, data, and video, requiring large amounts of spectrum to meet the bandwidth-intensive nature of such applications. Complicating the situation is that portions of the spectrum have characteristics that make them particularly well suited for specific types of applications. The frequencies that most engineers consider ideal for mobile communications, for example, are located between about 0.5 and 3 GHz—frequencies that are rapidly becoming congested.

Technology advances are expanding usable capacity, but it is unclear if such advances will be able to keep up with rising demand for services in the longer term.[50] Unlike wireline systems, which can add capacity or serve more users by laying more wires, the capacity of the spectrum is limited by current technology. For any given set of frequencies, the spectrum can only serve a limited number of users and cannot be expanded. Technology advances such as more efficient modulation, cellular architectures, narrower channels, digital compression, and use of higher frequencies can reduce overcrowding—by extending the usable spectrum and increasing efficiency and capacity—but demand for radio frequencies has historically outstripped supply. The long-term implication of rising demand is that spectrum will continue to be in short supply.

Faced with rapidly rising demands, Congress, the executive branch, and the FCC all have taken important steps to ensure that the wireless indus-

---

[48] Andrew M. Seybold, *Using Wireless Communications in Business* (New York, NY: Van Nostrand Reinhold, 1994).

[49] For a more complete discussion of the spectrum needs of various radio services, see U.S. Department of Commerce, *U.S. National Spectrum Requirements*, op. cit., footnote 16.

[50] For a more optimistic assessment of the ability of technology advances to stay ahead of demand, see Robert J. Matheson, "Spectrum Stretching: Adjusting to an Age of Plenty," National Telecommunications and Information Administration, April 1995. The author argues that technologies such as digital compression and frequency reuse can increase spectrum efficiency—and capacity—dramatically.

try has access to adequate spectrum. In 1993, for example, Congress required NTIA to identify and transfer 200 MHz of spectrum to private use.[51] In response, NTIA released a preliminary report in February 1994 identifying 50 MHz that could be transferred immediately and a final report in March 1995 that identified an additional 185 MHz for transfer.[52] The FCC, in cooperation with NTIA, recently proposed making 18 GHz in 12 bands available for the development of new commercial technologies. These would include licensed and unlicensed applications such as vehicle radar systems and extremely high-bandwidth applications, including two-way video and multimedia computer communications.[53] The FCC has also recently began auctioning frequencies for new mobile telephone services (PCS—see chapter 3) and has completed or launched a number of proceedings specifically aimed at bringing more spectrum resources to wireless data applications (see chapter 4). Although sufficient for the short term, it is too soon to tell if more spectrum will be needed for these applications in the long term.

Such actions, however, treat only parts of the problem, and policymakers will continue to struggle to match the supply of spectrum with demand. The ways in which spectrum is allocated and managed in the United States may need to be changed to respond to a new, more mobile world. To plan for the future and avoid piecemeal, reactionary decisionmaking, a national vision for long-term spectrum use is needed. Coordinated and focused spectrum planning—combining the efforts of both NTIA and the FCC—has been legislated several times (most recently in the Omni-

bus Budget Reconciliation Act of 1993), but has never been accomplished. The FCC and NTIA have not worked cooperatively to build a comprehensive framework for radiocommunications policy, although the FCC does have a liaison who coordinates policy at the staff level with NTIA. **The lack of a unified vision of future spectrum use could undermine long-term planning efforts and development of spectrum policy (including priority-setting), and may hamper development of innovative wireless technologies.**

The federal government has not maintained an aggressive approach to long-range spectrum planning for practical as well as ideological reasons. Practically, allocating spectrum for needs that have not been identified is difficult, and ideologically, such a planned approach was seen as too closely resembling "industrial policy," which past Administrations have tried to avoid. Furthermore, management of private sector spectrum in the United States has long relied on petitions by prospective users to determine uses rather than a priori planning. As a result, policymaking has tended to concentrate on specific portions of the radio spectrum without always addressing how individual decisions might interact. However, as the number and kind of wireless systems and users have grown and the technologies and services have begun to merge, the need for a more integrated policymaking framework has become necessary because multiple systems can now deliver essentially the same service.

Developing a practical and effective approach to long-term spectrum planning will be challenging. Planning for needs and technologies that do

[51] "The Omnibus Budget Reconciliation Act of 1993," Public Law 103-66, Aug. 10, 1993. Title VI deals with telecommunications issues.

[52] U.S. Department of Commerce, National Telecommunications and Information Administration, *Preliminary Spectrum Reallocation Report*, NTIA Special Publication 94-27, February 1994; U.S. Department of Commerce, National Telecommunications and Information Administration, *Spectrum Reallocation Final Report*, NTIA Special Publication 95-32, February 1995.

[53] The bands are located between 47 and 153 Ghz. These frequencies historically have been limited primarily to military and scientific purposes, and are generally only lightly used. Sixteen of the 18 GHz specified will be shared with government users. General Motors, Ford, and Chrysler have submitted comments to the FCC on vehicle radar systems they have already begun to develop. "Notes on the FCC 40 GHz Plus Proposal," *Telcom Highlights International*, vol. 16, No. 47, Nov. 23, 1994. "FCC Identifies Spectrum Above 40 GHz for Commercial Use, New Technologies," *Telecommunications Reports Wireless News*, vol. 4, No. 22, Nov. 3, 1994.

not yet exist is nearly impossible, and would not necessarily lead to efficient use of the spectrum. The tradeoffs between encouraging efficiency and promoting development of new technologies must be carefully weighed as a part of determining future radiocommunication policy. It may be possible to craft policies and regulatory efforts that encourage both, but it will be necessary to carefully balance the needs for efficiency with the demand for new technologies and services.

In any case, even better spectrum planning will not guarantee that a market for the planned service will actually develop or that the services/systems planned will become economically viable. The 12-GHz band of frequencies, for example, was planned more than a decade ago to provide television programming services directly from satellites to homes. Initial efforts to launch a service failed, however, and DBS systems are only now beginning commercial service. The history of DBS shows both the difficulties and ultimate success of one government planning effort. DBS frequencies went unused for many years as proponents struggled to launch operating systems, but without that early allocation, companies might not have developed new technologies so quickly. In addition, without early government action, companies might still be fighting for spectrum and customers might still be waiting for service. This case clearly illustrates the inherent uncertainties in planning for future, undefined spectrum applications.

In 1991, NTIA issued a report on improving spectrum management, and implemented some of the recommendations. However, some of its most fundamental conclusions for improving U.S. spectrum allocation and assignment processes were never put into practice.[54] It may be time to revisit some of these options. NTIA also recently completed a major study that identifies the spectrum requirements of most radio services for the next 10 years—an important first step in improving spectrum planning.[55]

The process of allocating spectrum, however, is only part of the problem. Until recently, spectrum was assigned to individual entities by the FCC on the basis of comparative hearings or lottery. In 1993, Congress authorized the FCC to use competitive bidding—auctions—to distribute some licenses.[56] Auctions are believed to be the most economically efficient way to assign licenses, while also raising money for the federal government. Given the financial success of the PCS auctions, which raised more than $7 billion, some analysts and policymakers have now begun to consider auctions as a way to assign spectrum for other services in the future. Despite their financial success, however, the longer term operational and economic effects of the auctions are still unknown.[57]

In any case, auctions may not be applicable to all radio service users. Federal, state, and local governments, for example, have a wide range of operations that support vital public interests such as national defense, air traffic control, public safety, and emergency preparedness functions. These types of services are not currently affected by auctions, and there would likely be a great deal of resistance to auctioning such spectrum. There are also a number of economic and public policy issues, in addition to administrative and practical questions, that would have to be addressed before such an approach could even be seriously considered.

Options

To ensure that adequate spectrum continues to be made available in the future, Congress could:

---

[54] U.S. Department of Commerce, *U.S. National Spectrum Requirements,* op. cit., footnote 16.

[55] Ibid.

[56] Omnibus Budget Reconciliation Act, op. cit., footnote 51.

[57] Many of the winners in the Interactive Video Data Service auction, for example, defaulted on their bids. This will slow the development and deployment of the service.

- mandate the transfer of additional spectrum from the federal government to the private sector. This effort would build on already-conducted NTIA studies of spectrum needs and reallocation.
- build on existing efforts to determine spectrum needs and existing planning, and enforce previous mandates for the FCC and NTIA to engage in cooperative long-term spectrum planning.
- establish research funds for development of high frequency (40 GHz and up) radio communication service, through the federal government and/or private sector initiatives.
- evaluate new methods for allocating and assigning spectrum, including the recommendations in earlier NTIA reports and the possibility of auctioning all future radio licenses. This may entail developing new rules for auctions.

■ Research is Needed

Research on the social, economic, and public policy implications of widespread use of wireless technologies is very limited, and research on the longer term effects and implications of wireless devices and systems is only at the conceptual stage. This situation is directly tied to the nascent state of the various segments of the wireless industry as a whole. Many of the technologies that will make the biggest impacts are not yet operating, and evaluating their social and economic effects is impossible. Even in the more mature wireless industries, research is sparse.[58]

One of the most important, and underappreciated, aspects of the development of wireless technologies is the problem of scale. Problems that seem trivial with only a relatively small number of users become magnified as the number of users grows. Some have commented that "society is not ready" for the many changes that ubiquitous

wireless communications will bring. One study estimates that 45 percent of the population will be using mobile communications devices (phones and/or laptop computers) by 2005.[59] And although some information and statistics have been collected on various aspects of mobility, there is little hard data that allow a good understanding of the characteristics of personal and professional mobility, and what implications they may have for the implementation and use of wireless services and for society. One example is 911 service. Despite the fact that only about 10 percent of urban customers have cellular phones, 911 operators receive, on average, eight reports for each traffic accident. As subscriber and penetration levels rise, 911 system administrators may be inundated with calls.

The most controversial area of research, and the one most in need of additional study, concerns the possible impacts radio communication systems could have on public health (see chapter 11). Some members of the public and a few scientists believe that radio waves can damage human cells. Research to date, however, has been inconclusive. No direct link has been found that radio waves are harmful, but it is still not possible to say with certainty whether the devices or antennas pose a risk to human health or how serious any risk may be. This issue is extremely emotional and polarized. Some people are convinced that wireless systems are dangerous and should be banned or severely limited. The wireless industry, however, believes that development of wireless technologies should continue because there is no conclusive evidence that either phones or antennas are harmful. Research is now being conducted, but much of it is sponsored by industry, either directly or indirectly, and it is unclear whether the public will be satisfied with the results. The federal government has played only a minor role in research on this

---

[58] The exception is broadcasting. There is a long history of economic, public policy, and social science research into all areas of radio and television broadcasting.

[59] Personal Communications Industry Association, "PCIA 1995 PCS Technologies Market Demand Forecast Update, 1994-2005," (Washington, DC: Personal Communications Industry Association, January 1995).

topic. Representatives from several government agencies, however, are involved in oversight and review of industry research.

Research on the economic structure of the various wireless industries and long-term outcomes of competition is even more limited.[60] In wireless voice and data services, for example, many new companies will enter the market over the next five years. Gathering accurate data on cost structures, revenues, and customer demand is only just beginning, and many companies will not divulge such information. Even industries that have been around for decades, such as broadcasting, will be affected. Both radio and television broadcasters are preparing for radical change as digital technologies replace analog, and as new competitors—some wireless (DBS, wireless cable, and cellular television) and some wireline (telephone companies)—enter the market for audio and video programming and services. The ultimate outcome of all these changes cannot yet be predicted, and the economic studies and modeling of such competition are just beginning.

Likewise, wireless telecommunications' contribution to productivity, economic growth, and employment is unclear. Industry studies indicate that wireless telecommunications account for significant productivity increases through better use of time, particularly for higher paid employees who spend time away from their offices. There are no credible data on additions to the gross domestic product or on future employment (either in the industry or in the economy generally) due to wireless telecommunications, though the cellular industry has experienced significant economic and job growth over the past decade.

Finally, the implications of wireless technologies for individuals, organizations, and society are only now emerging; they are likely to involve increased personal and business efficiency, but also increased stress and concern about health effects, monitoring, and privacy. Wireless technologies are likely to play a role in the continuing evolution of new organizational and social forms, including their geographic dispersion and functional disaggregation. The widespread deployment of mobile communication technologies also portends a change in the average wireless user—from mobile professional/field service representative to mass market consumer.[61] Again, the effects of this change are unknown.

Technical research and development is the exception to OTA's finding on the state of research. Research and development of new radio technologies and services is moving quickly. Some industry representatives, especially those representing larger companies, see no need for government support of technology research. Whether this position is shared by all technology developers is uncertain. The satellite industry, for example, has put together a list of topics they would like the federal government to help them in exploring.

Options

To increase understanding of the many economic, social, and regulatory issues surrounding the integration of wireless technologies into the NII, and establish a basis for informed policymaking, Congress could:

- monitor the development of various industry segments and social issues, including privacy, security, and especially health effects to determine if future congressional action may be necessary.
- establish funds to promote research into these issues. Congress already funds research in a

---

[60] OTA contracted for two studies—one to examine the basic economics and one to analyze the evolving structure—of the wireless industry. Both authors noted the lack of empirical data available on the various segments of the wireless industry, and the lack of appropriate models for studying wireless economics. Egan, for example, notes that "...based on publicly available data (including that from investment houses in their efforts to calculate prospective market penetration rates and net cash flows to establish valuation benchmarks for the investor community) indications are that the state of the art in engineering economics and financial modeling of network systems is not very far along." See Egan, op. cit., footnote 32, p. 43, and Woroch, op. cit., footnote 32.

[61] Frieden, op. cit., footnote 47.

number of related fields, such as transportation, labor statistics, and public health that could be expanded to cover wireless topics. Alternatively, a portion of the funds received from spectrum auctions could be designated for this purpose.

▮ State and Local Government Roles are Unclear

States have a significant interest and role in protecting their residents from services that are priced too high or that offer poor quality. Municipalities have an important, historically-defined role in local zoning matters and protection of public rights-of-way. However, the federal government, primarily the FCC, also has a legal role to play in advancing the communications systems of the country. **Since the Communications Act of 1934 was passed, state, local, and federal authorities have been struggling among themselves and in court to define the boundaries of their rights and responsibilities.**

Current proposed legislation will not end the debate. Bills under consideration in Congress generally prohibit states from enacting laws that "may prohibit or have the effect of prohibiting the ability of any entity to provide any interstate or intrastate telecommunications services."[62] The bills, however, also permit states and local governments to impose requirements for universal service, protect the public safety, and manage public rights-of-way. Specific cases will no doubt arise where the two policies will clash. In the case of wireless technologies and systems, there are several areas of conflict between federal and local policy goals.

State and local governments currently regulate wireless services only lightly. Broadcasting is mostly free of local regulation. Half the states once regulated cellular in one form or another, and another 20 had laws stipulating that the state regulatory commission must forebear from regulation.[63] As a result of new regulations governing CMRS, however, no state will be allowed to regulate wireless rates or enact laws that stifle entry by new providers.[64] Satellite providers have been struggling against local ordinances and taxes for many years (see chapter 8).

In the future, however, state regulation of telecommunications services in general may have significant, if indirect, effects on new wireless services, especially those used as a substitute for local wired telephone service. Importantly, the states will retain regulatory jurisdiction over the terms and conditions regarding wireless companies interconnection with local telephone companies. States are also likely to have a significant role in helping to define universal service obligations and subsidy schemes, both of which could significantly affect new wireless carriers.

Currently, the most controversial battle between federal and local policies involves zoning and land use. Wireless companies need to erect antennas and towers to provide their services. Some municipalities, however, in response to citizen concerns about public health and property values, have enacted zoning laws or other prohibitions that can make it difficult to put up a tower. Such regulations have delayed or halted construction of radio towers already licensed by the FCC. More local governments are expected to enact similar prohibitions as the number of antennas and towers proliferates with the spread of cellular and the introduction of PCS and ESMR services (see chapter 8).[65] Industry associations have asked the FCC to preempt such regulations, maintaining

---

[62] S. 652, op. cit., footnote 34.

[63] Woroch, op. cit. footnote 32.

[64] Eight states applied under the law to continue to regulate cellular/wireless rates, but the FCC denied all the petitions.

[65] Some rules set height limits, while others ban towers altogether in residential areas. See "City Zoning Rule Limits Radio Tower Height," *Telecommunications Reports*, vol. 61, No. 3, Jan. 23, 1995.

that new services will be slowed or even precluded.[66] The FCC has not yet ruled on this issue, and **the question of which should take precedence—federal laws that encourage the development of public communications systems or local control over land—remains unanswered.**

Options

Aside from specific issues relating to preemption, Congress may wish to establish an overarching framework to guide future policymaking. Establishing a cooperative relationship between federal and state regulators will be critical if the NII is to develop as quickly as possible. To determine the proper relationship between federal and state regulatory authority in a new competitive era, Congress could:

- make explicit its views on federal preemption regarding NII and wireless issues, indicating which authority should take precedence.

- hold a series of hearings in Washington and around the country or form a commission to gather input from all parties involved in federal-state telecommunications issues. As part of this broader effort, Congress could also establish more formal mechanisms for resolving federal/state/local disputes in telecommunications policymaking.

---

[66] The Electromagnetic Energy Association and the Cellular Telecommunications Industry Association have filed petitions for rulemaking on the issues. "FCC Asked To Preempt States' RF, Radio Tower Rules," *Telecommunications Reports*, vol. 61, No. 1, Jan. 9, 1995.

Mobility and the Implications of Wireless Technologies | 2

The need for mobility underlies many applications of wireless technology.[1] It is the single feature of wireless systems that other telecommunications technologies cannot replicate. Wireless technologies permit users to access communications networks while they are "on the go," and also make it easier for individuals to stay connected as they move around. However, the concept of "mobility" and its implications for the deployment of wireless technologies are poorly understood.

Individuals and businesses already use a number of wireless technologies, including cellular telephones, pagers, and various wireless data services, but over the next five to 10 years, a variety of new mobile communication systems—personal communications service (PCS), enhanced specialized mobile radio (ESMR), satellite-based telephony, and higher-bandwidth wireless data communications systems—will begin operation (see chapters 3, 4, and 5). To consider the potential success or failure of these new technologies and the implications of their widespread use, it is critical to understand the underlying forces that might motivate people and businesses to use them. Is society becoming more mobile? How does a technology deployed at scale challenge policy-making in this area? What are the potential social implications of widespread deployment of wireless telecommunications? An analytic framework used to address these issues places mobility,

[1] Some of the material in this chapter is based on Philip Aspden and James Katz, Bell Communications Research, Morristown, NJ, "Mobility and Communications: Analytical Trends and Conceptual Models," contractor report prepared for the Office of Technology Assessment, U.S. Congress, Washington, DC, Jan. 20, 1995.

the unique property of wireless, at the center. This chapter attempts to anticipate the answers to some of these questions.

FINDINGS

- **The concept of *mobility* is rudimentary and unfocused. Although some information and statistics have been collected, there is little hard data that allow a thorough understanding of the various characteristics of personal and professional mobility and their implications for the implementation and use of wireless services.** Mobility is an enduring social quality that affects people in both their personal and business lives. People move about in their private lives every day to shop, visit friends or relatives, or run errands. Understanding the patterns of what they do, how they do it, and what information they could use in the process will be crucial to understanding how wireless technologies may play a role in peoples' lives.

- Worker mobility is particularly significant because businesses tend to lead the way in the use of telecommunications technologies, and are expected to be the earliest and heaviest users of wireless technologies as well. Considering the ways in which a job can be "mobile," **OTA conservatively estimates that nearly 50 million workers (44 percent of the workforce) are mobile in some way today and the percentage of the workforce that can be classified as mobile is increasing.**

- From the technology side, research on how people use or would use wireless devices is sketchy. Some marketing surveys have been conducted, but they do not offer a very compelling or complete picture of how the average consumer might use communication or information resources in a mobile or portable setting. Prospectively, it is difficult to ask people to evaluate a technology or service they have never used and may not completely understand. A better grasp of mobility will improve product development and inform policymaking regarding new wireless technologies and systems.

Academic researchers have been studying "mobility" for years, but they tend to concentrate on one particular element only—work patterns, time spent commuting, or time management, for example. Researchers in different disciplines do not generally collaborate or communicate, and it appears difficult for them to conceptualize how their work may fit together. Little work has been done to bring together the disparate elements that define mobility, and no theoretical framework exists for studying mobility as a unified concept. **More research on all aspects of mobility—and their relation to telecommunications technologies, including wireless—is needed.**

- Because mobility is so poorly understood and the wireless technologies that will enhance the mobility of both people and machines are not yet widely used—even the penetration rate for cellular telephones is less than 10 percent—assessing the implications of this widespread use is difficult. **The impacts of wireless technologies on individuals, organizations, and society are only now emerging; they are likely to involve increased personal and business efficiency, as well as increased stress and concern about monitoring and privacy. Wireless technologies are likely to play a role in the continuing evolution of new organizational and social forms, potentially reinforcing geographic dispersion and functional dissolution.**

SELECTED EXAMPLES OF MOBILE WIRELESS SERVICES

Other than using wireless telecommunications for traditional telephony, what might people do with wireless? People or organizations often use new technologies to perform old functions. Over time, however, a technology deployed for one purpose may be used for something quite different than its designers intended. Many new applications of wireless technologies are still only at the developmental stage, and have yet to pervade the public's consciousness. A brief description of some of the current and projected uses of mobile communications may provide an idea of the broad scope and scale of their potential applications. The following examples are drawn from existing commercial or demonstration projects.[2]

■ Inventory Management

The emphasis on lowering costs and increasing efficiency in business operations has focused on improving inventory management, total quality management, and just-in-time manufacturing. Wireless technologies can help where mobility is a key feature of the process. For example, in warehouses, knowing more about the location of specific items can greatly reduce costs. Equipment to track inventory is becoming more popular. For example, Rexham, a box manufacturer in Charlotte, North Carolina, has revamped its quality control function with a wireless bar code system that provides up-to-the-minute information about job status and product information. Sensors can be attached to items so they can be tracked and found when needed. Using such a system, one chemical company experienced a 100 percent payback in six months and reduced its accounting personnel for this function from 12 to one.

■ Electronic Newspapers

Newspapers have been one of the most transportable sources of information. A few companies are now developing the capability to deliver newspapers via wireless systems directly to customer's laptop computers. One publisher has developed a prototype electronic newspaper whose screen resembles the front page of a newspaper. The user can touch a picture or headline to receive additional information in any form: video, sound, or text. Designed to overcome some of the shortcomings of traditional print newspapers—bulkiness and limited circulation—newspapers distributed using wireless could use digital cellular networks or

new PCS systems to reach readers efficiently, wherever they are. Prospective users might include travelers who want to read hometown newspapers or business executives who want to purchase electronic publications while aboard trains or airplanes.

■ Classroom Networking

Duke University recently participated in an experiment using a wireless local area network (LAN) to connect engineering students' laptop computers to one another and to the instructor's computer.[3] The system consists of an infrared transceiver attached to each student's computer that sends and receives messages to and from every other computer. Transmitting at 230 kbps, the infrared system is completely transportable, and an ad hoc network can be established in about 20 minutes in the library, lab, or dormitory.

Tying together the computers allowed the comparing of notes, facilitated collaboration on group projects, and allowed the professor to project one student's computer screen onto a large screen in front of the entire class for discussion. In addition to allowing students to work easily with one another, the system also sends the instructor's comments directly to every student's computer, perhaps communicating ideas more effectively. Based on this experiment, Duke University is considering implementing similar systems in other classes.

■ Real Estate Marketing

In the last few decades, real estate agents have come to rely heavily on computerized databases such as the Multiple Listing Service. However, while on the road, agents are out of touch with these databases, and must make frequent trips back to their offices to use them. To help agents save time and eliminate unproductive travel, sev-

---

[2] For a broader set of examples, and an analysis of the emerging uses of wireless technology, see Virginia Polytechnic Institute and State University, Center for Wireless Telecommunications, Blacksburg, VA, "A Survey of Emerging Applications of Wireless Technologies," contractor report prepared for the Office of Technology Assessment, U.S. Congress, Washington, DC, Sept. 15, 1994.

[3] Gary Hughes, "Wireless Network Goes to School," *Wireless for the Corporate User*, vol. 3, No. 3, 1994.

eral wireless services have been developed. For example, a low-power FM transmitter can be placed on a property for sale which will broadcast messages that identify key characteristics of the property. An agent driving past the house can tune in to hear the details. Another system uses a personal digital assistant (PDA) and a cellular digital packet data (CDPD) radio modem to provide real estate agents access to a multimedia database of homes. The data consist of pictures of houses, maps of residential areas, and detailed statistics that can be searched by price, location, number of rooms, etc. Rather than having to return to the office to search the listings again, a revised list of appropriate offerings could be accessed from any location.

▮ Field Service

To handle more calls, provide faster and more accurate inventory control, and reduce the time spent sending dispatch instructions, Coast Plumbing of Solana Beach, CA, implemented a data communications system that integrated dispatch, billing, and inventory functions.[4] The system connects a portable computer in each of the company's 20 trucks to a host computer in the office. The system delivers text dispatches to the plumber's portable computer, which displays the customer's name, address, and the reason for the call. The system also allows the technicians to check on part availability, access prior service history, and look up prices. Once a job is finished, the system automatically transmits billing and inventory information back to the host computer, which then updates parts lists and customer accounts. Coast also uses specialized mobile radio for voice and data communications.

After implementing the system, Coast Plumbing increased the number of calls per day handled by each plumber, dramatically reduced the amount of time spent on physical inventory, improved customer satisfaction, and streamlined administrative processes. The company estimates that the system saves it more than $10,000 per month in total costs.

▮ Disaster Recovery and Assistance

The success of emergency relief and recovery efforts relies on the ability of workers to communicate effectively, efficiently, and securely. Wireless is uniquely suited to these applications because: 1) disasters typically do not affect wireless communications links, especially satellite links, and 2) the rapid deployment of a communications system for mobile field workers is more efficacious with a wireless system. The users of these systems include insurance companies; emergency relief workers; federal, state and municipal disaster agencies; emergency medical personnel; and other suppliers of necessary services. They will typically need communications in the field to report assessments of damage, call for reallocation of resources, predict additional consequences, and file insurance claims or pay such claims electronically. For example, one company provided Iowa's 99 counties with backup protection during the flood of 1993 with a portable 18 GHz digital microwave system. The system was engineered, manufactured, delivered, and installed in just four days.

▮ Intelligent Transportation Systems

Intelligent Transportation Systems (ITSs)[5] apply information and communication technologies to surface transportation systems to reduce traffic

---

[4] Deborah Kirtland, *Wireless for the Corporate User,* vol. 3, No. 2, 1994, p. 53.

[5] ITS was formerly called Intelligent Vehicle Highway Systems (IVHS), but was changed to ITS to include public transit and other transportation modes.

congestion, improve safety, make public transit options more attractive to commuters, and decrease transportation-related environmental impacts.[6] Interest in ITS stems from the realization on the part of transportation experts that building more roads and/or expanding existing ones is often too costly and only marginally effective in reducing congestion, and does little to alleviate safety and environmental problems. ITS could make more efficient use of the current transportation infrastructure, improve safety, and allow public transportation to be more responsive to passenger demands. To transmit information to mobile units (automobiles, buses, and trains, etc.) from a fixed location, and vice versa, wireless technology of some kind is necessary.

In 1991, Congress passed the Intermodal Surface Transportation Efficiency Act (ISTEA[7]), which committed $659 million over six years for ITS projects. ISTEA also mandated the U.S. Department of Transportation (DOT) to establish the ITS Architecture Development Program. This program brings together DOT, a public/private consortium called ITS America, and various private transportation and communication companies for the purpose of forming a framework to develop an integrated, interoperable ITS in the United States.[8] More recently, the Federal Communications Commission (FCC) has allocated 26 MHz in the 902-928 MHz band for what it terms Transportation Infrastructure Radio Service, or TIRS.

Among the areas being developed are advanced traveler information systems that will inform people on the best way to get to their destinations; advanced traffic management systems that will gather and distribute data on traffic congestion and alter the timing of control signals to move traffic more efficiently; automatic toll collection; parking and security applications; and automated vehicle control. To date, most ITS efforts have focused on providing route guidance to travelers and fleet monitoring and control to transportation companies. Systems in Japan, Europe, and the United States rely on the Global Positioning System (GPS) for vehicle location, often used in conjunction with dead reckoning.[9] Some systems also employ terrestrial-based wireless data systems to relay traffic conditions to travelers in their cars. Both scenarios involve sophisticated mobile units for the users that can cost as much as $8,000.[10] One system marketed by Oldsmobile offers drivers stored information about local points of interest, such as restaurants, with the option to receive updated traffic, weather, and special event information via a wireless link.

More complex ITS proposals will require more sophisticated technology, both in-vehicle and in the public transportation infrastructure, than existing systems now offer. For example, some plans call for a radar-equipped vehicle that will sense the distance between it and the car in front and automatically apply the brakes if the gap is too small. Some plans for these Advanced Vehicle Control Systems (AVCS) may also incorporate sophisticated sensing equipment in the roadway, which would work in conjunction with systems in the ve-

---

6 National Research Council, "Primer on Intelligent Vehicle Highway Systems," Transportation Research Circular 412, Washington, DC, August 1994. See also U.S. Congress, Office of Technology Assessment, *Intelligent Transportation Systems for Metropolitan America—Background Paper*, background paper for OTA's Project on the Technological Shaping of Metropolitan America (in progress).

7 Public Law 102-240; Dec. 18, 1991.

8 U.S. Department of Transportation and ITS America, ITS Architecture Development Program; Phase I, Summary Report, Washington, DC, November 1994.

9 Dead reckoning is a technique by which vehicle location can be calculated and matched to on-board maps by calculating the distance traveled from a specific starting point.

10 W. Clay Collier and Richard J. Weiland, "Smart Cars, Smart Highways," *IEEE Spectrum*, Apr. 4, 1994, pp. 27-33.

hicles to automatically track the vehicle down the road at a constant speed toward the driver's destination.

CHARACTERISTICS OF MOBILITY

At the root of interest in wireless telecommunications is its ability to accommodate the physical mobility of people and things. However, our understanding of mobility is intuitive and poorly characterized analytically.[11] Some of the broad outlines of mobility are sketched here to provide a sounder basis for analyzing mobility and its implications. Key questions are: Why are people mobile? What are the features and forces that have shaped people's mobility patterns? What are the key trends in mobility? How do wireless technologies fit with mobile activities? What are the consequences of mobility?

Mobility has a number of dimensions that give it different meanings for different people. For example, some mobility is local, as in a hospital where nurses and doctors are constantly on the move, but within well-defined boundaries. Some mobility is long-distance, as with cross-country trucking or cellular roaming to cities outside the home service area. Some people are mobile but do not communicate en route, such as executives travelling to meetings in distant cities. Others communicate en route over long distances, such as salesmen who need to get up-to-date information before meeting their next prospect. While each is mobile, the wireless telecommunications technologies each would likely use may be quite different. Using the single term "mobility" masks its multiple dimensions and deprives it of analytic precision. Data on mobility characteristics, as described below, do not exist at present.

From the examples of applications given above and data on past and projected demand, the following characteristics or drivers of mobile access can be inferred:

- **People want to be mobile because they can increase their control and reduce uncertainty** in the conduct of their business or personal affairs. People move to see and do things remotely so that they can control their activities or gather information that reduces their uncertainty.
- **People want to communicate while moving**, or while in transit. They want flexibility in deploying and redeploying assets, services, etc. In many situations, people cannot predict their communications requirements, either for type of service or its location. All of these needs are met by a variety of wireless technologies, at a low cost, depending on the application, the data rate, and security requirements.
- **People want to communicate and get information** *immediately.* When they are traveling or away from wired telecommunications links, the urge to be connected is strong. Although one can usually travel to a place that has communications resources, the time pressures of today's society and business world dictate that those who have easiest access to communications resources have a competitive advantage.

Typologies like those in table 2-1 could be used to develop research programs and data sources on mobility and communications that could assist both policymakers and business planners. In particular, this framework could help determine the potential scale of wireless communications. Decisionmakers would then know whether particular public wireless communications systems are likely to be confined to small populations of workers or users, or are likely to be applicable to large segments of society. They could also provide information on the impact these technologies may have on individuals, organizations, and society at large.

---

[11] One attempt is that of the Cross-Industry Working Team, Corporation for National Research Initiatives, "Nomadicity: Characteristics, Issues and Applications," March 1995.

TABLE 2-1: Typology of Mobility and Communications

Mobility characteristics:

- *mobility extensiveness* (how far: global, national, regional, local, or home/office)
- *mobility intensiveness* (how much mobility is required for an activity)
- *mode of transport*
  self-propulsion (walking or biking)
  limited occupancy vehicle (private automobile, truck, small boat, or small airplane)
  public transport (bus, airplane, or ship)
- *variety of routes undertaken*
  standardized
  externally directed (defined by third party)
  spontaneous

Activities or information might be categorized by:

- *function*
  data-gathering and entry
  data analysis
  execution or control of activity or function
- *time factors*
  real time
  asynchronous
- *information type*
  symbols
  audio
  text images
  still picture images
  moving picture images
- *delivery paths*
  point-to-point
  point-to-multipoint
  dispatch
- *location*
  information obtained while in transit
  information obtained by visiting many locations
  information unrelated to location

SOURCE: Office of Technology Assessment, 1995.

JOBS AND MOBILITY

Although the use of wireless technologies is likely to be pervasive, their greatest impacts may be in working environments and on jobs. Jobs previously fixed may become mobile with new technologies, altering a wide range of business practices. Unfortunately, no government or private agency collects data on mobility in employment, nor are there measures on the degree of mobility typically associated with particular job classifications. Private studies tend to focus on particular market segments, such as white-collar office workers and executives on the road.

OTA made a preliminary estimate of high and low degrees of mobility in the work force to illustrate the argument made here. Based on Bureau of Labor Statistics data, and using rough estimates of the mobility requirements of jobs across the whole U.S. economy, OTA estimates that 34 million people are somewhat mobile and 15 million are highly mobile for significant parts of their working day, for a total of about 44 percent of the U.S.

TABLE 2-2: Occupations and Mobility Classification, 1994				
Occupation	Total employed (millions)	High mobility workers	Moderate mobility workers	Percent mobile
Executive, administrative, and managerial	**15.4**	**0.8**	**3.8**	**25.8**
Service deliverers	**12.2**	**5.7**	**5.5**	**51.3**
Engineers, architects, surveyors	1.9		0.9	
Social, recreation, religious workers	1.1	0.5	0.5	
Sales representatives, financial and business services	2.3	1.2	1.2	
Sales representatives, except retail	1.5	0.8	0.8	
Adjusters and investigators	1.4	0.7	0.7	
Protective services	2.2	1.6	0.5	
Mechanics, appliances, equipment	1.8	0.9	0.9	
Campus/building-wide workers	**8.8**	**0.0**	**8.0**	**91.3**
Health diagnosing occupations	0.9		0.9	
Health assessment and treating occupations	2.6		2.6	
Teachers, college and university	0.8		0.8	
Health technologists and technicians	1.5		0.8	
Cleaning/building service occupations	3.0		3.0	
Workers who move people or goods	**6.0**	**4.2**	**1.7**	**33.2**
Mail and message distributing	1.0	0.5	0.5	
Transportation, material moving occupations	5.0	3.8	1.3	
Outdoor workers	**8.3**	**4.2**	**4.2**	**54.2**
Construction trades	5.0	2.5	2.5	
Farming, forestry and fishing	3.3	1.7	1.7	
Others	**68.7**	**0.4**	**11.0**	**16.4**
Total	**119.3**	**15.2**	**34.3**	**44.0**

SOURCE: U.S. Department of Labor, Bureau of Labor Statistics, "Employment and Earnings," January 1994, and Office of Technology Assessment, 1995.

work force (see table 2-2). Workers may be grouped in five clusters, as indicated in the table. Because the criteria used in arriving at these estimates were extremely conservative, it seems likely this is an under- rather than over-estimate of the true amount of mobility in the work force.[12]

There is also some evidence that the number of mobile workers is increasing as a proportion of total workers. Using the classifications above—and past, current, and projected employment levels in each subcomponent—the pool from which the future mobile workforce will be drawn can be illustrated (see table 2-3). Such estimates show that these job clusters will increase 51 percent over the 1983-2005 period, somewhat faster than the projected growth rate for all jobs, which is 42 percent. Executive and managerial jobs are projected to increase fastest (73 percent), followed by campus/ building-wide workers (60 percent), and service deliverers to homes and businesses (54 percent). Finally, recent projections of occupational employment levels suggest that many of the fastest-growing job categories are mobile, further supporting the argument that mobility is likely to

---

[12] DYG, Inc., "The Growing Emergence of Mobile Workers," report prepared for Cowles Business Media, 1994, p. 2.

TABLE 2-3: Estimated Total Number Of Workers In Mobile Occupations, 1983-2005 (millions)

Occupations	1983	1994	2005
Executive, administrative and managerial workers	10.8	15.4	18.6
Service deliverers to homes and businesses	9.9	12.2	15.1
Campus/building-wide workers	7.1	8.8	11.4
Workers who move people or goods	5.0	6.0	6.9
Outdoor workers	8.0	8.3	9.4
Total	**40.8**	**50.7**	**61.4**

SOURCE: Office of Technology Assessment and Bureau of Labor Statistics, 1995.

be more important in the future.[13] Several other more limited studies have come to similar conclusions.[14]

IMPLICATIONS OF INCREASED MOBILITY

Having established that wireless telecommunications are here to stay, though still not ubiquitous in business and society, what are the implications of increased mobility for individuals, organizations, and society? How will wireless telecommunications affect peoples' personal and business lives? Much remains unknown, and issues are just beginning to be identified.

■ Implications for Individuals

Increasingly, communication will be made to a person, not a place. The potential for more complete integration of people with each other and information sources may be increased considerably with widespread deployment of wireless telecommunications. Networks may center on people rather than on physical connections, which could have both positive and negative effects on individuals.

Increased Contactability

The most striking impact of wireless communications systems may be the ability to make and receive phone calls from any location at any time, enabling users to be constantly "in touch." A recent Bellcore survey on telephone use asked people their opinions on their need to be reachable (see table 2-4). Nearly 50 percent agreed with the statement that "my responsibilities require me to be 'easily reachable,'" even on holidays. About 20 percent of those surveyed disagreed with the idea that they need to be readily contactable.

TABLE 2-4: Opinions on the Need To Be Reachable

	Strongly agree	Agree	Neutral	Disagree	Strongly disagree	No answer
My responsibilities require me to be "easily" reachable	13.4	35.4	26	18.9	2.9	3.5
People need to contact me about important matters	12.5	35.9	27.7	17.5	3.3	3.1
There are often times when I urgently need to get through to another person	8.3	36.5	29.7	20.0	2.2	3.3
I "stay in touch" even when I am on holiday	8.9	40	21.3	20.8	5.9	3.1

SOURCE: Bell Communications Research, "The Telephone: Making It Work Better For You," Bellcore national postal survey, 1993.

[13] Bureau of Labor Statistics, *Monthly Labor Review*, November 1993, cited in Peter Francese, "Cellular Customers," *American Demographics*, vol. 16, No. 8, August 1994, p. 56.

[14] William F. Ablondi and Thomas R. Elliott, "Mobile Professional Market Segmentation Study," BIS Strategic Decisions, Norwell, MA, pp. 1-3, and Alison L. Sprout, "Moving Into the Virtual Office," *Fortune*, May 2, 1994, p. 103.

TABLE 2-5: Personal Life Improvement Due to Cellular Telephone Use

Cellular telephones have:	Percent of cellular telephone users agreeing strongly or agreeing somewhat in:	
	1991	1993
Increased your flexibility	86	94
Increased your efficiency	79	75
Helped you make the most of your personal time	76	81
Added a significant amount of time to your day	68	62

Interviewing for the Motorola survey was conducted by telephone between December 1990 and January 1991, and between March and April 1993. In 1991, the nationally representative sample size was 650, and in 1993 it was 660 people. In the 1993 survey, 63 percent of the sample was male, 37 percent was female.

SOURCE: The Gallup Organization, "The Motorola Cellular Impact Survey: Evaluating 10 Years of Cellular Ownership in America," Princeton, NJ, 1993.

Thus, wireless may enable people to remain in continuous contact. In some cases, this higher level of connectedness could reduce the sense of alienation that plagues many people who are out of physical contact with others. People may feel secure in dense networks of communications with people they can rely on, and be able to conduct many activities with considerable remote control. Survey research on current cellular users shows that cellular telephone users feel positive about the technology with respect to its ability to help them maintain contacts in both business and private life. A survey of users in 1991 and 1993 yielded the results in table 2-5.

In general, this survey reports that users believe that cellular phones help them make better use of their personal time. About half of respondents said they feel better connected to their families because of their cellular telephones. A significant fraction said they couldn't do without their cellular telephone (46 percent), and most believe that the phone has been a good value for the money (85 percent). Clearly, this data should be viewed with some skepticism. Respondents were all paying cellular customers who may justify their purchase of cellular services by alleging benefits. People who had tried and ultimately rejected cellular services were not polled as to their attitudes.

Although many users clearly value the ability to communicate more easily, there are drawbacks. The same device that allows users to call out enables other people to call in—potentially reducing privacy and control over one's time.[15]

> Paradoxically, the most important aspect of the mobile telephone may be the ability to reach others with it and to be reachable anywhere, which implies both absolute mobility *and* the opposite of mobility as traditionally understood! The owner of a mobile telephone may be highly mobile, but is always "at home," always "there," as long as he or she carries a [personal phone], thus making simultaneously possible a freely floating, highly mobile society and a very traditional, immobile social and spatial structure.[16]

The tension between accessibility and privacy is easy to underestimate, because people may choose to use communications technologies that help them perform certain tasks, but that may also bind them in unwanted ways. People generally seek wide communications access to others, and they want to be able to receive messages quickly and reliably. Yet people seldom want to be universally accessible to others; they want to limit access by certain people, they want control over when they receive calls, and they want to choose with

---

[15] It should be noted that the same concerns were also raised when telephones were first introduced almost 100 years ago. It may take some years before new social/business protocols regarding mobile telephone use are internalized by society. For users who are bothered by this prospect, the phone can always be turned off—unless an employer expects it to be on.

[16] J. P. Roos, "300,000 Yuppies? Mobile Telephones in Finland," *Telecommunications Policy*, August 1993, p. 458.

whom they communicate.[17] There also may be disadvantages of greater accessibility to others. While it may be possible to reach a specific person, wherever he or she is, that person may be reached in an unexpected context. Thus, a call to an office worker may reach him or her away from desk and files—in the company cafeteria or in the bathroom, for example. The physical context of communication is important to the business that can be conducted, a context that previously was provided by the knowledge that calls reach people in fixed places.

The inability to control incoming calls, which may come at inconvenient times, may also be resented.[18] There may be no easy way to avoid such demands: if wireless telecommunications use becomes the norm, then turning a phone off completely may signal to a caller "This person is purposefully limiting availability." For some people, increased personal psychological stress could result from loss of control over when and where people can contact you.[19]

Personal Monitoring and Privacy

Another implication of increased contactability—when coupled with remote sensing equipment and databases—is the potential for personal passive monitoring. Both benefits and threats to personal freedoms could occur. The benefits include systems that offer remote monitoring of health care delivery devices and those that protect personal

security. Large firms and government agencies have used such dedicated systems for years; however, due to their expense and size, they have been unavailable to private users or small firms until relatively recently. The Advanced Research Projects Agency (ARPA), for example, has been funding wireless telecommunications research for several years; one of its main interests is in equipping soldiers with battlefield personal monitors to relay vital signs and injury information to medical centers to provide timely and accurate responses to the wounded.[20] Similar systems are now being developed for consumer use. "Electronic house arrest" has been enforced with electronic monitoring devices that signal when a convict leaves his or her home.[21] Marathon runners have recently been issued shoelaces with wireless microchips embedded in them to prevent them from deviating from the prescribed course.[22]

Information about movement and activities also could be obtained and used in ways that violate privacy. Relations between employers and employees could deteriorate if some employees feel burdened or controlled by employer or job requirements. Their supervisors, however, may welcome the ability to monitor activities even during off hours. People may feel that there is no way to escape the control of others if they have no choice but to be equipped with wireless technologies as part of their jobs.[23] Chapter 10 discusses the privacy of location information in more detail.

---

[17] James E. Katz, "Caller ID, Privacy and Social Processes," *Telecommunications Policy*, vol. 14, No. 5, October 1990, pp. 372-411, and James E. Katz, "Controlling Access: Demographic Characteristics of Unlisted/Nonpublished Subscribers," Bellcore Technical Memorandum, Morristown, NJ, 1993.

[18] Ibid.

[19] Michael Ventura, "Trapped In the Time Machine," *The Washington Post*, Feb. 12, 1995, pp. C1, C4, excerpted from "The Age of Interruption," *The Family Therapy Networker*, vol. 19, No. 1, Jan. 1, 1995, pp. 18-25, 28-31.

[20] Randy Katz, wireless project manager, Advanced Research Projects Agency, interview, June 9, 1994.

[21] Joseph Hoshen, Jim Sennott, and Max Winkler, "Keeping Tabs on Criminals," *IEEE Spectrum*, February 1995, pp. 26-32, details location monitoring technologies in use and under development for nonincarceration alternatives to imprisonment. All the techniques use wireless systems, and newer systems provide highly accurate and continuous monitoring of location. Future systems are envisioned that will be able to actively restrain people who violate their conditions of parole, such as sounding an alarm or causing an electric shock or other restraining action.

[22] Shiv Sharma, "Sports Diary," *Manchester Guardian Weekly*, Mar. 5, 1995, p. 31.

[23] For literary treatment of these ideas, see George Orwell's *1984* (San Diego: Harcourt, Brace, Jovanovich, 1984), and Jerzy Kosinski's *The Painted Bird* (Boston: Houghton Mifflin, 1976).

The specter of gradually ceding the right to be left alone is of great concern to many. The time people spend alone in their automobiles is increasingly a thing of the past as commuters now conduct business with cellular telephones, car faxes, laptop computers, and books, journals and office materials on tape.[24] The feeling of personal solitude may well be eroded when a person can make a cellular or satellite phone call from every area of the planet, or know where someone is located to within 10 meters.

William Safire, the *New York Times* columnist and former speechwriter for President Richard Nixon, has commented:

I have fended off the threat of intrusive wireless communication almost from its inception. At the Moscow Summit in 1972, President Nixon's chief of staff, H. R. Haldeman, introduced us to the new "beeperphone." Through this amazing paging device, worn on the hip, the nation's chief executive could instantly track down any of his score of assistants anywhere in the capital of the rival superpower at any moment.

Being on the end of an electronic leash did not appeal to me; indeed, its big-brother aspect struck me as more representative of the Soviet society than our own...

Think again about the rush to total intouchedness. The telecommunications that produced telemarketing can produce telefugitives. No slack can be cut in the wireless wire; a society with no place to hide produces people with no

secrets worth keeping and individuals with no minds of their own.[25]

Personal Safety

Widespread deployment of wireless telecommunications systems may also lead to an increased sense of personal safety and security, because people can call for help regardless of where they are, and can report accidents and other incidents in situations where assistance may be required. In a Gallup survey, a total of 91 percent of respondents believed that having a cellular telephone made them safer and more secure, and 90 percent said they would be more willing to lend a helping hand to a stranger because they can call for help (table 2-6).[26] Accidents along major roadways in big cities result in an average of eight calls to 911.[27]

Using communications technology for safety purposes may be a double-edged sword, however. Mobile communications are extensively used by criminals as well as law-abiding citizens, and mobility can make criminals more effective and threatening.[28] Indeed, the demand for altered phones is fueled in large part by people who want to use cellular telephones to commit crimes.

■ Implications for Organizations

Wireless telecommunications technologies will find a significant role in the workplace and in organizations. Strategies to deal with the new possibilities of wireless work are being experimented

[24] Rajiv Chadrasekaran, "For Some Area Commuters, Work Begins Behind the Wheel," *The Washington Post*, Aug. 9, 1994, pp. 1, 10.

[25] William Safire, "Stay Out of Touch," *The New York Times*, Nov. 1, 1993, p. A1.

[26] The Gallup Organization, "The Motorola Cellular Impact Survey: Evaluating 10 Years of Cellular Ownership in America," Princeton, NJ, 1993, p. 11.

[27] This common use of cellular telephones has driven much of the secondary growth in cellular subscriptions. Cellular companies have recognized this and offer payment plans targeted to this segment of the market. However, easy accident reporting sometimes creates problems, because police must decide which of the calls gives the correct information, location, etc. People vary widely in the accuracy of their reporting.

[28] "Chicago Council Considers Measure Limiting Pay Phones," *Telecommunications Reports*, Sept. 19, 1994, pp. 7-8.

TABLE 2-6: Cellular Telephone Use and Personal Safety		
Have you ever used your cellular telephone to call:	**Percent responding "yes"**	
	1991	**1993**
For roadside assistance for your own disabled vehicle	31	38
For roadside assistance for someone else's disabled vehicle	7	13
For assistance for your own medical or health emergency	7	13
For assistance for someone else's medical or health emergency	23	28
The police to warn of hazardous road conditions (e.g., collapsed roadway, downed trees, weaving driver, or icy road)	24	28

SOURCE: The Gallup Organization, "The Motorola Cellular Impact Survey: Evaluating 10 years of Cellular Ownership in America," Princeton, NJ, 1993, p.11.

with by a small number of early adopters.[29] Interest is growing among policymakers as well; the National Research Council recently released a report addressing some of these issues.[30]

New telecommunications technologies, principally computer networks, but also mobile computing and wireless telecommunications systems, make it easier to decentralize or deconcentrate central office operations and introduce new spatial relationships among workers. These changes are still not widespread, but there is some evidence that new organizational relationships—such as subcontracting, teaming and contingent organizational forms, and greater demands for flexible response to changing market conditions—are facilitated by use of new telecommunications technologies.[31] These are not solely due to wireless telecommunications technologies, but it seems certain that these technologies will play a role in such restructuring.

Another force driving the restructuring of physical organizations is the cost of space and facilities. Firms with large numbers of mobile employees see the cost of a private office as a drain on company revenues and some are experimenting with alternative arrangements that take advantage of mobile technologies. In occupations where workers spend a lot of time out of the office, such

[29] Early adopters were different from later adopters. Researchers are careful not to extrapolate projected usage patterns too far from this early adopter group. See DYG, Inc., op. cit., footnote 12, p. 5.

[30] National Research Council, *Research Recommendations To Facilitate Distributed Work* (Washington, DC: National Academy Press, 1994), p. 37. The study was requested by the Department of Energy in 1993.

[31] There is a growing literature in this area. Aspects of this development were addressed in U.S. Congress, Office of Technology Assessment, *Electronic Enterprises: Looking to the Future*, TCT-600 (Washington, DC: U.S. Government Printing Office, May 1994). See also Robert G. Eccles and Richard L. Nolan, "A Framework for the Design of the Emerging Global Organizational Structure," in *Globalization, Technology, and Competition: The Fusion of Computers and Telecommunications in the 1990s*, Stephen P. Bradley, Jerry A. Hausman and Richard L. Nolan (eds.) (Boston, MA: Harvard Business School Press, 1993), pp. 57-80; Tom Malone, J. Yates, and R. I. Benjamin, "Electronic Markets and Electronic Hierarchies: Effects of Information Technology on Market Structure and Corporate Strategies," *Communications of the ACM*, vol. 30, No. 6, June 1987, pp. 484-497; Ajit Kambil, "Information Technology and Vertical Integration: Evidence from the Manufacturing Sector," in Steve S. Wildman and Margaret Guerin-Calvert, *Electronic Services Networks: A Business and Public Policy Challenge* (New York, NY: Praeger, 1991); Stuart Smith, David Transfield, Hohn Gbessant, Paul Levy, and Clive Ley, "Factory 2000: Design for the Factory of the Future," *International Studies of Management and Organization*, vol. 22, No. 4, pp. 61-68. Examples focusing on wireless include: Mel Mandell, "Office of the Future?" *Across the Board*, October 1994, pp. 45-47; Alison L. Sprout, "Moving into the Virtual Office," *Fortune*, May 2, 1994, p. 103; Kirk Johnson, "Evolution of the Workplace Alters Office Relationships," *New York Times*, Oct. 5, 1994, pp. B1, B3.

as insurance adjusters or management consultants, alternatives could produce cost savings.[32] Nonterritorial or just-in-time offices are organized as shared facilities; work stations are allocated on a first-come, first-served basis or are shared with specific people. Moving to these newer organizational forms reduces the importance of place and increases the importance of communications links and networks. Further deconcentration may be facilitated by the widespread availability of wireless telecommunications. Long-term productivity benefits are as yet unknown because the short-term real estate savings, which can be significant,[33] may mask the effect of mobile office designs on work performance and employee development.[34]

Mobile work may result in more individual autonomy for workers because they will increasingly be able to work outside of traditional office settings.[35] Managers will have less visual assurance of job performance, and will have to rely more on other, perhaps performance-based, measures of job fulfillment.[36] Many supervisors ask how they can be sure their employees are working when they are not in the office.

On the other hand, such wireless-facilitated mobile work may also increase workers' isolation. They may have less face-to-face contact with co-workers and spend significant amounts of time away from their families.[37] Employee stress and burnout may increase in companies that adopt mobile office concepts. In some cases, employees are responsible for some of the costs associated with working on the road and for their home base; having to pay these costs themselves could undermine morale. Mobile workers typically work longer and harder than their office-located counterparts. Reconciling the desire to get more work out of employees with the need to keep morale high poses some dilemmas for firms.

Productivity and Efficiency

In business, the ability to be in touch with others through wireless telecommunications may be a real benefit to those who spend time away from telecommunications systems unwillingly, such as road-bound sales representatives, nurses on the move, or soldiers in the field. There is growing evidence that wireless devices drastically cut the time required to locate people in offices and hospitals.[38] Stockbrokers find it increasingly difficult to be out of touch with the global securities and financial markets because a gap in their trading day can mean large shifts in market positions and

---

[32] IBM has cut real estate costs by 50 percent for its marketing and sales costs in the New York-New Jersey area by moving to a converted warehouse in Cranford, New Jersey. Office space decreased by 75 percent, and only 200 of the 700 employees have permanent desks. Ira Sager, "The Few, the True, the Blue," *Business Week*, May 30, 1994, pp. 124-126. In one comparative study of such new office facilities, consulting firms Anderson Consulting in San Francisco and Ernst & Young in London reduced their need for space by 68 and 32 percent, respectively, saving $137,000 and $383,000 per year in gross space costs for 70 and 96 people. Franklin Becker, Bethany Davis, and William Sims, "Using the Performance Profile To Assess Shared Offices," *Facilities Management Journal*, May/June, 1991, pp. 13-29.

[33] Mel Mandell, "Office of the Future?" *Across the Board*, October 1994, pp. 45-47.

[34] Sue Shellenbarger, "Overwork, Low Morale Vex the Mobile Office," *The Wall Street Journal*, Aug. 19, 1994, pp. B-1, B-4.

[35] DYG, Inc., op. cit., footnote 12, lays out many of the characteristics of this type of worker, often called the untethered worker, the mobile worker or the self-contained worker. See also Mark Weiser, citation in National Research Council, op. cit., footnote 30.

[36] For example, see National Research Council, op., cit., footnote 30, pp. 12-13. This point is echoed frequently in the business press and wireless telecommunications trade press.

[37] Kirk Johnson, "New Breed of High-Tech Nomads: Mobile Computer Carrying Workers Transform Companies," *The New York Times*, Feb. 8, 1994, pp. B1, B5.

[38] One study found that the time: 1) required to locate a nurse fell from 28 minutes to 20 seconds with a wireless, office-based telephone system, 2) a nurse waited by a phone for a returned page fell from 52 minutes to less than 2 minutes, and 3) callers were put on hold fell from 62 minutes to 36 minutes. "Effects of Communication Delays on Hospitals: SpectraLink Workflow Study Results, August 1993," SpectraLink Company document, n.d.

TABLE 2-7: Business Performance Due to Cellular Telephone Use		
	Percent of cellular telephone users agreeing strongly or agreeing somewhat in:	
Cellular telephones have:	**1991**	**1993**
Increased your flexibility	91	97
Increased your efficiency	87	91
Enhanced communications which has made your life less stressful	83	82
Made you more productive at work	81	84
Made you more competitive	67	73
Added a significant amount of time to your day	78	80
Made you more successful in business	70	74

SOURCE: The Gallup Organization, "The Motorola Cellular Impact Survey: Evaluating 10 Years of Cellular Ownership in America," Princeton, NJ, 1993.

values. Stock quote devices such as QuoTrek, Quotam, and Metriplex deliver up-to-date information via digital broadcasting facilities or FM side bands generated by radio stations.[39] Users, at least, believe that wireless technologies improve their performance (table 2-7).

Wireless telecommunications may increase productivity for workers who can perform parts of their jobs in the "dead time" while in transit between places, as noted above. One study assessed employees' ability to recapture time spent away from the office by using cellular telephones.[40] Table 2-8 gives the annual productivity gains for broad job categories.[41]

Larger amounts of time recaptured in this model yield greater productivity gains. Thus, if a sales

TABLE 2-8: Annual Productivity Gains Attributable to Cellular Telephone Use				
Occupation	Annual income (dollars)	Average hours lost per week away from office	Time spent making cellular calls, hours per week	Annual productivity gains per employee,* (dollars per year)
President or chief executive officer	100,000	12.4	1.2	2,220
Sales or other revenue-generating employee	65,000	18.8	1.6	1,200
Middle management/director/ supervisor	65,000	9.6	1.2	780
Field service person/technician	60,000	15.5	1.3	680
Technical/R&D	45,000	7.4	1.1	-60
Administrative/secretarial	30,000	3.3	0.7	-550
Entry level	25,000	3.7	0.7	-680

SOURCE: "Cellular Use and Cost Management in Business," study prepared for PacTel Cellular by Yankelovich Partners, Newport Beach, CA, 1993, pp. 15-18.

---

[39] Jay Mathews, "Getting a Grip on the Markets," *Washington Post*, May 20, 1994, pp. F1, F3.

[40] "Cellular Use and Cost Management in Business," study prepared for PacTel Cellular by Yankelovich Partners, Newport Beach, CA, 1993.

[41] Senior executives in the large-sample survey reported they were away from their offices 149 minutes per day, and that they used cellular telephones about 10 percent of this time. The study then calculated the annual productivity gain by multiplying time recaptured by the average wage rates for various job classifications.

TABLE 2-9: Survey Respondents' Estimates of Time Added Per Day Due To Cellular Telephone Use		
Time added per day	1991 (percent)	1993 (percent)
A half-hour or less	31	38
Between a half-hour and an hour	16	17
1 hour	23	24
2 hours	17	11
3 hours	4	3
4 hours or more	3	1
Don't know/not sure	6	3
Lost time/very little	NA	3
Mean average (hours)	1.06	0.92

SOURCE: The Gallup Organization, "The Motorola Cellular Impact Survey: Evaluating 10 Years of Cellular Ownership in America," Princeton, NJ, 1993.

TABLE 2-10: Survey Respondents' Estimates of Productivity Increases Due To Cellular Telephone Use		
Productivity improvement	1991 (percent)	1993 (percent)
Zero	9	16
10	17	11
20	15	18
30	15	12
40	6	6
50	10	11
60	4	3
70	6	6
80	6	9
90	2	3
100	4	1
Don't know/not sure	6	4
Mean (percent)	36	34

SOURCE: The Gallup Organization, "The Motorola Cellular Impact Survey :Evaluating 10 Years of Cellular Ownership in America," Princeton, NJ, 1993.

representative recaptured 20 percent of time away from the office, the productivity gain would be about $3,540. The negative figures for technical, R&D, administrative, and entry level categories indicate that the productivity gain due to recaptured time does not cover the cost of wireless service and equipment.

Another survey reported on the time added to a person's day, which averages about one hour, and on productivity, which averages about 35 percent (tables 2-9 and 2-10).[42]

Although these numbers are suggestive of the positive effects mobile wireless technologies could have on productivity and efficiency, few studies of the deployment of these technologies have been undertaken. The National Research Council report on distributed work notes that such work can enhance productivity, but it also suggests that sociological and organizational studies of distributed work will be needed to ensure that distributed work can be carried out to serve the needs of individuals and organizations effective-

ly.[43] In addition, productivity improvements due to communications and computing technologies are difficult to measure.[44] Quantitative research is needed to determine the effects of wireless telecommunications on productivity.

▮ Implications for Society

Universal Service

One of the promises of wireless systems is that they can provide communication and information services to citizens who cannot access them via wireline models or who cannot afford them. While about 94 percent (240 million) of the U.S. population currently have telephone service, 6 percent (15.3 million) do not. A number of underserved populations could benefit from the use of wireless technologies (see chapter 9).

For example, there are four to five million migrant farmworkers without a permanent residence

---

[42] The Gallup Organization, op. cit., footnote 26. These figures have declined from 1991 to 1993, probably because more cost-conscious users, the so-called "second-tier" users, have subscribed.

[43] National Research Council, op. cit, footnote 30, p. 37.

[44] U.S. Congress, op. cit., footnote 31, pp. 51-52. See also, Richard A. Kuehn, "Enhanced Technology Doesn't Always Enhance Productivity," *Business Communications Review*, vol. 24, No. 4, April 1994, p. 83.

living and working in the United States,[45] and possibly two million homeless people living in shelters or on the street.[46] These people are inherently mobile, and thus have the most difficult time gaining access to reliable and affordable communications services.[47] There is currently no effort at the federal level to address their need for telecommunications access.

Recently, social agencies have begun to provide homeless people with voicemail boxes to facilitate their efforts to find employment, and to stay in touch with support services, families, and others.[48] Advocates for the homeless say that optimally, people should have access to immediate communications, such as might be provided by wireless, which would help assure better safety, services and employment prospects, all key concerns for the homeless. Failing personal telephone service, voicemail is an attractive alternative.

Land Use and Transportation Effects

Regional sprawl may also be associated with wireless telecommunications—easier mobile communications together with easy transportation may exacerbate travel patterns already in place due in part to earlier development of transportation and communications infrastructures.[49] Past telecommunications development facilitated (though probably did not cause) the growth and power of major cities and urban cores, while at the same time enabling production to be coordinated in factories located outside the cities. In many cases, the dispersal of production into outlying areas promoted the relocation of people to those areas as well. While this migration was not caused directly by either telecommunications or transportation system improvements, it is unlikely that such changes would have been so great without them.

The effect of wireless telecommunications on travel behavior and land use has not been widely studied, but preliminary work suggests that it may contribute to urban and suburban sprawl.[50] Already there is evidence that car offices are used increasingly by mobile professionals and services.[51] This minimizes the need for costly office overhead, but presumably increases the time

---

[45] U.S. Department of Health and Human Services, Public Health Service, Health Resources and Services Administration, *An Atlas of State Profiles Which Estimate Number of Migrant and Seasonal Farmworkers and Members of Their Families* (Washington, DC: U.S. Government Printing Office, March 1990); National Advisory Council on Migrant Health, *1993 Recommendations of the National Advisory Council on Migrant Health,* (Rockville, MD: National Advisory Council on Migrant Health, May 1993). Migrant workers are difficult to identify, because of their mobility and language differences from the majority population. Various federal agencies have different definitions and counting methods. See Valerie A. Wilk, *The Occupational Health of Migrant and Seasonal Farmworkers in the United States,* (Washington, DC: Farmworker Justice Fund, Inc., 1986), pp. 11-12.

[46] National Coalition for the Homeless, "How Many People Are Homeless in the U.S. and Recent Increases in Homelessness," information sheet, issue no. 5 (Washington, DC: Homelessness Information Exchange, National Coalition for the Homeless, January 1994). The Census Bureau estimates, conservatively, that there are about 250,000 homeless people in the United States. U.S. Department of Commerce, Bureau of the Census, *Statistical Abstract of the United States, 1994* (Washington, DC: U.S. Government Printing Office, September 1994), table 84, p. 69.

[47] See Chantal de Gounay, "L'âge du citoyen nomade," *Esprit* Paris, France, no. 11, November 1992, pp. 113-126, for a discussion of contemporary nomadism and culture in advanced industrial societies.

[48] N.R. Kleinfield, "For Homeless, Free Voice Mail Can Be a Key to a Normal Life," *The New York Times,* Jan. 30, 1995, pp. B1, B6; "Hold My Calls," *Newsweek,* Mar. 30, 1992, p. 9; "No Home: Please Hold," *The Economist,* Dec. 18, 1993, p. 29.

[49] Ithiel de Sola Pool et al., "Foresight and Hindsight: The Case of the Telephone," in *The Social Impact of the Telephone,* Ithiel de Sola Pool (ed.), (Cambridge, MA: The MIT Press, 1977), pp. 127-158.

[50] Youngbin Yim, Institute of Transportation Studies, University of California, Berkeley, interview, Jan. 24, 1995. See also Youngbin Yim, Adib Kanafani, and Jean-Luc Ygnace, "Expanding Usage of Cellular Phones: User Profile and Transportation Issues," PATH Research Report, UCB-ITS-PRR-91-19, (Berkeley, CA: University of California, Institute of Transportation Studies, Program on Advanced Technology for the Highway, December 1991).

[51] See, for example, Sue Ellen Christian, "It's Not a Car, It's a Mobile Office," *The Washington Post,* Aug. 8, 1994, pp. 17, 21.

spent on the road, and enables greater deconcentration from central facilities. Much more
work will be required to determine what factors
account for urban form and land use, including
sprawl, and what role wireless telecommunications technologies may play.[52]

---

[52] An ongoing OTA study is examining questions of information and other technologies and urban form. See U. S. Congress, Office of
Technology Assessment, *Technological Reshaping of Metropolitan America,* (in progress).

Part B:

Wireless Technologies and Applications

Developing a framework for discussing and analyzing wireless technologies in the context of the National Information Infrastructure (NII) poses many challenges. Historically, systems and services were classified and regulated in terms of the technologies used to transmit or deliver them—broadcast, telephone, cable, satellite, cellular, microwave, and so on. Such distinctions, however, are less meaningful now because the diffusion of digital technology and the convergence of services have blurred the categories. Other categorization schemes have been suggested based on: 1) technology drivers, 2) differences in the type of service delivered (mobile or fixed access), 3) broadband or narrowband, and 4) level of interactivity.

To present the technologies and their applications in the most intuitive and understandable way, the Office of Technology Assessment uses a scheme that divides wireless technologies and applications along functional, service-oriented lines: voice, data, and broadcast and high-bandwidth applications. This scheme is not perfect; there will be overlap among categories and between systems, particularly as technology continues to advance. Some, but not all, distinctions between the categories, for example, will likely disappear as different systems begin delivering similar services and information. However, many systems are likely to remain much the same well into the future. Different consumer and business needs and costs will drive users to make many tradeoffs—between cost and coverage and speed and cost, for example—allowing many different types of systems and services to survive and prosper. Chapters 3, 4, and 5 discuss the technologies being developed to provide wireless voice, data, and video/broadband services, respectively.

- Voice Technologies and Applications
- Wireless Data
- Broadcast and High-Bandwidth Services

Voice Technologies and Applications | 3

Much of the media attention surrounding wireless technologies has focused on mobile telephone services, primarily cellular telephony and new Personal Communication Services (PCS). Industry representatives and analysts have pointed to the high growth rates of cellular service as evidence of pent-up demand for mobile voice services. In response to this perceived demand, existing wireless carriers and new companies are planning to greatly expand the capacity and variety of wireless voice services they provide. The first part of this chapter examines the systems—both existing and under development—that will offer mobile voice communication services. Mobile data services, often provided by the same physical systems, are discussed in the following chapter.

In addition to providing *mobile* services, wireless technologies can also be used in *fixed* applications—to provide telephone (and data) service to homes and businesses.[1] Radio-based technologies may serve some households more efficiently or easily than traditional wireline technologies, and, in particular, wireless may be less expensive than wireline in remote areas, where long copper loops are expensive to install and maintain. However, wireless may play a role even in urban areas because it may allow new competitors to enter the market for local telephone services. With a few transmitters, new entrants can provide local exchange service to a neighborhood, avoiding the expense of re-creating the incumbent local exchange carrier's extensive copper network.

---

[1] "Fixed" refers to the fact that the user's equipment is physically connected to a specific location.

Fixed wireless services are discussed in the second half of this chapter. Many of the issues and implications of deploying wireless voice services in the National Information Infrastructure (NII)—such as interconnection, health concerns, and standards—are discussed more extensively in later chapters.

FINDINGS

- **The regulatory distinction between mobile and fixed wireless services, while based on valid historical, technical, and regulatory reasoning, is becoming increasingly unclear and should be revisited.** The wireless technologies that will be used to provide mobile telephone services and fixed services are very similar. In fact, it is possible to serve both fixed and mobile users with the same network. Current regulations, however, continue to treat fixed and mobile voice services differently, based on technical limitations that no longer exist and the protection from competition that regulators afforded the local telephone companies. Under current rules, the treatment of various mobile service providers—including cellular and PCS—regarding services provided to fixed locations remains inconsistent and unclear. The Federal Communications Commission (FCC) may need to clarify the conditions under which wireless providers can provide fixed service. Without action on this issue, wireless will be unable to compete effectively in the market for local telephone service.

- **OTA finds that the amount of spectrum dedicated to terrestrial mobile voice services is currently adequate, but additional allocations may be required over the long term.** Over the last three years, the FCC has allocated a large amount of spectrum for terrestrial mobile services. This should provide adequate capacity for current mobile voice services until

after the turn of the century. However, if current voice systems plan to upgrade their services to provide high-speed data, video, and multimedia applications, current spectrum allocations may be inadequate in the long term. If high-bandwidth services take off, additional spectrum may be needed.

The need for additional spectrum for commercial mobile satellite services, however, is less clear. U.S. satellite companies have long maintained that international and domestic frequency allocations are inadequate—limiting the services that can be provided and the number of companies that can compete in the market. The U.S. government has vigorously pursued additional spectrum allocations in international fora for a number of years, an effort that will continue at the 1995 World Radiocommunications Conference. However, at least five companies are poised to enter the satellite voice communications market over the next five years, and more firms may try to join in. Given the number of companies planning to offer satellite-delivered voice communications services and the uncertainty of the demand for such services, it is far from certain that the market will be able to support these firms.[2] Such spectrum needs should be carefully evaluated against other uses of the spectrum.

Public safety users have long fought for more spectrum, but their needs continue to be unmet. Congestion of public safety radio spectrum is common, and users report that it can seriously impact the usefulness of public safety radio systems. The growing use of data, images, and even video in law enforcement will severely tax public safety radio frequencies. A recent congressionally mandated FCC study of public safety spectrum needs has been criticized by the public safety community for seriously underestimating their needs. A more

---

[2] For a more complete discussion of the marketing and technical challenges facing mobile satellite companies, see U.S. Congress, Office of Technology Assessment, *The 1992 World Administrative Radio Conference: Technology and Policy Implications,* OTA-TCT-549 (Washington, DC: U.S. Government Printing Office, May 1993).

indepth evaluation of these needs is required, including an analysis of technology trends that could either alleviate the problems or exacerbate them.

- The emergence of competition in the market for mobile voice communications is likely to benefit consumers by lowering prices, encouraging higher quality and reliability, and promoting innovation that could lead to a wide range of new services. However, **new competitors to the incumbent cellular service providers will face technical and economic challenges that may ultimately result in the benefits of competition being less than proponents predict.** For example, although a given geographic area could potentially have up to 10 competing mobile telephone providers, it is unlikely that the more sparsely populated, rural areas of the nation will see this level of competition. These areas will not have enough prospective customers to support a large number of service providers. The long-term effect may be that in some areas, competition will not be sustainable and the benefits promised do not materialize. Although such shakeouts are a normal byproduct of competition, their longer term effects on prices and the diversity of services remain uncertain. If and when wireless communications systems become carriers of last resort, the effects of these long-term market structure concerns will be magnified.

MOBILE VOICE SERVICES

For most of the history of telecommunications, users have only been able to communicate to and from fixed locations—wherever the copper wires could reach. In the past few years, however, advances in wireless technologies have made it pos-

sible to imagine a future in which communication can take place anytime and anywhere. Mobile phone services, which allow users on the move to make and receive calls much as they would with an ordinary wireline phone, will play an increasingly important role in the NII. Within a decade, according to some projections, there could be almost 100 million mobile phones in use.[3] New wireless technologies may lead to a shift in the nature of communications, away from today's model of place-to-place communications to one based on person-to-person communications.

■ The Evolution of Mobile Telephone Service

Mobile telephone service began in 1946,[4] but subscribership grew very slowly. Because the FCC allocated only a small amount of spectrum to mobile telephony, systems were limited in the number of users they could support. Demand for service quickly outstripped capacity, leading to poor service at busy times of the day. Users often would have to try several times before their call went through. In some cities, carriers had to restrict the number of subscribers in order to maintain a reasonable level of service. For example, in 1978, the mobile telephone system in New York served only 525 customers, and there were 3,700 customers on the waiting list.[5] Even with restrictions on the number of subscribers, over half of the calls attempted did not go through.[6]

Wireless telephone service entered a new era when the first cellular telephone system began operating in Chicago in 1983. The FCC allocated much more spectrum to the cellular operators than it had previously allocated to mobile telephone services. In addition, the use of cellular technolo-

---

[3] Personal Communications Industry Association, "1994 PCS Market Demand Forecast" (Washington, DC: Personal Communications Industry Association, January 1995); Personal Communications Industry Association, "PCIA 1995 PCS Technologies Market Demand Forecast Update, 1994-2005," (Washington, DC: Personal Communications Industry Association, January 1995).

[4] The first system was in St. Louis. In less than a year, mobile telephone service was being offered in more than 25 cities. For a discussion of the early history of mobile communications, see George Calhoun, *Digital Cellular Radio* (Norwood, MA: Artech House, 1988).

[5] William C.Y. Lee, *Mobile Cellular Telecommunications* (New York, NY: McGraw Hill, 1995), p. 2.

[6] Ibid., p. 3.

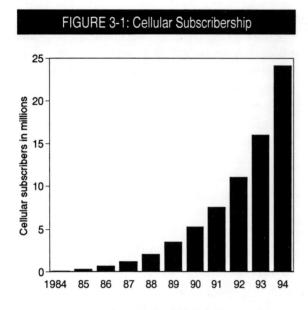

FIGURE 3-1: Cellular Subscribership

SOURCE: Office of Technology Assessment, 1995.

gy allowed system operators to use their spectrum more efficiently. Subscribership grew steadily in the 1980s, as businesses and professionals recognized the advantages of being able to stay in touch at all times (see figure 3-1); in the early 1990s, subscriber growth reached 40 per cent per year.[7] As prices have decreased, however, the profile of the typical cellular user has changed. Cellular carriers have begun to tap the broader consumer market and attract customers who use their phones for personal, rather than business, calling. There are now over 24 million users of cellular service.[8]

Many believe that the high rate of growth in mobile telephony will continue for the foreseeable future. In part, these projections are based on the fact that cellular penetration is still only 10 percent of the potential market.[9] However, future growth will also be driven by technological advances that enable a more functional, lower cost service. Handsets are becoming smaller, lighter, and less expensive, continuing their evolution

from bulky car phones to small portables. In addition, the transition from today's analog wireless technology to digital technology will allow wireless systems to support many more users at a lower cost per user. The combination of affordable service and small handsets has allowed service providers to envision a future in which tens of millions of users take pocket phones with them everywhere they go.

The projected growth in demand for mobile telephone service led the FCC to allocate a large amount of additional spectrum to mobile telephony in 1994. This new spectrum will be shared by up to six additional wireless operators in each market. The FCC refers to these new licensees as *Personal Communications Service* providers, reflecting the new vision of mobile communications systems targeted to users with pocket phones rather than car phones. PCS providers will compete with the existing cellular operators, driving the cost of mobile telephony down even further, and also will explore new niche services. Some PCS providers plan to offer service by the end of 1995, but most of the new operators will not begin service until the end of 1996 or early 1997. Additional competition will be provided by the Specialized Mobile Radio (SMR) operators, who are beginning to transform their dispatch systems into true mobile phone services by deploying a new generation of technology.

New technologies are also expanding the reach of mobile communications services. For example, network operators are increasingly providing in-building coverage in arenas, train stations, and public buildings. In addition, the deployment of a new generation of satellite systems will allow users to communicate wherever they are in the world—on ships, on airplanes, and in remote areas that could never support a terrestrial wireless service such as cellular. In the future, a single phone may be able to act as a cordless phone in the

---

[7] Cellular Telecommunications Industry Association, *Industry Data Survey,* December 1994.

[8] Ibid.

[9] Ibid.

FIGURE 3-2: Generic Wireless Communication Architecture

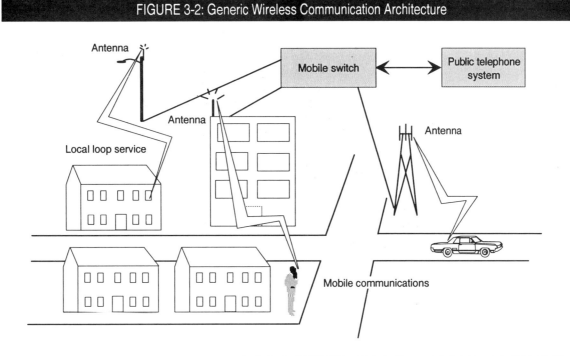

SOURCE: Office of Technology Assessment, 1995.

home, a cellular phone in the city, and a satellite phone when traveling in remote areas. Seamless systems that integrate all of these functions will help realize the vision of "anytime" and "anywhere" personal communications services.

∎ Services and Users

Three mobile phone services—cellular, PCS, and SMR—use terrestrial wireless technologies, relying on antennas mounted on buildings and towers to provide radio coverage in cities and along highways (see figure 3-2). These terrestrial systems will be complemented by mobile satellite services, which can provide mobile telephone service in areas where terrestrial systems are not viable. It is difficult to draw distinctions between the three terrestrial mobile telephone services because the technology they use and the services they provide are similar. However, they differ to some extent in their history, industry structure, and target market.

Cellular Telephony

Cellular is the best known and most established mobile telephone service, drawing its name from a system design concept that allows for efficient use of the spectrum. At first, cellular operators designed their networks to provide a car phone service. Over the past decade, however, technological advances have allowed the manufacture of small portable phones that weigh only a few ounces. As a result, cellular systems increasingly are being designed to provide good coverage for pedestrian users as well, both indoors and on city streets. Cellular systems are operational in most cities and larger towns, and along most major highways as well.

The cost of becoming a cellular user has declined substantially over the past decade, due primarily to the impact of economies of scale on the cost of the phone. The apparent cost of the handsets is further reduced by subsidies that the cellu-

As cellular telephone technology has advanced, phones have become progressively smaller and lighter, as seen here with a car-mounted cellular phone (top left), a transportable phone (bottom left), and a pocket-sized phone (above).

lar carriers use to attract new customers. Customers then pay a basic monthly rate, as well as a per-minute *airtime* charge. Airtime charges vary by company and by time of day, but typical rates during the day range from 30 to 40 cents per minute, with lower rates in the evening and on weekends. Carriers also offer a range of calling plans, targeted at different users, that include some "free" minutes. As lower volume customers who use their phones primarily for personal calling or in case of emergencies have signed up, the average monthly bill for cellular service has declined from $96.83 in 1987 to $56.21 in 1994.[10]

Cellular users can choose between two providers in each market. One of the carriers, the *B-side* or *wireline* carrier, is a subsidiary of the local telephone company; the other carrier, the *A-side* or *nonwireline* carrier, is independent—although many A-side carriers have been acquired by or entered into agreements with telephone companies operating out of their home territories. In creating the cellular industry, the FCC divided the country into 734 markets and assigned licenses separately for each market. As a result, ownership in the industry has been highly fragmented. Over the past several years, however, there has been considerable consolidation as carriers have acquired or merged with other carriers.[11] In part, industry consolidation has been driven by the need to assemble capital for the PCS auctions, but it also allows car-

[10] Ibid.

[11] See, for example, John J. Keller and Leslie Cauley, "Fear of Being Left Out of a Wireless Future Spurs Frantic Alliances," *The Wall Street Journal*, Oct. 25, 1994, p. A1.

```
┌──────────────────────────────────────────────────────────────────────────────┐
│                              BOX 3-1: Roaming                                  │
├──────────────────────────────────────────────────────────────────────────────┤
```

When the Federal Communications Commission created the cellular industry, it divided the nation into many small license areas. Because users did not want to be restricted to using their home system, the cellular industry has worked to make it possible for users to make and receive calls while traveling outside the home area. This is called *roaming*.

One basic requirement for roaming—that a user's phone be compatible with all cellular systems—was met when the FCC instructed all cellular carriers to use the same technology, the Advanced Mobile Phone System (AMPS). In the future, however, compatibility may not be guaranteed. The FCC has not specified a standard technology that all carriers have to deploy as they upgrade to digital, the next generation of cellular technology. These standards issues are discussed in detail in chapter 6.

Roaming also requires that the home and visited systems be able to exchange messages about the roamers. Before it allows a roamer to make a call, the visited system checks with the roamer's home system to determine if they are a valid user or a fraud risk. The visited system also tells the roamer's home system where its customer is located. The home system is then able to forward any incoming calls to the visited system, allowing users to receive calls wherever they are located. To exchange messages about roamers, the cellular industry has set up roaming networks using leased lines and special computer communications systems.

Roaming has become easier over the past five years, but can still be problematic. Not all carriers have deployed the most advanced roaming technology. In some cases, roamers have to give a credit card number before they can make a call or have to dial a special code in order to activate call delivery every time they enter a different service area. In addition, carriers often impose a daily fee on roamers, as well as per-minute charges much higher than their home airtime rates, although carriers have begun to compete with each other on roaming charges. Roaming is generally easiest among properties owned by the same carrier, or among carriers that have agreed to an alliance that includes a common brand name.

SOURCE: Office of Technology Assessment, 1995.

riers to offer a larger service area than their competitor, an important selling point.

The recent consolidation is the latest effort by the cellular industry to overcome the fragmented licensing structure imposed by the FCC. In the early years of cellular, subscribers were limited to service within their home market. But users soon demanded the ability to make calls when they traveled in other cities or to continue calls when they drove into a neighboring license area. Users needed to be able to temporarily use another operator's system, which is known as *roaming* (see box 3-1). Cellular carriers have worked together to develop the technologies and business relationships that allow users to make and receive calls outside their home service area, but roaming is still not always seamless. Users may have difficulty placing calls, and calls to them may require callers to know where they are and dial access codes. Moreover, users incur substantially higher airtime charges when roaming, often $1 per minute or more.

Personal Communications Services

In 1993, the FCC reallocated 120 megahertz (MHz) of spectrum for PCS.[12] This spectrum is between 1850 and 1990 MHz, often referred to as the *2 gigahertz band* (cellular, on the other hand, is

---

[12] Federal Communications Commission, *Second Report and Order, Amendment of the Commission's Rules To Establish New Personal Communications Services*, GEN Docket No. 90-314, 8 FCC Rcd 7700 (1993).

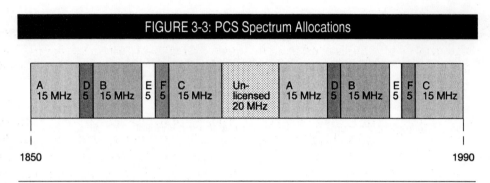

FIGURE 3-3: PCS Spectrum Allocations

SOURCE: Office of Technology Assessment, 1995.

in the *800 megahertz band*). The 120 MHz will be divided among six licensees in each market—three will get 30 MHz and three will get 10 MHz (see figure 3-3). The 30 MHz blocks are comparable in size to the 25 MHz blocks assigned to the cellular carriers, while the 10 MHz blocks could either be used for niche services or aggregated with other PCS or cellular spectrum. Depending on whether the 10 MHz blocks are used for a stand-alone service or aggregated with other spectrum, there will be between three and six PCS carriers in each market.

The FCC has defined PCS broadly as a "family of services" that will serve a variety of communications needs.[13] In practice, the term PCS is used less to define a particular wireless service and more as a label for the operators that will be using the new 2 GHz allocation. At first, it was believed that PCS providers would offer a service somewhat distinct from that offered by the cellular operators. According to this concept, PCS would be a lower cost service than cellular, but would not offer the same functionality. It would be an *enhanced cordless phone* or *low-tier* service that would not support vehicular-speed mobility, but would still allow pedestrian users to make and receive calls. Because the system would not be required to support vehicular-speed mobility, the

handsets could be simpler and therefore smaller, lighter, and less expensive.

Over the last several years, however, proposed PCS services have begun to look more like those offered by the cellular carriers. One reason is that potential licensees have come to believe that there is greater demand for a service that can be used in both the pedestrian and vehicular environments. Moreover, even *high tier* cellular-type handsets are becoming smaller and less expensive. As a result, it now appears that the main impact of the new PCS spectrum will be in providing competition to the two existing cellular carriers.

In 1994, the FCC granted *pioneer's preference* status to three companies that the Commission believed had done significant work in experimenting with new PCS technologies.[14] These licensees—in Los Angeles, New York, and Washington—have already begun constructing their networks and may be offering service by the end of 1995. The other PCS licenses are being assigned this year by auction. The first round of auctions, for two of the 30 MHz licenses, ended in March, and the remaining licenses will be auctioned later in 1995. The first of the systems built by an auction winner is not expected to be operational until the end of 1996.

---

[13] Ibid., p. 7713.

[14] Federal Communications Commission, *Tentative Decision and Memorandum Opinion and Order, Amendment of the Commission's Rules to Establish New Personal Communications Services,* GEN Docket No. 90-314, Nov. 6, 1992. The FCC granted pioneer's preference licenses to American Personal Communications (APC) (for the Washington market), Cox Enterprises (Los Angeles), and Omnipoint Communications (New York).

In the first round of auctions, three entities acquired many of the licenses. One consortium that acquired a large number of licenses consisted of Sprint and three major cable companies, Cox, TCI, and Comcast; they hope to attract customers with a package of long distance, local wireline, and local wireless service. AT&T, which also acquired many of the PCS licenses, hopes to use PCS spectrum to fill in the gaps between its cellular licenses, creating a nationwide wireless network. The third active participant in the auctions, a consortium of four cellular companies, Bell Atlantic Mobile, NYNEX Mobile, AirTouch, and US West New Vector, is pursuing the same strategy. They will offer their customers "dual-band" phones that work at the cellular frequencies where the carrier has cellular licenses and at the PCS frequencies where the carrier has PCS licenses.

Specialized Mobile Radio

Until the early 1980s, the primary use of wireless systems was for business and public safety dispatch communications. In dispatch communications, brief messages with a duration of less than a minute are exchanged between a control center and mobile users in the field. Dispatch systems are widely used by police and fire departments, taxicabs, delivery services, and construction companies. Because dispatch systems are used primarily for the internal communications needs of an organization, and are generally not interconnected with the landline public switched network, the FCC classifies them as private mobile radio services.

In some cases, organizations operate their own dispatch system. In others, they obtain service from a third party, known as a *Specialized Mobile Radio* (SMR) provider. Instead of each business operating its own dispatch radio system, the SMR carrier operates the system and sells dispatch service to several different businesses. Taxicabs, plumbing companies, and limousine services are good examples of customers that use SMR dispatch service. The SMR service was established by the FCC in 1974.

Although dispatch service is the traditional mainstay of SMR carriers, some SMR systems, especially those in rural areas, provide interconnected mobile telephone service. The spectrum inefficiency of SMR technology is not as critical in rural areas, allowing it to compete with cellular. In addition, cellular service came last to rural areas, many years after the cellular networks in the cities began operating; the last of the cellular Rural Service Areas licensed by the FCC did not get service until 1992. As of December 1993, about 425,000 of 1.5 million SMR handsets could be used for interconnected service.[15]

In the future, mobile telephone service may become an even more important part of SMR service. Driving this development is Nextel, a company that began buying many small- and medium-sized SMR operators in the late 1980s. With these acquisitions, Nextel has been able to acquire licenses and systems throughout the nation. While it still has less spectrum than a PCS or cellular licensee, Nextel believes that it has enough to deploy a digital technology known as Enhanced Specialized Mobile Radio (ESMR), which replaces the traditional single SMR antenna with a cellular architecture—allowing ESMR systems to use the spectrum more efficiently than traditional SMR technology, and potentially allowing Nextel to provide a true mobile telephone service.

The deployment of ESMR technology was expected to transform Nextel into a competitor to the cellular carriers for mass market mobile phone service. However, there have been reports that ESMR sacrifices voice quality to achieve reasonable capacity in the limited SMR spectrum.[16]

---

[15] Federal Communications Commission, *Second Report and Order, Implementation of Sections 3(n) and 332 of the Communications Act,* GN Docket 93-252, Mar. 7, 1994, p. 59, at footnote 294.

[16] For a discussion of this issue, see Judith S. Lockwood, "Considering Nextel? What Wireless Users Need To Know," *Wireless,* March/April 1995, vol. 4, No. 2, p. 30.

NEXTEL

Nextel's Fully Integrated Digital Portable Flip Phone allows subscribers to place and receive voice calls, as well as receive messages, numeric pages, voice mail alerts, and text messages.

Nextel recently announced that it plans to scale back its plans to compete broadly with cellular in order to target business users, providing them with an integrated unit that combines telephone, paging, and dispatch capability in a single handset. Nextel will also try to capitalize on the fact that it owns licenses throughout the nation, allowing it to provide seamless roaming more readily than the cellular operators. There are currently about 10,000 ESMR customers[17] in Los Angeles and San Francisco, but Nextel plans to activate

service in other large markets throughout 1995 and 1996.

Mobile Satellite Services

Terrestrial wireless services such as cellular are not economical in sparsely populated areas because there are not enough users to justify the cost of building a tower every few miles. Although there is at least one cellular licensee in each of the 734 license areas defined by the FCC, the licensees typically do not provide coverage to every square mile. In rural areas, especially west of the Mississippi, there are large areas where the only cellular coverage is along interstate highways. Satellite services can fill in the gaps in areas where terrestrial systems are not viable and help realize the vision of communications services available everywhere in the nation.

Geostationary satellite systems[18]

Limited satellite telephone service has been available for several years through the International Maritime Satellite Organization (Inmarsat). The Inmarsat system was originally established to provide communications to ships, but now also provides land mobile communication. The phones are bulky, briefcase-sized units that weigh about 25 pounds and cost between $15,000 and $20,000. The service is expensive at $4.95 per minute. But Inmarsat provides telephone service almost everywhere in the world and has been widely used for disaster relief, news-gathering, and businesses such as oil exploration and mining. There are about 10,000 land mobile terminals operational in the Inmarsat system worldwide, accounting for about one-third of Inmarsat's customers.[19]

Later this year, American Mobile Satellite Corp. (AMSC) is expected to begin providing a more advanced mobile satellite service in the

---

[17] Because an ESMR customer is typically a business, each customer averages about 10 phones. Ibid.

[18] Geostationary satellites orbit the Earth 22,300 miles above the equator. At this altitude, they orbit the Earth at the same speed that the planet rotates. As a result, they appear fixed at a specific point in the sky, allowing satellite dishes on the ground to easily communicate with them.

[19] Jack Oslund, Director, External Affairs, Comsat Mobile Communications, letter to the Office of Technology Assessment, U.S. Congress, Washington, DC, Aug. 2, 1994.

United States. AMSC, a consortium of large telecommunications firms, was formed in 1987, and is currently the only company in the United States authorized to provide mobile satellite services.[20] AMSC plans to market its service as an extension of terrestrial cellular telephone systems, primarily targeting the mobile user market, although offering some fixed services as well. AMSC will offer a car phone service with the transmitter installed in the trunk of the car, but because of the large amount of power needed for the signal to reach the satellite, handheld portable phones cannot be developed for the system. The car phones will have dual-mode capability—connecting users to the cellular network in areas where there is coverage, and switching to AMSC's satellite in remote areas beyond the reach of cellular. A total of 140 cellular carriers have signed on to market AMSC phones and service to their customers. In addition to traditional cellular phone users, such as business travelers, AMSC's service is expected to appeal to trucking companies, owners of corporate or general aviation aircraft, as well as remote populations currently without phone service. AMSC expects to offer its service for $25 a month, plus about $1 per minute of usage.[21]

Low-Earth orbiting (LEO) satellite systems

A new generation of mobile satellite services is expected to become operational in the late 1990s. Instead of using a small number of geostationary satellites like those employed in the Inmarsat and AMSC systems, these new systems will consist of a constellation of many smaller satellites in low-

Earth orbit (LEO). Because the satellites orbit close to the Earth, LEO systems permit the use of a low-power handheld device about the same size as a portable cellular phone. There are two types of LEO systems. The so-called *little LEOs* are designed for low-speed data services only (see chapter 4), while the *big LEOs* are designed to provide both voice and data services.

Several companies have proposed big LEO systems (see box 3-2). Like the AMSC system, they will use a dual-mode phone that switches between cellular and satellite coverage as necessary. Handset costs will range from $500 to $3,000, and service costs are projected to range between $0.40 and $3.00 per minute. Unlike AMSC, big LEOs will offer global coverage, providing users with the convenience of service from a single provider anywhere in the world. Potential markets include international tourists, business travelers, relief organizations, and government agencies.

LEO proponents have overcome many hurdles. The first step was to obtain an international spectrum allocation at the 1992 World Administrative Radio Conference (WARC-92).[22] The five U.S. applicants who had sought approval to deploy big LEO mobile systems then had to work out a plan for sharing the small amount of available spectrum. Finally, in January 1995, the FCC granted licenses to three of the five applicants to begin construction of their systems.[23] The licensees still must obtain licenses to operate in other countries and assemble enough capital to deploy their systems, which will cost between $1.5 billion and $4

[20] LEO systems have been given authority to construct, but not yet to operate their services. AMSC's singular status, and its consortium composition, is the result of an FCC decision to grant only one license for geostationary mobile satellite service due the limited amount of spectrum available at the time. Public investors now control roughly 34.6 percent of AMSC followed by Hughes Communications (27.2 percent), Singapore Telecommunications (13.6 percent), McCaw (now AT&T) (12.5 percent), Mobile Telecommunications Technologies Corp. (7 percent), and others (5.1 percent). American Mobile Satellite Corp., *1993 Annual Report.*

[21] AMSC is already offering commercial service to trucking companies with a leased satellite.

[22] See Office of Technology Assessment, op. cit., footnote 2.

[23] FCC licenses were issued to Iridium, Inc., TRW, and Loral Qualcomm. Systems proposed by Mobile Communications Holdings, Inc. and Constellation Communications did not receive licenses. Action on these applications was deferred until January 1996 to allow the firms to show their financial qualifications. "FCC Clears Global Satellite Projects of Motorola, TRW, Loral, Qualcomm," *The Wall Street Journal*, Feb. 1, 1995, p. A4.

BOX 3-2: Proposed Big LEO Satellite Systems

In late 1990 and early 1991, five companies applied to the FCC to provide mobile communications services using low-Earth orbiting (LEO) satellites. Three systems—Iridium, Globalstar, and Odyssey—were later granted permission to construct, although final operating authority was withheld until international allocations for the links between the satellites and the gateways ("feeder" links) are agreed to and sufficient spectrum is available. The systems are now being built. Ellipso and Constellation were denied construction licenses until they could provide better financial qualifications, and have until January 1996 to do so. A sixth system, Inmarsat-P, has applied for a license in the United Kingdom.

Iridium (Motorola Satellite Communications, Inc.)

The Iridium system will consist of a constellation of 66 LEO satellites and 15 to 20 Earth-based gateways that connect users to the public switched telephone network. Iridium investors will own and operate the gateways and be responsible for obtaining national licenses for operation of subscriber handsets, spectrum utilization, transborder agreements, PSTN interconnection, service provision arrangements, and distribution agreements. The networked satellites will orbit the Earth on six different planes of 11 satellites each. They will travel longitudinally, ringing the planet from pole to pole, at an altitude of 770 kilometers and completing a full orbit in 100 minutes. The Iridium satellites will be capable of passing a telephone call directly from satellite to satellite—the only big LEO system to do so—making each satellite a small orbiting switch, and making the Iridium system the most technically complex.

Iridium plans to use dual-mode satellite/cellular handsets that will allow subscribers to use the terrestrial cellular infrastructure when available or the satellite network when the user is in an area not served by cellular. Handsets will cost up to $3,000 and calls will average $3 per minute. The system will use a combination of time division multiple access (TDMA) and frequency division multiple access (FDMA) schemes. Commercial service is expected to become available in 1998 with the company projecting a market of 1.5 million users by the year 2000.

The system is expected to cost $3.37 billion for design, production, and launch, plus $2.8 billion for operation and maintenance over the first five years of operation. Investments in Iridium totaled $1.57 billion as of February 1995, with Motorola, Inc. committed to meeting the construction costs and operating expenses necessary for system deployment. Motorola, Inc. is the largest investor with 27 percent of Iridium Inc.'s stock. Iridium's second largest investor is a consortium of 17 Japanese companies that invested about $235 million, led by DDI Corp., Japan's second-largest telecommunications company. Other investors include Vebacom GmbH, the telecommunications arm of German energy conglomerate Veba AG; Korea Mobile; Sprint; STET, Italy's PTT; Bell Canada; Raytheon; Lockheed and others.

Globalstar (Loral/Qualcomm L.P.)

The Globalstar system design calls for a network of 48 satellites located 750 nautical miles above the Earth that will relay global digital voice and data traffic from fixed and mobile handsets to a terrestrial gateway—there are no intersatellite links. Satellites have a 1,500-mile-wide footprint, and will be organized in eight planes with six satellites in each plane and provide "global" coverage between 70 degrees latitude north and south. The system will use code division multiple access (CDMA) transmission modulation. Globalstar predicts a handset priced initially at $700, and services will cost 30 cents per minute plus 10 cents per minute for interconnection. Monthly service charges will be $8 to $10. Service is scheduled to begin in 1998 with a company-projected market of 2.7 million users by the year 2002.

(continued)

BOX 3-2: Proposed Big LEO Satellite Systems (Cont'd.)

Loral/Qualcomm estimates the cost of the system at $1.554 billion, including system deployment and first-year operating costs. Globalstar, L.P., an international partnership founded by Loral Corp. and Qualcomm, Inc., invested $275 million in an initial financing round in March 1994. Funds totaling $492 million had been raised as of February 1995, including commitments from AirTouch Communications, Inc.; Alcatel N.V. and France Telecom of France; Vodafone plc of the United Kingdom; DACOM Corp. and Hyundai Electronics Industries Co. Ltd. of South Korea; Daimler Benz Aerospace AG of Germany; Finmeccanica of Italy; and the international Space Systems/Loral aerospace consortium.

Odyssey (TRW, Inc. and Teleglobe)

Unlike the Iridium or Globalstar systems, the Odyssey system is technically a medium-Earth orbiting system. Twelve satellites, equally divided into three orbital planes at an altitude of 10,354 kilometers, will provide global digital voice and data communications by linking mobile handsets with ground-based cellular and terrestrial networks via 10 or 11 earth stations, using CDMA/FDMA modulation schemes. No inter-satellite communications are planned. Handsets are expected to be priced at less than $500, and service will cost approximately 65 cents per minute, plus 10 cents per minute interconnection fees and a monthly charge of $24.

Odyssey will be established as a limited partnership, with TRW and Teleglobe serving as the founding general partners and jointly managing the project. TRW, Inc. estimates $1.8 billion to construct, launch, and operate the system for one year. Teleglobe and TRW will provide 5 percent and 10 percent of the equity, respectively. They are seeking financing for the remaining 85 percent, most of which is expected to be in equity and the balance a combination of debt and vendor financing. TRW said it has sufficient current assets and operating income to finance the project, and submitted a declaration during the licensing process committing TRW to expend the funds necessary to construct, launch, and operate the Odyssey system.

Ellipso (Ellipsat/Mobile Communications Holding, Inc.)

Ellipso plans to provide global digital voice and data services to mobile or handheld terminals through two constellations of medium-Earth elliptical orbit satellites designed to maximize service to the Earth's populated land masses. The Borealis subconstellation of 10 satellites would service northern latitudes and operate in two elliptical orbits of five satellites each with apogees of 7,846 kilometers. The six-satellite Concordia subconstellation would cover tropical and southern latitudes and operate in a single circular equatorial orbit at 8,068 kilometers. Like Globalstar and Odyssey, the satellite will serve as relays between users and gateways on Earth—no intersatellite links are planned. User terminals are expected to cost approximately $1,000 within one to two years of service initiation and 50 cents a minute for usage. They will use CDMA technology.

The system will cost $564 million to construct, launch, and operate for the initial year. MCHI said in its statement of financial qualifications that it would rely on internal support from its shareholders, vendor financing (including committed funds from Ariansespace in the form of convertible debentures), equity investments, and other committed funds to cover the expected system costs. MCHI shareholders include Barclays de Zoete Wedd Ltd. of London, Westinghouse Electric Corp., and Fairchild Space and Defense Co. Cable & Wireless plc of the United Kingdom recently acquired 50,000 shares or 2 percent of its stock with an option to acquire an additional 600,000 shares.

(continued)

BOX 3-2: Proposed Big LEO Satellite Systems (Cont'd.)

ECCO (Equatorial Constellation Communications)

Initially, 12 satellites will orbit in a single ring around the equator. The complete constellation would add seven planes of six satellites each (five operational and one spare) for a total of 54 satellites in orbit (46 operational, eight spares). The system is designed to provide mobile and fixed-site voice, data, facsimile, and position location services in more than 100 countries in Central and South America, Southeast Asia, India, Africa and the Middle East.

Constellation Communications, Inc. filed the original license at the FCC, but recently, Constellation, Bell Atlantic Enterprises International, and Telecommunicacoes Brasileiras S.A. ("Telebras") signed a Memorandum of Understanding as a framework for discussing the creation of a joint venture to own and operate a LEO satellite system. Constellation Communications, Inc. submitted commitment letters and balance sheets for its newly disclosed equity investors, Bell Atlantic and E-Systems, Inc. Constellation also said that Telecommunicacoes Brasilerias S.A. (Telebras) of Brazil intends to take an equity stake in the project later. Constellation estimates that constructing and launching the total system will cost $1.695 billion and that $26.4 million will be required to cover the first year's operating costs.

Inmarsat-P (ICO Global Communications Limited—a consortium including Inmarsat and 38 Inmarsat signatories)

Inmarsat-P, sometimes referred to as Project-21, would employ 10 or 12 satellites in intermediate circular orbits (10,355 km). Each satellite would have the capacity for 4,000 circuits and an expected lifetime of 10 years. Inmarsat handsets are expected to cost between $1,000 to $1,500 and calls will cost $2 per minute. Inmarsat has started the licensing process in the United Kingdom and hopes to begin offering service in 1999, with the system fully operational by the year 2000.

The cost to construct, launch, and operate the system for one year is expected to be $2.8 billion. $1.4 billion in initial financing was committed by 39 signatories to Inmarsat including a commitment of $150 million by Inmarsat as an organization. The Inmarsat Council has indicated that Inmarsat and its affiliates will maintain at least 70 percent ownership. Additional pledges of $900 million were turned away and the remaining $1.4 billion will be financed through equity and debt. The U.S. investor is Comsat Corp., the U.S. government's signatory to Inmarsat. In Europe, the biggest investors are Deutsche Telekom AG's mobile-phone unit and the Swiss, Spanish and Dutch state phone companies. Other major investors are: the Beijing Maritime & Shipping Co., an arm of the Chinese Ministry of Transport; Japan's main international phone carrier, KDD, Ltd.; India's international phone company; and Singapore Telecom.

SOURCE: Office of Technology Assessment, 1995

billion. Many analysts do not believe that there are enough customers to support all of the proposed systems.[24]

Technology

Advances in technology underlie the vision of small and light handsets and low-cost wireless service. Developments in semiconductor and microprocessor technology allow the functionality of a mobile phone to be squeezed into a small package. New technologies also permit power-efficient systems that can use a smaller battery, usually the heaviest part of the handset. But the most important development in wireless telephony is

[24] For a more complete discussion of the challenges facing the LEOs , see OTA, op. cit., footnote 2.

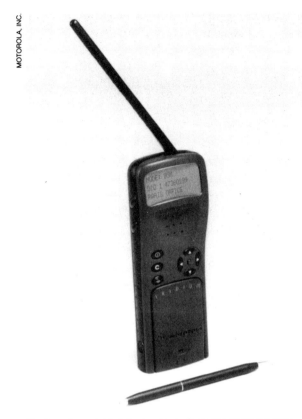

Dual-mode portable telephones, such as this prototype from Iridium, will first attempt to connect to a local cellular system. If no system can be accessed, the telephone will then use the satellite system to complete the call.

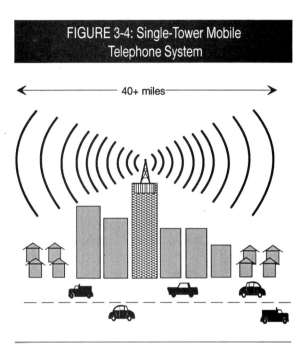

SOURCE: Office of Technology Assessment, 1995.

the evolution to digital transmission, which will allow network operators to serve three to 10 times as many users as today's analog systems with the same amount of network equipment. This capacity increase will translate into a substantially reduced cost to serve each user.

Terrestrial Wireless Technology

Terrestrial wireless systems provide radio coverage to their service area with antennas mounted on towers or on buildings. Until the early 1980s, terrestrial mobile telephone systems used a single, high-power transmitter on a tall tower or skyscraper to cover a metropolitan area. Any user within the signal's range, usually up to about 40 miles away, could get service. This single-tower architecture is still used for most SMR systems today (see figure 3-4), and is adequate when the predominant type of communication is short dispatch messages.

The cellular concept

Modern terrestrial systems use a *cellular* architecture that provides coverage with many low-power transmitters. Cellular technology provides the foundation for a mass market service by allowing a large number of users to share the limited spectrum more efficiently than the single-tower approach. Because it has now been proven over a decade of service, cellular technology will no longer be used only by the cellular carriers. In 1991, the FCC allowed Fleet Call (now Nextel) to deploy its ESMR technology by shifting from one high-power broadcast tower to a cellular architecture in six of the largest U.S. markets. New PCS providers will also use a cellular architecture.

Each of the low-power transmitters in a cellular system provides coverage to an area a few miles across, known as a *cell* (see figure 3-5). Cells are often drawn as circles or hexagons, but real-world cells are irregular in shape because buildings and trees obstruct the radio waves. By deploying enough transmitters or *base stations*, cellular operators provide continuous coverage wherever their customers are likely to be. Because users often pass through several cells as they travel

Wireless antennas come in many shapes and sizes, from large, conspicuous monopole designs, to practically invisible building mounted panel antennas.

through a city, a cellular system has to automatically *hand off* the call from base station to base station. As the user nears the edge of a cell, the system reassigns the user to a new cell by determining which of the other base stations in the area can provide the strongest signal.

The cellular architecture makes efficient use of the spectrum and increases system capacity. In a conventional single-tower system, each channel can only be used by one customer at any one time. By contrast, a cellular system allows a channel used in one cell to be *reused* by a different user in

PCS antennas (left) and roof top base stations (right) are smaller than their cellular counterparts.

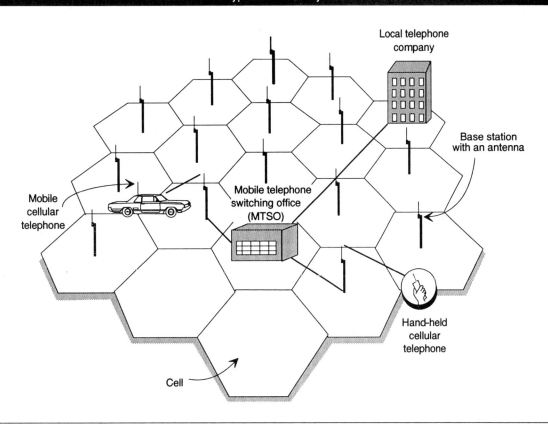

FIGURE 3-5: A Typical Cellular System Architecture

SOURCE: Office of Technology Assessment, 1995.

another cell, as long as there is enough separation between the cells to minimize interference (see figure 3-6). Network operators can further increase system capacity by splitting large cells into several smaller ones. The greater the number of cells, the greater the number of users who can use a channel at the same time. In typical systems, cells at highway interchanges or in downtown areas are less than a mile in diameter, while in areas where the traffic is light they may be up to 20 miles across (see figure 3-7).

The heart of a cellular system is the Mobile Telephone Switching Office (MTSO), which is connected by microwave or landline links to all of the base stations. It is also connected via a high speed digital link to the public switched telephone network. The user's voice signal is transmitted from the phone through the air to a base station, back to the MTSO, and then through the landline

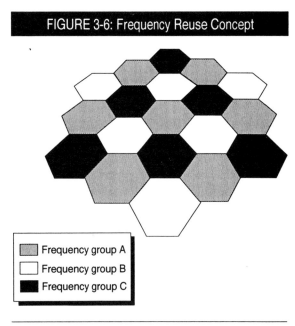

FIGURE 3-6: Frequency Reuse Concept

Frequency group A
Frequency group B
Frequency group C

SOURCE: Office of Technology Assessment, 1995.

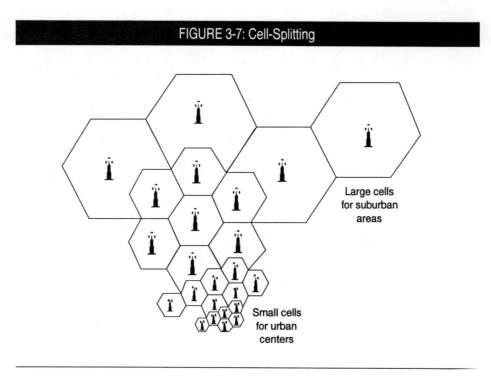

FIGURE 3-7: Cell-Splitting

Large cells
for suburban
areas

Small cells
for urban
centers

SOURCE: George Calhoun, *Digital Cellular Radio* (Norwood, MA: Artech House, Inc., 1988), p. 43.

network to its destination. The MTSO is responsible for managing the assignment of radio channels to users. When the user dials a number and presses the "send" button on their phone, the MTSO checks to see if there is a channel available and then assigns the channel. During the call, the MTSO monitors the signal strength to see if it should initiate a handoff to a nearby cell.

Digital transmission

Although the cellular concept is the foundation of terrestrial wireless technology, it is the transition to digital transmission that is most responsible for the vision of low-cost personal wireless services. Today's cellular technology, known as the Advanced Mobile Phone Service (AMPS) system, is based on an analog frequency modulation (FM) transmission scheme and dates from the mid-1970s. AMPS systems in large cities are starting to reach capacity limits that cannot be overcome by further cell-splitting. Digital systems are being deployed to provide higher capacity.

One way that digital systems increase capacity is by making extensive use of voice compression technologies. Once a voice signal has been transformed into digital form, complex mathematical manipulations can be used to reduce the amount of information that needs to be sent for good-quality speech. Reducing the amount of information that needs to be sent also reduces the amount of spectrum needed for each user, allowing more users to share the spectrum. For voice quality that is comparable to an AMPS system, at least three times as many users can be accommodated by a digital cellular network with the same number of base stations. Because voice compression technology will continue to improve, future systems will be able to achieve even greater increases in capacity.

The deployment of digital systems in the United States has been slowed by battles over standards. For the first generation of cellular technology, the FCC selected AMPS as a national standard and required all operators to use it. But with digital cellular, the FCC has left technology

---

BOX 3-3: TDMA and CDMA

In a cellular system, many users make calls at the same time in each cell. Clearly, it is necessary that these users' transmissions not interfere with each other. One solution to this "multiple access" problem is to ensure that each user transmits on a separate frequency or "channel." When a user initiates a call, the system tells the user's phone which frequency to tune to, much as a radio listener tunes to a particular station. This approach, known as Frequency Division Multiple Access (FDMA), is used in today's analog cellular systems.

Digital cellular systems could also use FDMA; the only difference would be that the information sent through the channel would be in digital, not analog form. However, it is more likely that digital cellular systems will use one of two alternate schemes, either Time Division Multiple Access (TDMA) or Code Division Multiple Access (CDMA). The TDMA and CDMA approaches differ from FDMA in that several users may share the same channel.

In a TDMA system, several users are assigned to a single channel, and they take turns. Each user's phone transmits a short burst of data, waits as the other users assigned to the channel transmit their data, sends another burst, and so on. At the receiver, the bursts are reassembled into a continuous signal and turned back into speech. In the U.S. TDMA system, three users share the same channel that the analog Advanced Mobile Phone Service (AMPS) system uses for a single user. Digital compression technology allows a user to send a good-quality speech signal even when using the channel only one-third of the time, tripling the capacity of the system.

CDMA systems use a much wider channel than TDMA or FDMA systems, and share it among a larger number of users. The users can all use the channel at the same time, but each user's transmission is uniquely coded. If the receiver knows the code, it can pick out a particular user's transmission from the combined signal. There is no strict limit to the number of users in each channel, but it becomes more difficult to pick out individual users' transmissions as the channel becomes crowded. Because CDMA uses a wider channel than FDMA or TDMA, it is sometimes referred to as a "spread spectrum" technology.

Both CDMA- and TDMA-based systems have been proposed for use as a replacement to AMPS. Over the past several years, there has been a debate about which approach is better. TDMA proponents have argued that TDMA is a more proven technology, whereas CDMA proponents have argued that their system will offer higher capacity—not just three times more users than AMPS, but 10 or 20 times more. TDMA-based digital cellular systems entered commercial service in 1992, and have several hundred thousand users. No CDMA systems are expected to be operational until late 1995 or early 1996.

The capacity estimates for CDMA are higher because it appears to overcome a fundamental challenge that faces designers of TDMA systems. In a TDMA system, the same channel cannot be used in adjacent cells because of excessive interference. A channel can only be safely "reused" in cells some distance away. As a result, only a fraction of the operator's spectrum can be used in each cell: typically one-seventh or less. CDMA systems, on the other hand, allow all of the spectrum to be used in each cell, increasing system capacity.

SOURCE: Office of Technology Assessment, 1995.

---

selection to the industry. Cellular industry standards committees have been unable to choose between two systems, one based on a technology called Time Division Multiple Access (TDMA) and the other based on Code Division Multiple Access (CDMA) (see box 3-3). Similarly, seven different technologies have been proposed for use in the new PCS band. There are more PCS than

Satellites in geosynchronous orbit 22,300 miles above Earth are being used to provide a variety of mobile communication services to aircraft, ships, and vehicles.

Mobile Satellite Technology

In a satellite telephone system, instead of sending the radio signal to a base station, the mobile phone beams the radio signal up to a satellite. In many ways, satellite systems are benefiting from the same technological advances as terrestrial systems. By using multiple "spot beams" in place of a single beam, satellite systems can exploit the same concept of frequency reuse that terrestrial systems use. They will also use digital voice compression to dramatically increase the number of users that can be served from each satellite, reducing the cost per user considerably. But the most significant new concept in satellite system design has been the development of nongeostationary LEO or Medium Earth Orbit (MEO) systems.

The Inmarsat and AMSC systems use satellites in geostationary orbit, 22,300 miles above the equator. At this altitude, the satellite appears to remain at a fixed point above the Earth. This simplifies system operation, but has several drawbacks. First, it takes a considerable amount of time for the signals to travel up to the satellite and back down to Earth, resulting in a noticeable and annoying delay. Second, because the satellite is so far above the Earth, considerable power is needed to transmit the signal up to the satellite, requiring bulky transmitters. Therefore, the AMSC system can only be used with car phones, not portables.

With a nongeostationary LEO or MEO system, on the other hand, the satellites orbit much closer to the Earth at altitudes between 500 and 7,000 miles. This reduces power requirements, allowing the use of handheld portables. Moreover, the delay incurred in sending the signal up to the satellite and back down to Earth is significantly reduced. But LEO systems are also more complex. While a geostationary system can provide global coverage with a small number of satellites, LEO systems plan to use constellations of 10 to 66 satellites (figure 3-8). The satellites move relative to the surface of the Earth, complicating system coor-

cellular technologies under consideration because the PCS band may be used for a wider variety of services. See chapter 6.

While ESMR and PCS operators will use digital technology from the beginning, cellular operators will have to deploy digital technology while also continuing to provide service to the millions of users who still have analog phones. Because it will take several years to convert every base station to digital, the new digital phones now being sold are also capable of analog transmission. The user can "fall back" to analog in areas where there is no digital service or where a different digital system has been deployed. Over time, network operators will add digital capability to more and more cell sites and continue to expand the amount of spectrum dedicated to digital service, while continuing to reserve some channels for users who still have analog-only phones. The transition to digital has only just begun and is expected to take about a decade.

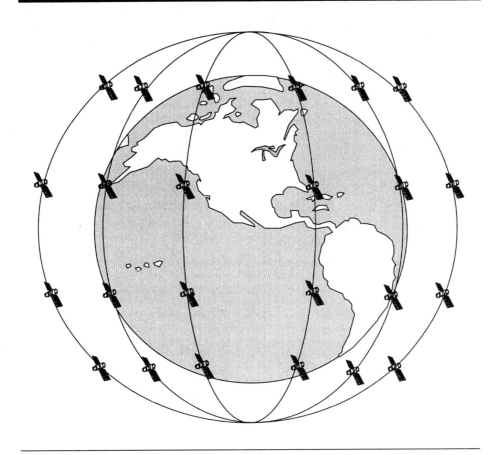

FIGURE 3-8: Low-Earth Orbit Satellite System

SOURCE: Office of Technology Assessment, 1995.

dination. Moreover, satellite lifetimes are significantly reduced, requiring replacement satellites to be launched continuously.

■ Regulatory Framework

The FCC is responsible for managing the spectrum used by commercial wireless services, and sets the rules regarding their licensing and operation. Historically, the FCC has regulated wireless services less than wireline services, believing that the wireless market is more competitive than the traditional monopoly wireline market. The FCC has relied on market forces to determine the prices

of wireless services, within the regulatory framework it established for each service.

As technology has advanced, the distinctions between different mobile telephone services have become less clear. Cellular, PCS, and ESMR will provide similar services. To streamline regulation of these existing and emerging services, Congress directed the FCC in the Omnibus Budget Reconciliation Act of 1993 (Public Law 103-66) to set up a new regulatory classification—Commercial Mobile Radio Services (CMRS)—that would allow the FCC to "treat like services alike," forbear from imposing some elements of common carrier regulation, and preempt state regulation of CMRS rates.

Spectrum Allocation, Licensing, and Market Structure

FCC spectrum allocation decisions play a key role in determining the degree of competition in the wireless industry. To a certain extent, the number of competitors can be increased by dividing the available spectrum among more carriers. When it created the cellular service, for example, the FCC first thought that it would only license one carrier, but then determined that competitive benefits would result from splitting the spectrum among two carriers. But because a mobile telephone network needs a minimum amount of spectrum to operate economically, additional competition usually requires that more spectrum be allocated. For example, by allocating 120 MHz to PCS, the FCC was able to create up to six new competitors in each market.

The FCC also influences market structure by specifying the size of the license areas. In most countries, wireless licenses are assigned on a nationwide basis. But the FCC chose to divide the United States into many small license areas, allowing a larger number of companies to take part in the industry. For cellular, the FCC divided the nation into 734 separate market areas—306 Metropolitan Statistical Areas (MSAs) and 428 Rural Service Areas (RSAs). PCS licenses, on the other hand, are being allocated on the basis of either 493 Basic Trading Areas (BTAs) or 51 Major Trading Areas (MTAs). Two of the 30 MHz PCS licenses are being allocated on the basis of MTAs, while the other PCS licenses will be allocated on the basis of BTAs. SMR service areas are defined by the propagation distance of signals transmitted from the operator's tower, but there are proposals to establish standardized service areas such as BTAs or MTAs.

Other FCC rules seek to maintain competition by limiting concentration of ownership. To ensure that the new PCS spectrum would be used to provide competition to the incumbent cellular carriers, the FCC did not permit cellular carriers to obtain more than 10 MHz of PCS spectrum in markets where they already owned cellular licenses. Similarly, local exchange carriers were also restricted to bidding on 10 MHz blocks of spectrum in their service areas, not the larger 30 MHz blocks. Finally, no carrier may have more than 45 MHz of cellular, PCS, and SMR spectrum in any given market.

The FCC's choice of a mechanism for assigning licenses also affects the structure of the industry. The first cellular licenses were assigned by comparative hearings where the Commission selected among applicants based on detailed proposals. But the task of allocating hundreds of licenses by this method overwhelmed the FCC. For the later cellular licenses, the Commission assigned the licenses by lottery. Most of these licenses were quickly sold to larger carriers.[25] To reduce the speculation on licenses and raise funds for the Treasury, Congress authorized the FCC to use auctions to assign licenses.[26] The first round of PCS auctions raised $7.7 billion.

Commercial Mobile Radio Services

PCS, cellular, and most SMR and mobile satellite services are regulated as CMRS carriers. The creation of the CMRS classification was a response to disparities in the regulation of SMR and cellular carriers, which became increasingly significant as the deployment of new technologies allowed SMR carriers to compete with cellular carriers. Beginning in August 1996, PCS and SMR carriers that provide mobile telephone service will be subject to the same rules as cellular carriers. The FCC has launched a series of proceedings to define the

---

[25] "Stanley Says 70% of Lottery Cellular Licenses Transferred," *Telecommunications Reports,* Apr. 26, 1993, p. 16.

[26] Public Law 103-66, section 309(j).

BOX 3-4: CMRS Proceedings

In the Second Report and Order that was issued in the proceeding that defined the rules governing CMRS carriers, docket No. 93-252, the FCC identified several issues that required further study:

- **The interconnection obligations of CMRS licensees**
The FCC issued a Notice of Inquiry as part of docket number 94-54 in June, 1994. In April, 1995, the Commission issued a Notice of Proposed Rule Making (NPRM) in which it tentatively concluded that it was premature to impose rules requiring CMRS carriers to interconnect with other CMRS carriers.

- **The imposition of equal access obligations on CMRS licensees**
The FCC issued an NPRM, in docket number 94-54, in which it tentatively concluded that cellular carriers should be required to give their customers a choice of long distance carriers. It also asked for comment on whether equal access rules should be imposed on other CMRS carriers. No order has been issued in this docket.

- **The reclassification of private radio licensees as CMRS**
The FCC has completed work on the technical and licensing rules that will apply to CMRS carriers, issuing a Third Report and Order and Fourth Report and Order in docket number 93-252 in early 1995. Among the issues addressed in this proceeding was the spectrum cap limiting CMRS carriers to 45 MHz of cellular, PCS, and SMR spectrum in any market.

- **Tariffing of local exchange carrier(LEC)/wireless interconnection**
The FCC issued an NPRM as part of docket number 94-54 in which it requested comment on whether LECs should be required to file tariffs specifying the rates charged for interconnection, or whether interconnection rates should be negotiated. No order has been issued in this docket.

- **Monitoring of competition in the cellular marketplace**
In the second report and order in docket 93-252, the FCC concluded that the cellular marketplace was not fully competitive and proposed collecting more information about competition in the cellular industry. The FCC has not acted on this issue.

- **Further forbearance from regulating certain types of CMRS carriers**
In the Second Report and Order in docket number 93-252, the FCC decided to forbear from applying some aspects of common carrier regulation to CMRS carriers. In docket number 94-33, the FCC issued an NPRM asking whether the regulation of some CMRS services should be relaxed further. No order has been issued in this docket.

- **Provision of dispatch service by CMRS carriers**
In March, 1995 the FCC issued an order permitting CMRS carriers to provide dispatch service.

SOURCE: Office of Technology Assessment, 1995.

---

regulations that will apply to CMRS providers, described in box 3-4.

CMRS carriers are less regulated than the wireline local exchange carriers because the local exchange carriers have a near-monopoly, while the wireless industry is competitive. There are two cellular carriers in every market, with the prospect of additional competition from SMR and PCS providers. In developing the new CMRS regulatory regime, Congress and the FCC determined that competition would, in most cases, be sufficient to protect consumers and keep prices reasonable. Although CMRS providers will be regulated as common carriers, subject to Title II of the Communications Act, Congress allowed the FCC to "forbear" from regulating interstate rates and requiring the filing of tariffs.[27]

More importantly, Congress preempted state regulation of intrastate rates.[28] While many states had already concluded that the cellular industry was sufficiently competitive that rate regulation was unnecessary, a few states still regulated cellular. Under the new law, states will only be able to regulate the price of cellular or any other CMRS service if they can demonstrate to the FCC that market conditions have failed to guarantee just and reasonable rates.[29] Eight states petitioned the FCC for the right to continue regulating cellular service, arguing that the industry would not become truly competitive until the PCS and ESMR providers were operational.[30] However, the FCC rejected these petitions in May 1995, freeing all wireless carriers of rate regulation by states.[31]

■ Issues and Implications
By allocating a large amount of additional spectrum for wireless telephony and creating the new CMRS framework, Congress and the FCC have established the foundations for a successful industry. Given the growth rates in cellular subscribership and the continuing development of low-cost wireless technology, the future of the wireless industry appears bright. However, there are several issues that will have an impact on the cost, utility, and availability of wireless services.

Spectrum Allocation for Commercial Services
The recent allocation of 120 MHz to PCS more than triples the amount of spectrum allocated to terrestrial commercial mobile telephone services. Combined with new, more efficient digital technologies, the current spectrum allocation should be sufficient to meet the demand for the next several years, even if subscribership continues to grow at a high rate.[32] However, if data, image, or video applications become important components of the service mix of commercial mobile radio services, additional spectrum may have to be found sooner than expected.

Mobile satellite systems may have more pressing spectrum needs. Currently, the five proposed U.S. LEO systems are required to share 33 MHz in the 1610 to 1626.5 MHz and 2483.5 to 2500 MHz frequency bands. If demand for mobile satellite service matches the expectations of its proponents, this allocation will be insufficient. The National Telecommunications and Information Administration (NTIA) has estimated that an additional 60 MHz of spectrum may be required for mobile satellite services over the next decade.[33] However, it is particularly hard to judge how much spectrum should be allocated to mobile satellite services. As yet, no mobile satellite services are operating on a wide scale, and demand remains unproven. Because these systems will generally not compete in the same markets as terrestrial services, demand estimates for these in-

---

[27] FCC, op. cit., footnote 15, pp. 68-70.

[28] Public Law 103-66, section 6002(c)(2)(A).

[29] Communications Act of 1934, 47 U.S.C., Section 332(c)(3)(B).

[30] Arizona, California, Connecticut, Hawaii, Louisiana, New York, Ohio, and Wyoming. Wyoming subsequently withdrew its petition.

[31] Each state's petition was handled in a separate proceeding. See, for example, Federal Communications Commission, *Report and Order, Petition of the Connecticut Department Public Utility Control to Retain Regulatory Control of the Rates of Wholesale Cellular Service Providers in the State of Connecticut*, PR Docket 94-106, May 19, 1995.

[32] NTIA has forecast that only 33 MHz of additional spectrum will be required for two-way commercial mobile radio services over the next decade. National Telecommunications and Information Administration, *U.S. National Spectrum Requirements* (Washington, DC: 1995), p. 33.

[33] Ibid., p. 57.

dustries provide only a rough guide to the potential for satellite-delivered services.

With current technology, the most desirable frequency bands for most mobile services are those below 3 GHz. At higher frequencies, radio waves are more subject to scattering by buildings, trees, and other obstructions. In addition, radios that operate above 3 GHz are more expensive to build. Unfortunately, there is very little unused spectrum below 3 GHz. Any future expansion in mobile services will require either technological advances that permit the economical use of higher frequency bands, or the reallocation of spectrum from other services.

Reallocating spectrum can be time-consuming and costly. Potential new users need frequencies for their proposed services, but incumbent users usually resist being forced to move to other frequency bands. Policymakers and regulators often have a hard time balancing the two competing sets of interests. Much of the spectrum now allocated to terrestrial mobile services—PCS, cellular, and SMR, for example—was once used for other purposes. PCS will operate in a band that is now being used by fixed microwave services (who will have to move their operations to higher frequencies), while the cellular and SMR services were allocated spectrum that had been previously been used for broadcast television's channels 70 to 83. To meet potential future needs for even more mobile spectrum, one plan being considered is reallocation of television frequencies as part of the transition to Advanced Television (ATV). In the case of mobile satellite services, a complicating factor is that they require global coordination and approval. At the WARC-92 conference, for example, additional spectrum was allocated to mobile satellite service in the 1970 to 2010 Mhz and 2160 to 2200 MHz frequency bands. However, in the United States, part of this spectrum is allocated to broadcast auxiliary services, and in the rest of the world it will not be available until 2005.[34]

Spectrum Allocation for Public Safety

Federal, state, and local public safety agencies, such as police and fire departments, may also have significant near-term spectrum needs. In the Budget Reconciliation Act of 1993, Congress mandated that the FCC submit to Congress a study of current and future needs of state and local government public safety agencies through the year 2010, and develop a plan to ensure that adequate frequencies are made available to public safety licensees. In its report, the Commission declined to allocate additional spectrum, but outlined the steps it would take to gather additional information and procedures to respond to emergency needs.[35] The report has been criticized by the public safety community for underestimating the urgency of their needs.[36]

The demand for public safety wireless communications has grown considerably in recent years. Part of the growth in demand is due to an increase in the number of public safety personnel. But a more significant factor is that future public safety communications systems will not only be used for voice communications, but will also have to accommodate increased use of imaging for mobile transmission of fingerprints, warrants, and mug shots. Image communications requires much more spectrum than ordinary voice communications. In comments submitted to the FCC, the Association of Public Safety Communications Officials (APCO) estimated that between 6 and 18 MHz of additional spectrum would be needed for public safety voice communications by 2010, but that 75 MHz would be needed for the new "wideband" applications.[37]

---

[34] OTA, op. cit., footnote 2.

[35] Federal Communications Commission, *Meeting State and Local Government Public Safety Agency Spectrum Needs Through the Year 2010,* Feb. 9, 1995.

[36] "Public Safety Officials Pan FCC's Spectrum Report," *Telecommunications Reports,* vol. 61, No. 7, Feb. 20, 1995.

[37] FCC, op. cit., footnote 35.

In allocating spectrum for public safety, several other issues have to be considered. First, public safety users are concerned that the use of auctions to assign commercial licenses will cause regulators to allocate any available spectrum to commercial services because an allocation to public safety would not provide revenues for the Treasury. Second, even if more spectrum is made available, there will be a continuing need for greater coordination of the radio systems operated by different public safety agencies. Currently, different public safety agencies use different frequency bands, preventing them from talking to each other in an emergency. A single frequency band, or a limited number of bands, could improve coordination of public safety activities.

The Development of Competition

Congress and the FCC are relying on competition, not regulation, to ensure that the price of mobile telephone service will be reasonable. The new PCS allocation will provide three to six new competitors to the cellular carriers, and the deployment of ESMR technology will provide additional competition. In the larger cities, there is the potential for robust competition on the basis of price and coverage. Most observers foresee four or five major competitors in the larger markets, with some niche players as well. However, many analysts believe that smaller cities and rural areas, where customers are fewer, will not be able to support as many competitors.

It should also be emphasized that most of the new PCS competitors will not have operational networks before the end of 1996. They will have to acquire hundreds of sites for base stations in each market and build their networks a time-consuming and expensive process. More importantly, they are unable to use some of the spectrum they have acquired at the auctions until its current occupants, fixed microwave users, are relocated to another band. Under the procedures established by the FCC, the PCS licensees will have to negotiate with the microwave licensees and pay the cost of their relocation. However, microwave licensees are not required to negotiate until 1997, and PCS licensees will not be able to request that the FCC involuntarily relocate a microwave user until 1998.

Finally, it is unclear how many competitors can be supported in the long term. The major players in the industry will have to spend billions of dollars to build out the new PCS networks. While the pioneers of cellular service had the luxury of building out their networks one cell site at a time, the PCS-band networks will have to enter the market with broad coverage and compete against established providers. The new PCS carriers, especially the licensees who qualified as small businesses or businesses headed by women or minorities, face a difficult challenge. Even the deep-pocketed cellular carriers, long distance carriers, and cable companies that are acquiring PCS licenses are risking large amounts of money on the assumption that demand will continue to grow and that they will survive potential price wars.

E-911 from Mobile Telephones

Emergency assistance available through 911 has been a significant driver of recent cellular telephone sales, and the industry promotes this with advertisements touting the benefits of mobile communications for personal safety and security. As a result, demand for 911 services from wireless users is growing with the rise in cellular subscribership. It has been estimated that 10 percent of 911 calls are from mobile users. The California Highway patrol reported that in January 1993 it fielded 80,000 emergency calls, of which 25,076 were from cellular telephones.[38]

However, while many wireless users can get access to 911 operators, they may not be able to fully

---

[38] Federal Communications Commission, *Notice of Proposed Rule Making, in the Matter of Revision of the Commission's Rule to Ensure Compatibility with Enhanced 911 Emergency Calling Systems*, FCC No. 94-237, CC Docket No. 94-102, RM- 8143, proposed rules, Sept. 19, 1994, para 9.

benefit from the Enhanced 911 or E-911 services available to wireline users. Systems equipped with special equipment providing E-911 services identify the location of the caller, even if he or she cannot speak. An Automatic Number Information database together with an Automatic Location Information database provide precise location information to the 911 public safety answering point (PSAP), from which the appropriate emergency service (police, fire, medical) is dispatched. About 90 percent of all wireline telephones have 911 services available, and of these about 76 percent also have E-911 capabilities.

The location of a cellular telephone user, however, currently cannot be automatically determined because cellular phones—unlike their wireline counterparts—are not linked to a specific location; they are designed to move around. As a result, unless the caller can provide clear and exact information or directions—which is often not the case—emergency assistance workers often do not know where to go when receiving a call from a mobile handset. It is unclear how many people understand that mobile phones do not offer the exact same services available from a wireline telephone. Some public safety officials believe that failing to integrate wireless systems into the E-911 framework undermines the $2 to $3 billion invested in 911 service since it was established as a nationwide goal 30 years ago.

To address this problem, the FCC has initiated a rulemaking to guide development of E-911 services and to ensure that location information will be available from all phones.[39] There are a number of technologies that may be useful in providing better location information, such as use of triangulation; Global Positioning Satellite (GPS) or LORAN systems; signal strength, angle and/or time delay measurements; antenna and cell site sectorization; and time synchronization.[40] Each

system has its strengths and weaknesses, and the FCC has invited comments on the technical, performance, and cost considerations of each of the candidate technologies, but appears inclined to set performance rather than technology standards for achieving accurate location identification. This proceeding has resulted in substantial debate about the importance of accurate location information technologies for wireless systems.

Other Issues

In addition to the issues discussed above, there are a number of critical issues that will affect how new and existing wireless voice technologies will be integrated into the NII and what effects and implications ubiquitous mobile services may have for individuals and businesses. These issues are only briefly discussed here, but are analyzed in more detail in later chapters.

Standards

For the first generation of cellular technology, the FCC specified a standard technology (AMPS) that had to be used by all carriers. This guaranteed that every cellular phone would work anywhere in the nation. But for digital cellular and PCS, the FCC has refrained from picking a standard. Industry standards committees have been unable to agree on a single standard, in part because manufacturers have an incentive to promote their own technology as a standard. Because there is no standard, each network operator will have to choose from among the contending digital technologies. There is considerable concern that different technologies will be deployed, making roaming impossible. These issues are discussed in detail in chapter 6.

Interconnection to local exchange carriers

The interconnection of wireless and wireline networks allows their users to call each other. FCC

---

[39] Ibid.

[40] These technologies were reviewed in C. J. Driscoll & Associates, "Survey of Location Technologies to Support Mobile 9-1-1," report prepared for California Department of General Services, Telecommunications Division and the Association of Public Safety Communications Officials, ed. 1.0, July 1994.

rules require that local exchange carriers (LECs) allow wireless carriers to interconnect with their network. But wireless carriers have to pay LECs a fee for every minute of traffic. In the past, state regulators have allowed these interconnection charges to be significantly above cost to provide additional revenues that support the LEC's low residential rates. The high cost of interconnection is becoming an increasingly important issue because it raises the price of wireless service. These issues are discussed in more detail in chapter 7.

Interconnection obligations of wireless carriers

Because LECs have a monopoly in their market, they are required to interconnect with wireless carriers, long distance carriers, and, increasingly, wireline local exchange competitors. Wireless carriers, on the other hand, do not control a bottleneck—there are at least two competitors in each market, with more to come. Therefore, they have not been required to interconnect with other wireless carriers or with all long distance carriers. For example, the FCC does not require wireless carriers to give their customers a choice of long distance carrier. An issue of growing importance is whether the interconnection obligations of wireless carriers should continue to be minimal, or if they should be modeled on those of the LECs. For further discussion, see chapter 7.

Local restrictions on antenna siting

A cell site consists of base station equipment and an antenna mounted on a building or tower. The cellular carriers have deployed about 18,000 cell sites to date, and it is expected that the wireless industry will have to deploy an additional 100,000 cell sites over the next decade. Some communities, however, are becoming concerned about possible health effects from electromagnetic radiation and the aesthetics of the towers. Increasingly, zoning regulations and other ordinances are being used to limit or halt the construction of new towers. The wireless industry contends that such restrictions will hamper their efforts to provide ubiquitous wireless service, and has petitioned the FCC to preempt all local restrictions on antenna siting. Zoning issues are discussed in more detail in chapter 8.

Privacy and security

The issues surrounding the confidentiality and security of wireless communications will become a more important issue as more consumers and businesses begin to use mobile/portable devices. Already, eavesdropping is a concern to many individuals and businesses who fear that important or sensitive personal or business information may fall into the wrong hands. Fraudulent use of wireless telephones is a particularly difficult problem, costing the industry and consumers an estimated $480 million a year. Finally, the use of wireless devices also raises questions related to the location of the user. Wireless technologies can be used to track people and things, but may also be used to hide ones' location. These issues are discussed in chapter 10.

Health effects

One of the most controversial issues surrounding the widespread use of wireless technologies involves any possible health effects caused by either the devices (cellular telephones, for example) or the transmitting antennas. Although research has been conducted, it has not been conclusive—it is not yet possible to say with certainty whether the devices/antennas do or do not pose a risk to human health or how serious any risk may be. In the face of this uncertainty, some researchers and members of the public believe that the safest course is to redesign, restrict, or even ban the use of wireless systems, while the industry believes it should be allowed to pursue its plans until there is convincing evidence that health problems are likely. This issue is intensely polarized and is already being played out in battles over cellular/PCS antenna siting and local zoning (see above). It is likely to become a more important political issue as citizens raise the issue with state and federal policymakers and regulators. This controversy is discussed in chapter 11.

Interference between devices

As the number of wireless devices used by consumers and businesses increases, there is a likelihood that interference will increase. Radio devices can cause interference to other wireless communication systems or to some electronic devices, giving rise to poor quality communications or malfunctions. Cellular phones, for example, may interfere with aircraft navigation systems and some medical equipment, including monitoring devices, pacemakers, and hearing aids. Electronic equipment, such as a computer, can also interfere with wireless communications unless it is properly shielded. These issues are discussed in chapter 12.

FIXED WIRELESS TECHNOLOGIES

In the United States, telephone service to houses and other fixed locations is generally provided over copper *loops* that run between the telephone company's *central office* and the customer's premises. But fixed service can also be provided with wireless technologies (see figure 3-2). Instead of being transmitted through copper wires, voice signals are transmitted from a radio tower or satellite to an antenna on the outside of the home. In the past, wireless was more expensive than copper, and would have required a prohibitive amount of spectrum to serve a large number of households. However, today's wireless technologies may be able to serve fixed users more efficiently. Spectrum allocations and current regulatory uncertainty, however, present obstacles to the widespread use of fixed wireless.

∎ Services and Users

One reason for the growing interest in "wireless local loops" is that they may now be comparable in cost to copper loops, due to the development of new spectrum-efficient technologies. In addition, wireless local loops allow telephone service to be rolled out quickly. Service can be provided to thousands of users as soon as the base stations are in place, without the need to install copper loops to each household. For this reason, wireless is now the technology of choice in developing countries that have little or no telephone service.[41]

Reducing the Cost of Telephone Service

In the United States, fixed wireless systems may be able to provide lower cost telephone service in some, especially rural, areas. One of the characteristics of wireline technology is that the cost to serve a household depends on its distance from the central office. It is much more expensive to provide telephone service in sparsely populated rural areas—where homes are typically far from the central office—than in cities. Wireless, on the other hand, has a cost structure that is largely distance-independent. The cost to serve a household is much the same whether it is close to the transmitter or far away.[42] In addition, radio waves can cross canyons or other difficult terrain that rule out wireline telephone service or make it extremely expensive.

Recognizing that wireless could reduce the cost of rural telephone service or provide it to unserved households, the FCC created a service called Basic Exchange Telecommunications Radio Service (BETRS) in 1988.[43] BETRS allows telephone companies to use a limited number of frequencies in the 450 MHz band to provide fixed wireless services in rural areas. Currently, no more than a few thousand households are served by wireless due to the small amount of available spectrum and the high cost of early BETRS equipment. However, with advances in technology, wireless may soon play a key role in delivering service to rural areas.

Wireless could also be used in suburban or urban areas. Carriers would like to take advantage of

---

[41] Terry Sweeney, "Lenders Backing Wireless Loops," *CommunicationsWeek International*, Dec. 12, 1994, p. 3.

[42] In the most extreme cases, even terrestrial wireless may be too expensive and satellite services may be used.

[43] Federal Communications Commission, *Report and Order, Basic Exchange Telecommunications Radio Service*, CC Docket No. 86-495, 3 FCC Rcd 215 (1988).

lower installation costs, reduced maintenance costs, and faster deployment of service in new housing developments.[44] In some cases, suburban areas are growing so quickly that the demands cannot be met with the existing network.[45] Even where wireline facilities exist, wireless may provide a less expensive solution if the copper loop is deteriorating and is affecting the reliability of telephone service. Furthermore, by using wireless loop technology, telephone companies would no longer have to dig up established yards and streets to replace facilities.

Local Exchange Competition

For many years, the expense of duplicating the incumbent's copper loop was seen as evidence that the local exchange market was a natural monopoly. Today, regulators who are trying to facilitate competitive entry have had to require the LEC to *unbundle* its network, allowing competitors to connect their switch to the existing loops. Only the cable companies, who already have a wire to many homes, can easily contemplate entering the local exchange market with wireline technologies.

Wireless technology may provide an alternative that will allow new local exchange competitors to enter the market. With wireless, a new entrant does not have to install a copper wire to each customer. Instead, it can deploy enough base stations to provide telephone service to every household in a city. This strategy would be especially attractive for long distance carriers, cellular carriers, and other companies that already have switches or other infrastructure in place. Cable companies, for example, could install base stations at various points on their network in order to provide local telephone service to a neighborhood. The cable network would be used to connect the base stations to a switch at the cable headend.

■ Technologies

In most of the world, the equipment used in wireless local loop applications is much the same as that used for mobile services. In some cases, mobile network operators use excess capacity to provide service to fixed users as well. In other cases, a modified version of mobile network equipment is deployed specifically for use in fixed applications. The major difference between fixed and mobile systems is that a fixed system does not require handoff capabilities. These modified cellular systems have been successfully deployed in local loop applications throughout Central Europe, in developing countries, and in other parts of the world where there is little or no wireline infrastructure.

In the United States, most of the wireless local loops are based on a technology specifically developed for use in the BETRS service in the 450 MHz band. However, newer wireless local loop technologies are being developed for use in the new 2 GHz PCS band. These "low-tier" wireless systems would provide service to fixed users, and would also allow limited mobility in the neighborhood around the user's home. Hand-offs between cells would be supported at walking speeds, but not at vehicular speeds. These low-mobility systems generally offer voice quality and data transmission capabilities that match or surpass those of a copper loop.

In suburban or urban areas, wireless local loop systems would consist of many "radio ports" or base stations mounted on telephone poles or street lights. Each radio port would serve an area with a radius of about 1,000 feet, which would allow each port to serve 35 to 40 homes.[46] Because it is difficult for radio waves to penetrate the walls of buildings, antennas would be mounted on the outside of customers' homes. The antennas are then connected by a wire to a phone inside the house. In

---

[44] See comments of Southwestern Bell (now SBC) in Federal Communications Commission, *First Report and Order, Allocation of Spectrum Below 5 GHz Transferred From Federal Government Use*, ET Docket 94-32 (1995).

[45] See comments of US West, ibid.

[46] Southwestern Bell comments, op. cit., footnote 44.

rural areas, where homes are further apart, more powerful transmitters are used and the antenna is mounted higher, allowing the signal to travel several miles to the customer's home.

▌ Regulation

In most states, fixed wireless service can only be provided by the incumbent, monopoly, local exchange carrier. The reason is that most state regulators have only allowed the monopoly telephone company to provide local exchange service. When the FCC created the BETRS service, it was careful to state that only the incumbent LEC or another carrier that had been certified by state regulators would be permitted to provide BETRS. The FCC also does not allow cellular, SMR, and PCS licensees to provide service to fixed users, except on an "incidental" or "ancillary" basis. If the FCC's rules had permitted these licensees to provide fixed service, this might have been seen as sanctioning local exchange competition.

▌ Issues and Implications

Limited Spectrum

Despite the promise of wireless local loops, almost no spectrum has been allocated for this application. The only spectrum used for fixed wireless is the 26 frequencies assigned to BETRS in the 450 MHz band. These are allocated on a co-primary basis, and are only available in rural areas. The FCC rules allow cellular spectrum to be used for BETRS, under certain circumstances, but, in practice, only the 450 MHz band has been used. LECs have been asking the FCC to allocate additional spectrum for fixed wireless, claiming that the small amount of spectrum available has limited wireless local loops to niche applications and prevented their use on a wider scale.[47]

Because of the limited amount of spectrum allocated specifically to fixed wireless applications, the FCC needs to clarify whether PCS or other mobile telephone spectrum can be used for wireless local loops. During the proceeding that created PCS, the FCC emphasized that the new spectrum was to be used for mobile services.[48] The FCC has since indicated that, in certain cases, it would be open to waiver requests from operators seeking to offer a fixed service with PCS spectrum.[49] However, this position was stated only in passing in an unrelated proceeding, and may only apply to rural areas.

Local Loop Competition

Many of the state rules limiting local exchange competition are gradually being dismantled. Moreover, legislation currently being debated in Congress would preempt state restrictions on entry by local exchange competitors. Many of these competitors are considering wireless as a vehicle to enter the market. However, because the FCC's rules on fixed PCS are unclear, it is uncertain whether a cable company could use a PCS license to compete with the LEC. This is another reason why FCC policies regarding the provision of fixed service by PCS, cellular, and SMR carriers need to be revisited.

Universal Service

Fixed wireless service may be able to advance the goal of universal service in the emerging NII. For many years, telephone penetration rates in rural areas lagged behind those in the cities. To promote universal service, regulators established a variety of mechanisms to direct billions of dollars in subsidies from low-cost urban users to high-cost rural users. This subsidy flow is now being threatened by the transition to a more competitive industry in

---

[47] United States Telephone Association comments, op. cit., footnote 44.

[48] "In ... allowing fixed use of PCS spectrum only on ancillary basis to the mobile service, we note that there is only a limited amount of spectrum available to meet the primary purpose of serving people on the move." FCC, op. cit., footnote 12, at 7712.

[49] FCC, op. cit., footnote 44, para. 20.

which prices are expected to move closer to cost. Wireless may provide a way to lower the cost of providing rural telephone service, making it affordable even if subsidies are reduced. The relationship between wireless and universal service is discussed further in chapter 9.

Wireless Data | 4

T he term *wireless data* describes a wide array of radio-based systems and services centered around pagers, portable computers, personal digital assistants (PDAs), and specialized applications for business. These wireless technologies enable users, who range from mobile professionals, to delivery drivers, and to factory and office workers to exchange electronic mail, send and retrieve documents, and query databases—all without plugging into a wire-based network. To date, however, growth in wireless data services has been low. Applications have been slow to develop, current speeds and capabilities cannot match those of wired services, and prices have been high. Like developments in other wireless technologies, there is great uncertainty regarding what applications the mass market wants, what it is willing to pay for, and what types of devices will match user needs.[1] Before the potential for wireless data services can be realized, service providers and manufacturers will have to overcome a number of technical, economic, and consumer-knowledge obstacles.

FINDINGS

- **The wireless data industry is at a nascent stage. Wireless data applications and systems will continue to grow, but at a slower pace than most analysts predict.** The acceptance

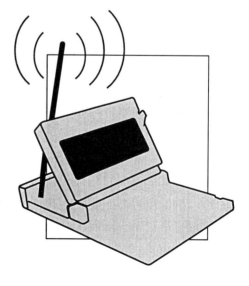

[1]Ken Dulaney, "Mobile Computing—Mobilizing the Organization," materials provided at Gartner Group presentation at the World Bank, Washington, DC, Feb. 10, 1994. Dulaney cites the example of Apple's Newton as a product that did not have a clear market or purpose, hence its low sales. He also notes that questions surrounding the ergonomics of portable computing devices—keypad and screen size, interface technologies, and capabilities—will only become clearer as users actually start to buy machines.

of wireless data by consumers and the general business market may be much lower than expected, especially in the short term. To date, use of wireless data technologies and systems remains concentrated in a small subset of business users—primarily in the fields of trucking, public safety, field service, and (taxi and courier) dispatch services. Current estimates of the total number of mobile data users range from 275,000 to 600,000.[2] Residential consumers, however, make little use of most wireless data communication technologies, and the prospects for significant growth in the consumer market are highly speculative and long term. Users have just begun to see the benefits wireless data can offer.

Despite this slow start, most industry analysts still expect wireless data to be one of the fastest growing sectors of the wireless industry. Applications and services are improving, and a host of new wireless data systems are expected to be introduced within the next five years. Some analysts predict the use of wireless data will grow as much as 30 percent per year, and many expect this growth to accelerate over the next decade.[3] OTA believes, however, that actual growth rates will be lower due to uncertain demand and technical difficulties in integrating wireless and wireline data networks and applications.

- **Technical challenges will continue to slow industry growth, but most analysts believe the problems will be solved as the technologies and the industry naturally mature.** Wireless data services lag those offered on wire-based networks, including the public telephone network and public/private computer

networks, in many respects. Current speeds offered over wireless networks, for example, are usually substantially less than those available using wireline technologies, and it is unclear how advanced networking applications and protocols, such as Asynchronous Transfer Mode (ATM), will be adapted for use in a wireless environment.

Interoperability between wireline and wireless networks and services is a continuing problem. Services and applications designed for wired media work less well (and sometimes not at all) using the often-noisy and congested airwaves. Interoperability problems also result from wireline data communication standards and protocols that generally have not incorporated wireless features and requirements. Finally, the multitude of new wireless data companies that has sprung up has also led to many companies selling proprietary products and services that do not work together. Companies have been started that integrate wireless services, and software is being developed that attempts to mask as many of the differences as possible.

Fundamentally, these problems exist because the development of wireless data technology is still in its early stages, but they also reflect frequency allocations that were made based on past applications—when needs and spectrum requirements were lower. The federal government has recently allocated more spectrum to wireless data services, and private companies are working to improve their products by making them easier to use and more interoperable with existing wireline networks and services.

---

[2] For individual services, estimates of subscribership vary, and, in most cases, are closely guarded. Some analysts suggest, for example, that RadioMail has only 1,000 paying customers, Ram Mobile Data between 3,000 and 15,000, and Ardis some 50,000, but with flat growth. "RadioMail Slashes Mobidems to $199," *Mobile Data Report*, vol. 6, No. 7, Apr. 11, 1994, p. 4. David Strom, consultant, presentation to OTA, Oct. 10, 1994.

[3] BellSouth, for example, predicts that 33 percent of its wireless revenues will be from data and that 25 million Americans will use wireless data services. BIS Strategic Decisions predicts wireless data revenues will be $10 billion/year by 2000. Andrew Kupfer, "Look, Ma! No Wires!" *Fortune*, Dec. 13, 1993, p. 147. Datacomm Research predicts the value of mobile hardware, software, and services will grow from $450 million in 1992 to $3.7 billion by the year 2002. Datacomm Research, *Portable Computers and Wireless Communications* (Wilmette, Illinois, 1993).

WIRELESS DATA SERVICES

■ Applications

Wireless data services use a mix of terrestrial and satellite-based technologies to meet a wide variety of local (in-building or campus settings), metropolitan, regional, national, and international communication needs. Most often, wireless data systems are designed to serve user needs for mobility or portability—*mobile data* is a widely used term—but many mobile systems and applications can also serve the data communication needs of users who do not move about (*fixed* users).[4] A number of wireless data applications, in fact, are being designed with fixed users in mind.

Traditionally, wireless data applications and services have been concentrated primarily in a few narrowly defined, *vertical*, business markets, including:

- *Field service* (dispatch, sales, repair, parts ordering, work order processing). Field technicians rely on wireless communication systems to get their next assignment, order parts, and check customer histories and accounts.
- *Fleet management* (dispatch, parcel tracking, vehicle location, and security). Wireless services are heavily targeted to trucking and other transportation industries. Wireless systems allow companies to dispatch trucks faster and more efficiently, track cargo, locate trucks, plan routes, and find stolen vehicles and merchandise.
- *Messaging* (paging, e-mail, short messages). Wireless systems also allow remote workers to stay in touch. A regional manager can be contacted at any store, sales personnel can be sent

updated product information, and doctors can be paged for emergencies.

For the past decade, the use of these services has been limited to a small group of business users with high mobility and connectivity needs—those who could afford the high prices of equipment and service. Package delivery companies such as UPS and Federal Express, for example, rely on wireless data to keep up-to-the-minute track of parcels and for dispatch services (see box 4-1).

Today, however, the kinds of people and companies who use wireless data products are changing and expanding. As the United States has moved into a more competitive international environment and a more service- and information-based economy, the use of computers in the workplace has increased. In addition, more workers are getting out of the office—but even within the office or factory setting, the value of being mobile (but in touch) is being recognized (see box 4-2). These changes are beginning to affect the consumer mass market and the more general, *horizontal*, business market for wireless data products and services.

Moving from specific (and specialized) applications to products designed for the general user, however, is proving difficult. Services designed for one company often do not translate well to another with different needs and expectations. However, some general applications have been identified, including computer network extension, Internet access, wireline replacement (point-of-sale terminals, alarm monitoring), personal services (computer services, online services, and other information services), and other data applications, such as medical monitoring equip-

---

[4] As noted in chapter 1, fixed use can be thought of as a subset of mobile use. Cellular phones, for example, work just as well (sometimes better) when one is standing still as when one is driving or walking. Intuitively, if a system can serve mobile users it can usually serve fixed users as well. Although some engineering concerns (power level, building penetration) may be different, in many cases the same system can serve both types of users.

BOX 4-1: United Parcel Service's Use of Cellular Technology

In 1992 United Parcel Service (UPS) began developing a nationwide, real-time, package tracking system, combining UPSnet—UPS's existing wire-based network—with cellular technology. To provide this service, UPS had to stitch together a network of over 70 large and small cellular carriers, including GTE Mobile Communications, AirTouch (formerly PacTel Cellular), McCaw Cellular, and SouthWestern Bell Mobile Systems. These companies also arranged to provide UPS with a single point of billing for air time. The project involved technical as well as logistical challenges for UPS, the cellular industry, and equipment manufacturers. In February 1993 UPS initiated the new service it calls TotalTrack.

When delivering a package, UPS drivers use a device called a DIAD (Delivery Information Acquisition Device) to scan the bar codes on the package's label. When the driver returns to the truck, he inserts the DIAD into the DIAD Vehicle Adapter (DVA). The DVA transfers the package information from the DIAD and transmits it to a cellular telephone tower via an in-vehicle cellular modem. The data are then routed through the cellular system to the wireline UPSnet, and on to the UPS Data Center in New Jersey. Here, package information is stored in a database where it can be accessed by a UPS customer service representative.

By implementing the TotalTrack system using the U.S. cellular infrastructure, UPS has been able to keep pace with Federal Express, which in the 1970s and 1980s built their own private wireless data network to provide real-time package tracking. On an average day, UPS will track roughly 6.3 million packages with TotalTrack, moving about 290 million bytes of data over the cellular network. This utilization of the current U.S. wireless infrastructure enabled UPS to meet the growing demands of its clients.

SOURCE: Office of Technology Assessment, 1995.

---

ment. At the leading edge, videoconferencing and video telephony products are being developed for laptop computers.[5]

Finally, an increasing number of wireless data applications will not involve people at all. Systems are being developed to locate stolen cars, track individual pieces of cargo on trucks and trains, and remotely monitor environmental conditions (tides, wind, snowfall) and industrial operations such as natural gas/oil wells or pipelines. These systems give companies more immediate information and closer control over their operations in locations where wireline technologies either will not work or are impractical.

As the needs of wireless data users beome better defined, new technologies and applications will be deployed. Service providers and equip-

ment manufacturers have entered a period of innovation and uncertainty as they seek to (re)design products and applications to appeal to a wider audience. The next several years are likely to be characterized by rapid product turnover, slim margins, and consumer confusion. The technologies and services that succeed and those that do not will only be determined as users buy, and the market reacts.

■ Factors Driving Demand

The most important factors fueling the demand for wireless, especially mobile, data communications services are: 1) the dramatic increase in sales of portable computers; 2) a growing familiarity and use of computer networks; and 3) a rising expectation of being able to access information anywhere,

---

[5]Current systems use telephone lines and V.34 (28.8 kbps) modems to deliver video at 7 to 10 frames per second (normal video runs at 30 fps), and cost from $1,000 to $1,500. The Personal Conferencing Specification now being developed will offer a standard for videoconferencing.

BOX 4-2: Wireless Technology in Restaurants

To speed order processing and improve customer service, some restaurants are implementing wireless technologies that allow wait staff to send customer orders directly from the table to the kitchen or the bar. One such system, the Squirrel Restaurant Management System, uses Fujitsu palmtop computers with PC card radio modems, and allows wait staff to transmit orders, call up drink and menu inventory, and process credit cards—all from the customer's table.

This system uses frequencies in the unlicensed 902 to 928 Mhz band to transmit signals from the hand-held unit to a system base station. The system achieves a burst data rate of up to 242 kilobits per second and has a range of 300 feet indoors and 800 feet outdoors. The Fujitsu PoqetPad sells for $2,285, including the radio modem. Additional costs include the Squirrel Restaurant Management software package and the restaurant base stations, each of which can accommodate five hand-held units.

In addition to speeding the delivery of the customer's order, such systems have enhanced accounting processes in many restaurants. Prior to implementing such a system, restaurant management would have to go through every order slip to track the number of salads, bottles of wine, etc. they had sold in a day/week/year. With the automated system, restaurants can have the wireless device send one copy of the order to a printer in the kitchen, and one to a main computer which keeps records of the sales. This makes tracking inventory and checking employee theft much simpler for restaurant accounting offices.

Although increasingly popular, these systems have encountered some problems. For example, one restaurant which implemented a wireless ordering system found that, without extensive training, wait staff spent too much time looking at the hand-held device while at the customer's table, and not enough time talking with the customer.

SOURCE: Jeff Tingley, "Wireless Pen Computing Serves Restaurant Industry," *Wireless for the Corporate* User, vol. 3 No. 1, 1994, p. 52.

---

anytime.[6] Today, worldwide notebook computer sales total almost 8 million units, accounting for 17 percent of the market for personal computers (see figure 4-1).[7] By 1998, at least one company predicts that sales of notebook computers will capture 22 percent of the total market. These figures suggest that workers in many jobs and who exhibit varying levels of mobility are using portable computers—no longer will they be confined to traveling professionals and executives. Most industry observers believe that the latent demand for mobile/portable computing is enormous, and that the development of mobile computing applications and software will lead to a corresponding increase in the demand for wireless connectivity (see box 4-3). This may be a reflection of the same trend that is fueling increasing cellular phone subscriptions by small businesses and even mass market consumers—the increasing desire and/or need to be connected to family, friends, the office, customers, or suppliers.

At the broadest level, wireless data applications are being driven by the increasing demands for mobility and by a need to access information immediately from any location. Almost 50 million workers have jobs that can be classified as mobile in some way (see chapter 2). For some, mobility is an inherent part of the job—a supervisor on a fac-

---

[6]Decision Resources, "Wireless Data Communications: Scenarios for Success," written by Clifford Bean of Arthur D. Little, Inc., cited in *Mobile Satellite News*, vol. 5, No. 18, Sept. 15,.1993, p. 4. For a discussion of the trends affecting the mobile computing industry, see Dulaney, op. cit., footnote 1.

[7]Paul Taylor, "Small, Light—and Powerful," *Financial Times*, May 3, 1995, p. 5.

MOTOROLA, INC.

Personal digital assistants (PDAs) allow users to take many of the functions of their office with them when they travel. This unit combines pen based computing with wireless electronic mail and fax capabilities.

tory floor, a sales representative with a large multistate territory, or a repair technician working in a metropolitan area. These people need to be in touch with colleagues, customers, and suppliers; access company files; and transmit status reports and updated information. Wired networks may not always be easily accessible or convenient.

For other workers, mobility is only an occasional part of the job. Professionals and white-collar workers often use computers and computer networks in their offices, but when they travel—to visit clients, attend a conference, or take a vacation—these resources stay behind. Increasingly, however, users are demanding access to the same capabilities when they travel as they have in their offices, including electronic mail (one of the most common uses of mobile data services), remote file access, and fax.

WIRELESS DATA SYSTEMS

The following sections describe the various wireless data systems according to the character of the information sent (one-way or two-way) as well as the distinctions created by past regulatory and

FIGURE 4-1: Worldwide Notebook PC Market

Worldwide notebook PC market (millions of units)

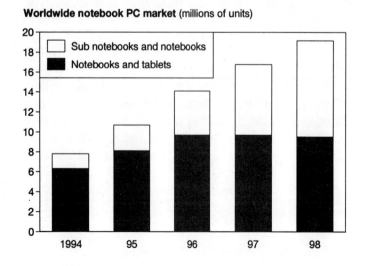

Sub notebooks and notebooks
Notebooks and tablets

1994   95   96   97   98

Notebook share of total PC market

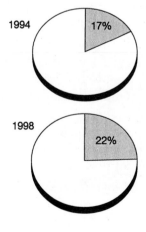

1994   17%

1998   22%

SOURCE: Dataquest.

BOX 4-3: Wireless Data Devices and Products

Personal Digital Assistants. In the last two years, products have been introduced that combine many of the functions of a personal computer with wireless communication capabilities, including e-mail, paging, faxing, and remote data access. Some also enable users to place and receive phone calls. These personal digital assistants, PDAs, include Apple's Newton, Tandy's Zoomer, Motorola's Envoy, and IBM/BellSouth's Simon. Prices range from $200 to 1,500, depending on features. Some of these devices now use cellular or private data networks to allow users to communicate.

By most estimates, the introduction of PDAs has been a disappointment. Although experts disagree on which factor was most important in their low sales (poor handwriting recognition, slow processing speeds, etc.), nearly all agree that lack (and/or the high price) of communications software was an important contributing factor. Apple's Newton, for example, could communicate with other Newtons, but adding the capability to communicate with the "outside world" cost more. The fact that communications is viewed as so important in their demise, however, may mean that future PDAs (now sometimes called personal communicators) with standardized (and affordable) communications capabilities for messaging, faxing, e-mail, and perhaps even voice will be more successful.

Other factors contributing to the slow start of PDAs include unreliable (due to poor quality of links) transmission, and high prices both for the units themselves and transmission and data services. The machines also use competing operating systems: Apple and Sharp use Newton, Tandy/Casio use Zoomer (software by Geoworks), Microsoft (with Compaq) has developed Winpad , and General Magic, whose backers include Sony, Motorola, ATT and Apple, and Phillips, has developed software called Magic Cap.

Pen-based computing. In contrast to the disappointing sales of PDAs, pen-based computers serving specific business uses—field technicians, delivery personnel, insurance caseworkers—have been relatively successful. Each of these vertical applications, however, usually will not work with the others. Special software is customized for each user; with different capabilities and ways of entering information. Many applications require the individual user to fill an electronic "form" that is designed to capture specific kinds of information—census data for example—that would not transfer to other businesses.

PCMCIA cards. Personal Computer Memory Card International Association (PCMCIA) cards, also known as PC cards, are credit card-size devices that plug into a special slot in a (laptop) computer and perform a range of functions—modem, LAN access, hard drive, even GPS capabilities. In modem and LAN access applications, PC cards can use cables to connect or radio waves. PC cards have had their share of problems—software incompatibilities, excessive memory and power requirements, and hardware connectivity—, but these problems seem to be subsiding as manufacturers and developers refine their designs and products.[1] However, wireless PC card adapters are still expensive; costing from $600 to $800 each.

SOURCE: Office of Technology Assessment, 1995.

---

[1]For an overview, see *PC Magazine*, Jan. 24, 1995, which has a series of articles on PCMCIA cards.

technological differences. Many of these systems can or will offer essentially the same service(s), but at different costs and with slightly different features and coverage areas.[8] Six types of systems are discussed: broadcast, two-way messaging, cellular data, wireless computing, unlicensed services, and satellite data services.

■ Broadcast Systems

Broadcast technologies are well suited to distributing data from one central location to many users (point-to-multipoint) in a given area, or to reach a user whose location is unknown or who is moving about. Although these technologies are one-way only, they often provide the lowest cost alternative for keeping in touch with family, business associates, and employees, and are increasingly being used by residential consumers as well as businesses.

Paging

Paging services represent the most basic form of wireless data delivery. Use of pagers has boomed in the past five years as prices have dropped 50 percent—to below $100 for basic models.[9] About 600 paging companies operate in the United States today, providing services to over 19 million people—making paging one of the most widely used wireless services. Paging systems provide service at all levels—local, regional, and national, and equipment and usage are usually quite inexpensive. Customers pay between $50 and $500 for a pager and between $15 and $100 per month for service.[10]

Paging companies provide a range of services. With tone-only pagers, the paging company transmits a signal to the user's pager, alerting them to call in for a message. With more advanced tone/

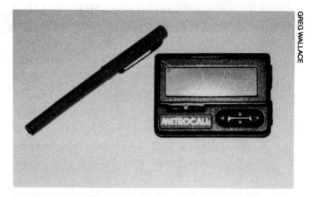

GREG WALLACE

Pagers, such as this alphanumeric unit, have become one of the most popular means for people to stay in touch via wireless.

voice or numeric pagers, the user receives a voice message or phone number on their pager. The most advanced units, alphanumeric pagers, can receive short text messages, e-mail (even from the Internet), voice mail notification, and information services such as traffic alerts or stock quotes. In 1993, numeric pagers accounted for 87 percent of the pagers in use, alphanumeric 7 percent, tone-only 4 percent, and tone/voice 2 percent (see figure 4-2).[11]

Paging companies are expanding their services to provide more sophisticated communication services. MobilComm, the country's second largest paging company, began sending messages to Newtons and other PDAs using a receiver that costs $200. Mtel is building a $150 million network that will allow users to acknowledge received pages beginning in 1995, and its Skytel service has already begun testing two-way communications. Recent Federal Communications Commission (FCC) auctions of narrowband personal communications service (narrowband PCS—see below) frequencies made additional spectrum available for advanced digital and two-way paging services. This new spectrum will en-

---

[8]For an overview of current products and services, see various articles in *Data Communications*, vol. 24, No. 4, Mar. 21, 1995.

[9]Lois Therrien, "Pagers Start to Deliver More than Phone Numbers," *Business Week*, Nov. 15, 1993.

[10]David Strom, "Reality Check on Wireless Data Services," *Business Communications Review*, May 1994, pp. 62-66. See also Data Communications, op. cit., footnote 8.

[11]EMCI, Inc., based on EMCI paging survey, January 1994.

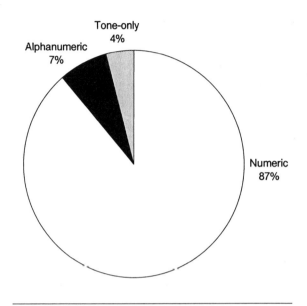

FIGURE 4-2: Pagers in Use by Type, 1993

Tone-only
4%

Alphanumeric
7%

Numeric
87%

SOURCE: Office of Technology Assessment, 1995, based on data from EMCI, Inc.

would have the paging chip be unprogrammed when purchased, allowing the purchaser to call their preferred service provider that would then program the chip over the phone.[13]

Regardless of the type of service provided, all paging systems use similar technologies and architectures to deliver service (see figure 4-3).[14] When a caller wishes to send a message to a paging customer, he calls the paging company, which then encodes the message with the paging customer's "address," called a cap code, and broadcasts it.[15] The subscriber's pager receives the transmission and alerts the user. To achieve the best possible coverage of an area, paging companies use a technique called *simulcasting* that transmits the same message from multiple transmitters at the same time. To extend the coverage of services, many companies establish agreements with other paging companies that allows their customers to use paging systems outside their home system. A few service providers have assembled nationwide networks using this approach. National paging services also use satellites to relay messages between local systems.[16] Because pagers are generally tuned to specific service providers, users cannot easily change carriers—unlike cellular phones, which can be easily reprogrammed.

able paging companies to offer a wide array of new information services, continuing the trend toward higher functionality.

By combining a computer or a PDA with a paging unit, users can receive data files, short messages, and other more advanced features. Some analysts expect that alphanumeric paging will become an integral part of portable computers before the end of the century, and that computer-based services will represent an increasing portion of the paging business.[12] In the future, paging devices may be reduced to a single computer chip and integrated into a wide range of computing and information devices. One idea now being developed

Radio Broadcast

Traditional AM and FM radio broadcasters are exploring ways to deliver information services using their broadcasting facilities. Some are experimenting with Radio Data System (RDS) technology that will transmit additional information—such as song title and artist, the station's call letters, and music format information—along with

---

[12]The Gartner Group, for example, predicts that 50 percent of all palmtop computers will have paging capabilities built in by 1998. See also T. Garber, "Special Report," *Radio Communications Report,* vol. 13, No. 10, May 23, 1994.

[13]Andrew M. Seybold, *Using Wireless Communications in Business* (New York, NY: Van Nostrand Reinhold, 1994).

[14]Paging companies are licensed in 25 kHz channels in four bands: Lowband, Highband, UHF, and 900 MHz. The recent FCC Narrowband PCS auction made available 1,300 KHz of additional spectrum in three bands between 900 and 941 MHz.

[15]Alphanumeric pagers are an exception. A computer with paging software and a modem, rather than just a telephone, is required to initiate an alphanumeric message. Telephone answering services (TAS) are available so that anyone with a telephone can call the TAS and leave a voice message with a representative, who then inputs the message through a computer to the paging company's encoding and controller station.

[16]D. Baker (ed.), *Comprehensive Guide to Paging* (Washington, DC: BIA Publications, Inc., 1992).

FIGURE 4-3: Generic Paging System

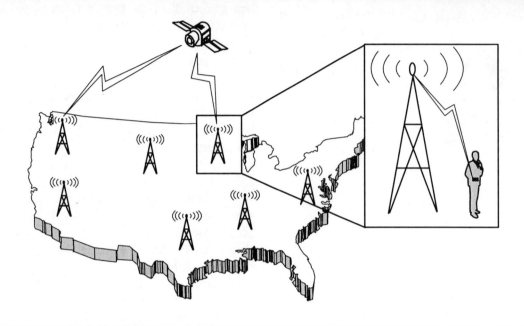

SOURCE: Office of Technology Assessment, 1995.

the regular programming.[17] Such systems have been used in Europe for many years, but the U.S. Radio Broadcast Data Service (RBDS) was only established in early 1993, and deployment of the technology has been extremely slow. By early 1994, some 100 stations were said to be using RBDS, but few RBDS-compatible radios are in use.[18]

Other radio data services being considered include travel advisories, local restaurant/hotel information, and advertising supplements. In FM radio, for example, broadcasters would like to use the FM subcarrier to transmit supplementary advertising information—school closings, stock quotes, and other information services—directly

to personal computers.[19] Such systems have been tested, but most efforts are only in the conceptual stage. Standardized (receiving and processing) technology for consumers has not been developed, and systems are not expected to be ready for widespread deployment until late 1995 or 1996. Speeds up to 19.2 kilobits per second (kbps) are expected to be available, and, like other broadcasting applications, these types of services are expected to serve both mobile and fixed users.

Television Broadcast
Using their existing equipment, television broadcast systems are capable of transmitting data in several ways. Over the years, a number of at-

---

[17]The system uses a *subcarrier* that is broadcast alongside the main radio signal and allows data to be sent at about 1.2 kbps. It does not interfere with the main radio programming. Reportedly, a higher data-rate standard is being developed by the National Radio Systems Committee of the National Association of Broadcasters—perhaps ready by 1995—that would carry information at speeds up to 20 times the existing standard. Bennett Z. Kobb, *Spectrum Guide* (Falls Church, VA: New Signals Press, 1994), p. 29.

[18]John Gatski, "RDS/RBDS Slowly Gains Acceptance," *Radio World*, vol. 18, No. 4, Feb. 23, 1994.

[19]Paul Farhi, "EZ Communications Forms Unit to Develop Radio Technology," *The Washington Post*, Dec. 4, 1994, p. D4.

tempts have been made to develop *videotext* services that send data—including stock quotes, newspapers, and other local information—in the vertical blanking interval (VBI), the black stripe at the top/bottom of a television picture. To decode the data, users had to have a set-top box that would capture the information, store it, and display it. None of these experiments were commercially successful. The VBI can also, theoretically, be used for applications such as paging and updating retail information (stolen credit card lists, for example), but there has been little demand for these services from businesses and most broadcasters are not providing them.

Other methods for transmitting data are also being developed. Recently, one company has developed a proprietary system that transmits high-speed data using the whole broadcast signal without interfering with the regular programming.[20] Similar to the sideband broadcasting applications being developed for radio broadcasting, other applications are being developed that use television secondary audio (SA) channels.[21] One system currently being tested uses audio channels transmitted via satellite to deliver current weather information and emergency weather and environmental alerts to personal computers located around the country.

The industry is also working with new companies to provide Interactive Video Data Services (IVDS). These systems would allow viewers to respond to polls, order merchandise, and play along with game shows by using a remote control and a set-top box connected, via special IVDS frequencies, to a local control center. Frequencies for IVDS were auctioned by the FCC in 1994, but services have not yet been deployed because of problems with the technology and availability of equipment. In addition, a number of the IVDS auction winners defaulted on their bidding commitments.

Two factors will seriously limit the implementation of data systems by broadcasters in the short run. First, most of the services developed so far are fairly low bandwidth, and demand has historically been low. Second, these systems are based on the existing analog technology currently used by television broadcasters. They will most likely not work with the digital broadcasting systems now being developed (see chapter 5). Once digital broadcasting technologies are implemented, broadcasters hope to use at least some of their spectrum to provide various information services. The terms under which such uses will be allowed have been a contentious issue for policymakers. Legislation now being debated in Congress generally allows broadcasters to provide "ancillary or supplementary services," subject to various licensing restrictions and payment of fees.[22] The definition of an "ancillary or supplementary" service remains unclear, however, and what services will be allowed remains uncertain.

▌ Two-Way Messaging

Two-way messaging services provide a variety of interactive low-speed data applications, and can serve fixed, portable, or mobile users. Many individuals use two-way services to send and receive electronic mail and access company data networks. Other services include remote meter reading, point-of-sale and credit card verification, and alarm monitoring. Some of these applications could be provided by wire-based systems, but the

---

[20]Presentation of Wave-Phore at the National Association of Broadcasters convention, Las Vegas, NV, April 1994.

[21]These channels are currently used to provide second-language translations for television programming, or, on some PBS stations, weather reports.

[22]U.S. Congress, Senate, *S. 652, Telecommunications Competition and Deregulation Act of 1995* (Washington, DC: U.S. Government Printing Office, 1995); U.S. Congress, House of Representatives, *H.R. 1555, Communications Act of 1995* (Washington, DC: U.S. Government Printing Office, 1995).

high cost of laying wire would likely be prohibitive, and, in many cases, it is easier to install a wireless system.

Packet Radio

The two-way data messaging industry is dominated by Ardis (backed by Motorola) and Ram Mobile Data (a joint venture of Ram Broadcasting and BellSouth Enterprises). These two providers offer specialized communication services primarily to companies, but are now trying to expand into more general markets (e.g., mobile professionals). Some analysts doubt that such a strategy will work, citing potential competition from both cellular carriers deploying cellular digital packet data (CDPD) and future narrowband PCS companies (see below).

Commercial messaging services are provided through terrestrial towers in each metropolitan area. Digital packet technology is used to send information over channels in the 800 MHz SMR frequency band. The Ram service is currently available in more than 250 metropolitan areas, while Ardis serves the nation's 400 largest metropolitan areas—coverage is not quite national. Both services are designed to deliver short (200 to 300 bytes) text messages, generally using specialized equipment. Ram operates at 8 kbps and Ardis is upgrading its network to offer speeds up to 19.2 kbps, but actual data throughput is usually about half these speeds. Each offers a range of pricing plans based on peak and off-peak times and different levels of use. Ram's prices range from $25 to $135 per month, and are based on the amount of data sent, while Ardis's range from $39 to $299 per month, and are based on the number of messages sent.[23]

Narrowband Personal Communications Service

In 1993 the FCC established a new category of wireless data services, narrowband PCS; allocated spectrum for it; and established the rules that would govern the systems' operations.[24] Following congressional mandates, in 1994 the FCC began auctioning narrowband PCS licenses. To date, 10 national and 30 regional licenses have been awarded; bringing in just over $1.1 billion.[25] A total of 3,554 licenses will be issued to companies that plan to offer new services as well as expand and augment existing networks and services. The first systems are expected to begin operation sometime in 1995.

The FCC defines narrowband PCS as a family of mobile and portable radio services that will provide a variety of advanced paging and messaging applications to individuals and businesses.[26] It promises low-cost, two-way data communication services that are expected to appeal initially to the traditional mobile data markets, such as field sales or (repair) service and fleet and courier dispatch.[27] Narrowband PCS licensees plan services that include: credit-card verification, locator services (for vehicle dispatch and tracking), voice paging, acknowledgment paging, and two-way exchange of short messages. These services will be delivered to user devices such as alphanumeric pagers,

---

[23]Joseph Palenchar, "Will Cellular Packets Lead the Way in Wireless?" *Mobile Office*, July 1994; *Data Communications*, op. cit., footnote 8.

[24]The service was allocated 3 MHz of spectrum at 901 to 902 MHz, 930 to 931 MHz, and 940 to 941 MHz, of which 1 MHz was held for future uses. Federal Communications Commission, *Amendment of the Commission's Rules to Establish New Narrowband Personal Communications Services*, First Report and Order, Gen. Docket 90-314, 8 FCC Rcd 7162 (1993).

[25]Licenses were divided among four types of service areas: 492 Basic Trading Areas and 51 Major Trading Areas (as defined by Rand McNally), five regional licenses, and 11 national licenses. Six companies paid $617,006,674 for 10 national licenses: Airtouch Communications Inc., Bellsouth Wireless, Inc., Destineer Corp (MTel), McCaw Cellular Communications, Inc., Pagenet, and Pagemart. Six licenses were auctioned for each of the five regions, with bids totaling $488,772,800. Robin Gareiss, "PCS: Making Sense of the New Services," *Data Communications*, October 1994, p. 49.

[26]Federal Communication Commission, op. cit., footnote 24.

[27]Gareiss, op. cit., footnote 25.

computers equipped with radio modems, portable fax machines, and portable computers.

Like other new wireless systems and services, the costs of building the systems and the prices that will be charged are closely guarded by the companies involved. One company, Pagemart, estimates the cost of building their system is about $200 per subscriber—more than a traditional paging system, but significantly less than the $750 that cellular companies say they spend on building their systems.[28] Prices are still being determined, but are expected to be higher than traditional paging services, but less than other wireless data services such as Ardis/Ram, CDPD, or cellular.

■ Cellular Data

Cellular telephone systems can also be used to send data, and represent an alternative to packet radio networks and future narrowband PCS. Cellular data transmission allows users to do everything they can do with a regular wireline modem— connect to office LANs, send and receive electronic mail or text files, access online services, and browse the Internet. Speeds remain slow; and operation is not as reliable as a wired modem, but cellular systems are rolling out new digital data services that should improve performance.

Circuit-Switched Cellular Data

The traditional method for sending data over a cellular system is much the same as sending data from a computer using the telephone lines. A computer is connected to a radio modem that dials the phone number and makes the connection using regular cellular channels. Radio modems add features to compensate for the different transmission characteristics of the airwaves, which are more prone to noise and interference than the wireline network.[29]

Alternatively, using a regular wireline modem, a user can connect his or her computer directly to a cellular phone through a data connection (RJ-11 jack) built into the phone (not all phones have such connections). This method is often less reliable, however, because wireline modems are designed for landline use and may not be able handle the differences in cellular phone networks and calling procedures; the phone may disconnect during cellular hand-offs, for example.[30] Such problems, combined with the interference and noise common in cellular voice calls, make cellular data calls less reliable than those made through the public telephone network.[31] Maximum speed is theoretically 9.6 kbps, but actual speeds are usually lower—2.4 or 4.8 kbps.

Current circuit-switched analog data applications, which currently account for about 3 percent of total cellular traffic, may grow in the next several years, but in the longer term, they will be discontinued.[32] Cellular providers are now deploying digital data technologies that use their existing networks (CDPD-see below), and eventually, they will completely replace their analog service with new digital services (see chapter 3).

---

[28]Ibid.

[29]Specially designed cellular modems offer advantages over regular landline modems for cellular use, but they generally require the same type of modem on both ends to work (a mobile worker with a cellular modem cannot just connect with anyone with a regular modem). To overcome this compatibility problem, some carriers have instituted modem *pools* that allow users with cellular modems to dial in to the pool and the carrier will serve as a go-between, translating the cellular modem signals into signals the modem being called can understand.

[30]Common cellular-network impairments include frequent cellular base-station hand-offs, dropouts, call interference, fading, echo, and other types of signal distortions. These problems require signal conditioning techniques not implemented in traditional landline modems.

[31]Datacomm Research reports that "even with special 'cellular modems,' one can expect call attempts to fail anywhere from 20 to 50 percent of the time," op. cit, footnote 3, p. 23.

[32]Palenchar, op. cit., footnote 23.

Cellular Digital Packet Data

To overcome the limitations and high cost of circuit-switched cellular data, a group of cellular carriers began to develop an alternative data transmission system called cellular digital packet data (CDPD) in 1992. Standards were agreed to in mid-1993. CDPD radio modems transmit data by breaking the information into digital "packets" and sending them over vacant channels on existing analog cellular systems—no one channel is dedicated to one data "conversation" as in circuit-switched service.[33] Although CDPD was originally envisioned as a dynamic system in which vacant channels would be identified "on the fly," in practice, system operators have set aside a certain number of channels dedicated to CDPD use in order to improve performance. CDPD systems can transmit at speeds up to 19.2 kbps, but actual throughput is closer to 9.6 kbps because of error-correction features added to increase reliability.

CDPD services are being designed to support a wide range of data applications. In addition to the mobile services used by professionals and field technicians, CDPD is also being developed for some fixed location applications, such as vending machines and remote utility installations like natural gas wellheads. CDPD can be used like a regular wireline modem to remotely connect to LANs, access databases, and exchange files, but it is especially useful for applications characterized by short "bursty" data. CDPD systems have been designed to favor shorter transmissions (less than 600 words) and have been optimized for users who send and receive many short messages (500 characters, or 50 to 75 words)—credit-card verification, real estate transactions, emergency services, dispatch, fleet management, package delivery and tracking, telemetry, two-way paging, Internet access, and electronic mail.[34]

Service packages currently range from $11 to $139 per month, and, like Ram and Ardis, are usually based on varying levels of usage that allow users to match their usage to their budgets. Following the technology, pricing structures favor shorter communications. Messages of up to 1,000 characters, for example, may cost as little as $0.17.[35] CDPD is expected to be more cost-effective than circuit-switched data services for short communications, while circuit-switched may be preferred for larger file transfers.

For cellular system operators, CDPD offers an important benefit; it allows them to upgrade their data capabilities without replacing their existing analog cellular infrastructure (antennas, transmitters, frequencies), and with the addition of very little additional equipment. This "overlay" approach may allow CDPD services to be rolled out faster and at less cost than competing services that have to be built from scratch like some of the new narrowband PCS services.[36] CDPD also offers performance advantages over circuit-switched cellular data applications, including better error correction; improved reliability; faster speeds; more flexible functions, including multicasting; and potentially lower costs.

One important advantage that CDPD has over most other wireless data technologies (except satellite services) is coverage. The potentially wide availability of CDPD—cellular services are currently available to about 95 percent of the population—would give it a distinct edge over existing wireless services such as Ardis and Ram, which

---

[33]Data are NOT sent in between pauses in conversations, but in the time between different conversations. When a voice conversation is assigned to a channel currently being used for data, the system will automatically find another vacant channel and switch the data communication so that no interference occurs. This is called "channel hopping." Research indicates that an average channel is unused as much as 30 percent of the time. John Gallant, "The CDPD Network," *EDN*, Oct. 13, 1994.

[34]Ibid.

[35] "Sending the same message via circuit-switched cellular could cost more than four times that amount because carriers bill for air time in one-minute increments, even if a transmission takes only a few seconds." Palenchar, op. cit., footnote 23.

[36]Chris Pawlowski and Peter McConnell, "CDPD Air Interface Basics," *Telephony*, Dec. 5, 1994.

typically provide coverage only in metropolitan areas. The relatively small number of potential users outside those coverage areas, however, may mean that this advantage is important only for users who need very broad coverage, such as trucking or package delivery companies. The interoperability of CDPD, however, has not yet been proven. Only a few carriers have struck CDPD roaming agreements, and the technical ability to connect different CDPD systems is only now being tested—true nationwide roaming may be years away.

There is still a great deal of uncertainty over CDPD's role in data transmission and how successful it is likely to be. CDPD standards were set in 1993, but deployment of CDPD capabilities has fallen well behind initial expectations due to technical difficulties. By mid-1995, only 19 systems were offering service, and another 22 were planning to begin operation by the end of the year.[37] Some analysts see CDPD as little more than an interim service that few people or businesses will use. Others, including the consortium of cellular companies that developed CDPD, believe it is the answer to publicly accessible wireless data services.[38] At least one forecast estimates that there will be 1.6 million CDPD users by 1998.[39] Given the slow deployment of CDPD, it is still unclear how successful it will be, or whether it will be quickly superseded by advanced digital cellular data applications.

Digital Cellular Data

Once cellular carriers switch to digital formats—time division multiple access (TDMA) and code division multiple access (CDMA)—new data formats will also become available. The data portions of these overall standards are being developed, but have not yet been finished, and no commercial data services are being offered. Cellular digital data applications will be deployed after the voice applications, which are already starting to appear. In the first implementations of TDMA, for example, existing analog channels continue to be set aside for analog and CDPD modem communications.

By contrast, more than two dozen Global System for Mobile communications (GSM—see chapter 3) systems around the world have already begun to offer data services.[40] However, only a few vendors are making GSM data equipment, and services are usually limited to the home system—roaming is not yet possible due to the lack of roaming agreements for data applications. Individual networks also must be upgraded to provide data services. Finally, compared to other wireless services, such as the international affiliates of Ram Mobile, GSM data communications can be more expensive. Vendors and analysts expect these initial problems to be solved quickly as more GSM systems are deployed and more users subscribe.

■ Wireless Computing

The use of wireless technologies by computer users is one of the areas projected for the strongest growth over the next several years, and a good number of companies have targeted mobile or wireless computing as a potential market for various kinds of wireless information services. This section will concentrate on the use of portable computers for general computer tasks—word

---

[37]Robin Gareiss, "Wireless Data: More Than Wishful Thinking," *Data Communications*, op. cit., footnote 8.

[38]Consortium members include: Ameritech Mobile Communications, Inc., Bell Atlantic Mobile Systems, Inc., GTE Mobilenet, Inc., Contel Cellular, Inc., McCaw Cellular Communications, Inc., Nynex Mobile Communications, Pactel Cellular, and Southwestern Bell Mobile Systems. CDPD service and product providers have also formed the CDPD Forum, Inc., a trade association composed of more than 80 companies involved in CDPD that will continue work on stardardization and interoperability.

[39]Report by BIS Strategic Decisions, cited in Pat Blake, "Wireless Data: The Silent Revolution," *Telephony*, Dec. 5, 1994.

[40]The following material comes from Elke Gronert and Peter Heywood, "GSM: A Wireless Cure for Cross-Border Data Chaos," *Data Communications*, op. cit., footnote 8.

processing, file transfer, and remote connection to computer local area networks (LANs).

Device Connectivity

At the simplest end, products are being developed that allow users to link wirelessly with their desktop computers while they are near, but not in, their offices. These products respond to studies of mobility on the job that indicate that most of the time people are out of their offices, they are still close by. The machines have varying levels of intelligence and storage built in, but allow users to remotely access their desktop computer—to read electronic mail or use any applications. The systems require modems at both ends and have a range of about 500 feet. Remote devices cost about $1,400, while radio modems for the desktop range from $600 to $700.

In addition to these products, infrared technologies—like those used in television remote controls and other consumer electronics devices—are also being developed that would allow portable computers, printers, and PDAs to communicate directly with one another. Infrared technology allows the ad hoc creation of low-speed networks (maximum data rate is currently 115 kbps, but speeds up to 10 Mbps are being developed)—at a meeting, for example—and direct device-to-device communication. Most PDAs, for example, already have infrared technology built in so they can communicate with each other, and one analyst estimates that 90 percent of all personal computers will have this capability by 1997.[41] In the future, proponents expect many kinds of devices to incorporate infrared communications capabilities, including public phones, computer printers, cash registers, and fax machines.

The advantages to infrared technology is that it is inexpensive (around $3 to $5 to equip a comput-

Portable personal computer makers are beginning to integrate wireless data communications capabilities, including remote local area networking, into their products. Users will soon be able to wirelessly connect to their office LAN from almost anywhere.

er with infrared, and $50 to $100 for an adapter, with prices expected to fall with increased volume), and potentially ubiquitous—companies from many countries have agreed to an international standard that will allow products to work around the world.[42] Computer hardware and software companies have already begun to build infrared communications capabilities into their products, and adapters that will connect to existing computers, printers, and telephones are expected to be on the market by mid 1995.[43] In the future, proponents expect infrared technologies to provide an inexpensive way to provide high-bandwidth communications over short distances—another way to access the resources of the NII.

Wireless Local Area Networks

LANs connect computers in a small area (in an office, for example) and allow them to share

---

[41]Dulaney, op. cit., footnote 1.

[42]The Infrared Data Association, which is composed of over 70 companies in the field, announced a set of infrared data standards in early 1994. John Romano, "Infrared Boosts the 'Personal Area Network,'" *CeBIT News*, Mar. 21/22, 1994.

[43]Materials provided in a briefing to the Office of Technology Assessment by the Infrared Data Association, no date.

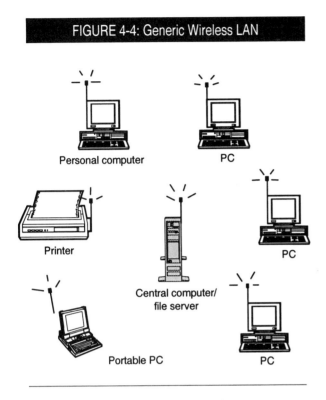

FIGURE 4-4: Generic Wireless LAN

Personal computer   PC

Printer

Central computer/
file server

PC

Portable PC   PC

SOURCE: Office of Technology Assessment, 1995.

memory, use a common printer, and exchange files and electronic mail (see figure 4-4). Wireless LANs substitute radio waves for the fiber optic or coaxial cables that connect most wire-based LANs. A computer equipped with a radio modem links to a central computer, called a server, which is also equipped with a modem or modems. Most wireless LAN radio modems also support direct device-to-device communication separate from the server.

Wireless LANs were originally designed to substitute for wireline LANs; to be used where wires were either too costly to install or where added flexibility (to move computers easily and/or quickly) was needed. For example, many older buildings are difficult to wire for computers (or even phone lines) because of their construction or the presence of hazardous materials such as asbestos. In these cases, wireless LANs may provide a cheaper solution. Box 4-4 compares wired and wireless LANs in school applications.

However, the market for such applications has not developed as expected. The primary problem, most analysts agree, is that wireless networks are significantly slower than wired LANs—1 to 2 Mbps on wireless versus 10 Mbps on most wired LANs. Wireless speeds are adequate for some applications—electronic mail and database queries, for example—but not for the higher-speed applications, such as image and graphics transfer, that are becoming increasingly popular. In addition, wireless LANs are often more expensive than their wired counterparts, with wireless modems costing up to $800 and access point equipment (that allows multiple computers to connect to the LAN remotely) costing up to $2,500 each.[44] As a result, wireless LANs have not proven popular simply as a replacement for wireline systems.

Currently, the wireless LAN industry is undergoing a transformation as vendors refine their products and marketing. Some see the concept of LAN extension—in-building mobility—or remote access to wired LANs as a more lucrative market. In fact, the market for wireless LANs has recently begun to improve. Commenters in a recent FCC proceeding provided sales figures demonstrating a rapidly expanding market for wireless LAN equipment—sales of $200 million for 1994 and expected sales as high as $2.5 billion by 1998.[45]

---

[44]By comparison, wired products cost from $150 to $500 for an adaptor and $500 to $1,500 for a multiple access hub. For a discussion of the speeds and prices of select systems, see David Newman and Kevin Tolly, "Wireless LANs: How Far? How Fast?" *Data Communications*, op. cit., footnote 8.

[45]Federal Communications Commission, *Allocation of Spectrum Below 5 GHz Transferred from Federal Use*, First Report and Order and Second Notice of Proposed Rulemaking, ET Docket 94-32, released Feb. 17, 1995, para 33.

BOX 4-4—School Networking and Wireless Technologies

Schools and school districts nationwide have been struggling for years to upgrade their communication and computer networks to keep pace with the rest of society. Facing tight budgets, many have found it difficult to afford the major capital investment of wiring classrooms, installing local area networks (LANs), and buying computers, let alone training teachers and administrators on the new technology.[1] Nevertheless, some schools and school districts have made computer networking a priority.

Although wireless LANs have been considered for many school applications, these systems have generally not been selected due to some combination of cost, reliability, and data-rate concerns.[2] As a result, wireless LANs are generally perceived as a second choice solution that is most appropriate for buildings that are hard to wire— historic buildings, those with asbestos, and buildings with insufficient room in the walls or ceilings for additional wiring—or for temporary school-building settings.[3]

As technology develops, however, wireless LANs may become a more competitive alternative to traditional wire-based LANs for school applications. In recent years, for example, wireless LANs have become more popular for business applications because of their enhanced security, higher throughput, and more competitive pricing relative to first generation wireless LANs.[4] However, as wireless technologies advance, so too do wire-based technologies. Some believe that 100 megabit/second wire-based LANs will soon become standard, dwarfing the throughput of even the fastest wireless alternatives.

School officials may wish to complement their existing wire-based LAN with wireless LAN technology. Many wireless LANs offer the flexibility to have numerous interconnected computers in a classroom one day and none the next. In addition, many wireless networks allow students to carry a portable PC or other device from classroom to classroom without sacrificing connectivity to the network. Other characteristics of wireless networks include: 1) implementation can be gradual (a school can purchase five transceivers for five computers, and increase the number as slowly or as quickly as demand warrants and the budget allows); 2) changes to the school are unnecessary (e.g., no asbestos removal or rewiring); and 3) installation takes days or weeks instead of the months required for a wire-based LAN. The table below provides a rough comparison of three wire-based LAN configurations for schools with three wireless alternatives. Because the installation cost of any LAN is dependent on the specific needs and circumstances of each user—which will vary greatly by site—the numbers presented below should only be considered as a crude illustration of the relative costs and merits of each system.

(continued)

---

[1]For an in-depth treatment of this subject, see U.S. Congress, Office of Technology Assessment, *Teachers and Technology: Making the Connection*, OTA-EHR-616 (Washington, DC: U.S. Government Printing Office, April 1995.

[2]Charles Procter, Florida Department of Education, Bureau of Educational Technology, personal communication, Mar. 9, 1995.
[3]Marty Heavey, Windata, Inc., personal communication, Mar. 28, 1995.
[4]Susan D. Carlson, "Wireless LANs Take on New Tasks," *Business Communications Review*, February 1995, pp. 36-41.

BOX 4-4—School Networking and Wireless Technologies (Cont'd.)

Comparison of Wire-based Local Area Network Costs and Capabilities
With Wireless Alternatives

Wire-based LANs	LAN Speed Internal	Internet Connection	Expansion Potential	LAN Cost	Drops	Per Drop Cost
MIT Model						
Low cost estimate	10 mbps	56 kbps	low	$37,100	67	$554
High cost estimate	<10 mbps	56 kbps	low	102,000	67	1,522
Central Kitsap, WA						
Low cost estimate	<10 mbps	512 kbps	high	357,500	350	650
High cost estimate	<10 mbps	512 kbps	high	412,500	350	750
Acton-Boxborough, MA	<10 Mbps	56 kbps	medium	25,393	82	310
Wireless Alternatives[1]						
Windata FreePort[2]						
High Mobility	<5.7 mbps	56 kbps	medium	136,495	67	2,037
Low Mobility	<5.7 mbps	56 kbps	medium	50,295	67	750
Proxim	<1.6 mbps	56 kbps	medium	52,385	67	782
Metricom[3]	10-40 kbps	28.8 kbps	high	13,400	67	200

NOTES: **LAN Cost** represents the one-time cost of installing the network (hardware costs and facility upgrades, including significant electrical upgrades with the wire-based LANs), but excludes computers and ongoing costs such as maintenance, usage fees, and personnel training. The wide discrepancy in the total LAN costs shown here represents different technology choices and also different-sized schools. For this reason, a better comparison can be drawn between the per-drop costs for the different LANs.

Per-Drop Cost is the cost of the LAN divided by the potential/intended number of users. This needs some qualifying in the case of the wireless alternatives because there are no "drops" per se, but rather wireless transceivers.

Expansion Potential refers to the ease (both financial and physical) with which additional users can be added to the various LAN architectures.

SOURCE: Russell I. Rothstein and Lee McKnight, MIT Research Program on Communications Policy, _Technology and Cost Models of K-12 Schools on the National Information Infrastructure,_ Feb. 10, 1995; Kent Quirk, Chairman, Citizen's Technology Advisory Committee, Acton, MA, personal communication, Mar. 29, 1995; Gordon Mooers, Coordinator Information Systems, Central Kitsap School District, Silverdale, WA, personal communication, Mar. 29, 1995; George Flammer, Metricom, personal communication, May 4, 1995; Windata, Inc.; Max Sullivan, Proxim, personal communication, May 16, 1995.

---

[5]The Windata and the Proxim systems are intended to complement an existing wireline LAN, thus, in addition to the wireless LAN components, a minimal wireline infrastructure is required, including a server ($4,000) and cabling to each wireless node ($520). These cost figures are taken from the MIT model and the Acton-Boxborough model, respectively.

[6]The Windata FreePort transceivers (the transmitters that provide communication from the PC to the rest of the wireless LAN) can support up to eight PCs. In the low mobility model, it is assumed that every computer will share the transceiver with seven others, thus reducing the amount of mobility realized for each user, and also reducing the cost dramatically. In the high mobility model, each computer has its own transceiver, thus increasing each user's mobility and the cost. We assume a total of four wireless hubs, at a cost of $7,450 per hub. Schools may require fewer, for example if all users are on one floor, then only one hub is needed.

[7]The Metricom Ricochet Network uses pole-top radios to relay wireless data from sender to receiver. These radios, which cost about $700 each, are owned and maintained by Metricom. Therefore, the only cost to the school is the $200 for the Metricom modem for each computer.

SOURCE: Office of Technology Assessment, 1995.

Several different kinds of wireless LANs are used today, which can be divided into three categories: infrared, narrowband, and unlicensed spread spectrum.[46]

Infrared

LANs using infrared signals are capable of transmitting data in fixed or portable LAN applications, although true mobility is hard to achieve. Infrared systems transmit information using both lasers (generally for point-to-point) and light-emitting diodes (LEDs—primarily for indoors).[47] These systems can operate at speeds of up to 10 Mbps, although throughput is much lower, and range is limited (60 to 150 feet). The technology works best with a direct line of sight between sender and receiver, but can also work by reflecting the signal off walls and ceilings—although not very well. Infrared signals, however, will not pass through walls or office partitions, limiting its usefulness for larger scale applications. Infrared data systems do not require licensing by the FCC and can be relatively inexpensive because they take advantage of production economies for other consumer electronic uses. The Institute of Electrical and Electronics Engineers (IEEE) is now developing standards for infrared LANs.

Licensed

A few companies have experimented with licensed spectrum to provide wireless LAN services. Motorola's Altair, for example, first introduced in 1992, operates at frequencies in the 18 GHz range and offers throughput at about 5.7 Mbps. One problem with licensed systems is that they are limited in the amount of spectrum they can use—only five channels in a 35-mile diameter area—and licensing is required.[48] In the case of Altair, Motorola controls the licenses. To avoid licensing and coordination problems and delays, most vendors have developed wireless LANs using unlicensed frequencies.

Unlicensed (Spread Spectrum)

Wireless LANs operate in the 900 MHz, 2.4 GHz, and 5.7 GHz bands (see discussion below on unlicensed data services). They offer speeds up to 5.3 Mbps, although actual throughput is usually 1 to 2 Mbps.[49] They use either direct sequence or frequency-hopping, spread spectrum transmission techniques (see appendix A). A number of wireless LAN products operate in the unlicensed bands, and the IEEE is currently developing industry standards for LANs as well as standards that will allow users' computers to communicate with each other directly—"ad hoc" or "peer-to-

---

[46]For further discussion of these systems, see Datacomm Research, op. cit., footnote 3.

[47]These systems currently do not work like most other radio systems—by modulating a radio wave. Instead, they simply turn the LED or laser on/off at high speeds to send digital streams of information—in the same fashion as digital fiber optic technology. Some companies, however, have begun to develop amplitude and frequency modulated systems. These systems could reduce interference and increase the range of infrared systems. High costs make the timeline for deploying such systems uncertain.

[48]Seybold, op. cit., footnote 13.

[49]Nathan Silberman, presentation to OTA staff, Sept. 16, 1994. Because these bands have been designated for "unlicensed" use by the FCC, neither manufacturers nor end users have to obtain a radio license from the FCC. The manufacturer is responsible only for ensuring that the product conforms to FCC technical rules and regulations, to prevent interference to other products.

peer" networking.[50] Development of products for the 2.4 GHz band has reportedly accelerated in anticipation of the IEEE standard for wireless LANs, the increasing congestion (see below) of the 902 to 928 MHz band, and the greater amount of bandwidth available compared to the 900 MHz band.

■ Unlicensed Data Services

One of the most rapidly developing and hotly contested areas of wireless data involves the use of spectrum that does not require the user to be licensed.[51] In 1985, the FCC opened up three bands for unlicensed uses (data and other types of communications) based on a set of regulations designed to minimize interference and encourage the development of new services.[52] Since then 130 companies have developed more than 200 systems and products for use in these bands—the 900 MHz band being the most popular—and more than 3 million devices are now in use by consumers and businesses.[53]

Unlicensed systems and devices are widely known as *Part 15* services because they operate according to Part 15 of the FCC's rules. Some of the services that operate under Part 15 include: automated utility readers, wireless LANs (see above), cordless phones, wireless audio speakers, home security systems, and some medical monitoring devices. In addition to these services, which are mostly self-contained or private, developers are also looking at the bands to provide more public services similar to those that now require a license—advanced paging and two-way messaging, for example—in order to avoid the expense (possibly exacerbated by auctions) and time (months or years) required to obtain a license. Zenith, for example, recently announced CruisePad, essentially a portable computer with a range of communication options, including remote LAN access operating in the 2.4 GHz band. Metricom uses a series of small (toaster-size) radios mounted on telephone or utility poles to create a microcellular, mesh network that provides metropolitan area coverage, and allows computers with appropriate modems to communicate with remote servers, send and receive e-mail, or access the Internet.[54] It serves utility monitoring, credit card verification, and personal communications functions.

In the past two years, the FCC has taken three actions to allocate more spectrum for unlicensed uses. First, as part of its broadband PCS proceeding, the FCC allocated the 1910-1930 MHz band

---

[50]The current standard for wireless LANs is 802.11, which specifies 1 Mbps or 2 Mbps. The European Telecommunications Standards Institute (ETSI) is developing a wireless LAN standard (expected to be completed in 1995) called Hiperlan that many in the United States feel is superior to the U.S. 802.11 standard. It allows wireless LANs to operate at speeds up to 22 Mbps over a range of 50 meters, and is capable of transmitting voice, data, and video in a user-to-user or broadcast mode. It does not require a license to operate. Hiperlan, however, is likely to be expensive and quite power-hungry, making portable applications difficult initially. To support these applications, and minimize interference, European countries have allocated a total of 350 MHz of frequencies at 5.2 and 17.1 GHz that will be dedicated to wireless LANs. Japan has also established two standards for wireless LANs, one operating at speeds less than 2 Mbps in the 2.4 GHz band, and the other supporting higher (greater than 10 Mbps) speeds operating near 18 GHz.

[51]In this case, unlicensed refers to the fact that neither the service provider, equipment manufacturer, nor the user must have a license. Cellular phone service, for example, is considered a licensed service because even though end-users do not need to be licensed, the company providing service does.

[52]The bands are 902 to 928 MHz, 2400 to 2483.5 MHz, and 5.725 to 5.875 GHz. See generally 47 CFR 15.247.

[53]"Review Could Lead to Auctions for Licenses in 902-928 MHz Bands," *Land Mobile Radio News*, vol. 48, No. 49, Dec. 16, 1994.

[54]The system operates in the 902 to 928 MHz band at 100 kbps total for each radio, which can be shared by several users. Shared use, however, brings down the bit rate available for each user. The system provides connection to the public telephone network, but does not allow handoffs; therefore it supports portable, but not fully mobile, communications. Metricom presentation to OTA staff, Sept. 14, 1994.

to unlicensed PCS—for both fixed and "nomadic" uses.[55] This allocation was designed to support a range of new data services centering around portable phones and computers, including wireless LANs. To reduce the potential of interference among users, the FCC adopted a "spectrum etiquette" that defines the technical rules that unlicensed PCS devices must meet to operate in the band. Systems cannot begin operating until the existing users of the band are moved, although exceptions will be permitted in areas where the unlicensed PCS system or devices can be coordinated with existing microwave system operators. It is not known how long the spectrum will take to clear or when such systems and devices will begin operation. For these reasons, this band is seen by industry as inadequate to meet short-term needs.

Second, in February 1995, fulfilling an earlier pledge to find more spectrum for unlicensed uses, the FCC reallocated 50 MHz of spectrum transferred from government uses by the National Telecommunications and Information Administration (NTIA).[56] Of that amount, 10 MHz is designated specifically for use by unlicensed radio services such as portable computers and wireless networks, and will be governed by Part 15 rules and the rules that govern data PCS applications. Part 15 users were allowed to continue to operate in another 15 MHz of the band already used for digital cordless telephones, wireless LANs, and inventory control systems. The FCC specifically indicated the benefits of this allocation for serving the needs of the NII: "The potential for open access to the information infrastructure offered by unlicensed PCS devices will provide benefits, not only to commercial users, but also to individuals and private users."[57] This allocation will be available immediately for use by unlicensed wireless data devices.

Finally, the FCC recently opened a proceeding into the possible uses of various frequency bands above 40 GHz by unlicensed (and licensed) services.[58] These frequencies would allow high-bandwidth communications to be transmitted, but only over very short distances (several miles at most). The FCC believes that data rates between 50 Mbps and 5,000 Mbps or more are possible, enabling systems to deliver extremely high-bandwidth services including high-speed data, high-resolution video and image transfer, and vehicle radar systems. The possible uses of these frequencies to provide NII access for consumers and backbone communications services for NII providers was explicitly recognized by the FCC.

■ Satellite Data Systems and Services

All the of the systems previously described use land-based towers to transmit information. Some systems—paging networks, for example—use satellites to connect local systems to form regional or national coverage areas. Satellites, however, have also been used by themselves for many years to transmit data and other types of information, primarily to fixed locations. The primary advantage of satellites is their ubiquitous coverage—the beam of one satellite can cover the whole United

---

[55]Of these frequencies, the bottom 10 MHz are reserved for *data PCS*. Material in this paragraph comes from Federal Communications Commission, *Amendment of the Commission's Rules to Establish New Personal Communications Services*, Memorandum Opinion and Order, GEN Docket 90-314, RM-7140, 7175, and 7618, released June 13, 1994; Federal Communications Commission, *Amendment of the Commission's Rules to Establish New Personal Communications Services*, Second Report and Order, GEN Docket 90-314, RM-7140, 7175, and 7618, released Oct. 22, 1993.

[56]Specifically, the bands allocated were 2390 to 2400 MHz, 2402 to 2417 MHz, and 4660 to 4685 MHz. These bands were the first transferred as part of a more general reallocation of government spectrum to private sector use mandated by the Omnibus Budget Reconciliation Act of 1993, Public Law No. 103-66, Aug. 10, 1993.

[57]Federal Communications Commission, op. cit., footnote 45.

[58]All information in this paragraph comes from Federal Communications Commission, *Amendment of Parts 2 and 15 of the Commission's Rules to Permit Use of Radio Frequencies Above 40 GHz for New Radio Applications*, ET Docket 94-124, released Nov. 8, 1994.

States. This easy national coverage also makes satellites uniquely suited to transmitting information to many sites that are far apart, and for transmitting to extremely remote areas that wire-based or terrestrial radio services cannot reach.

Several companies are now developing products and services that will take advantage of satellites' unique capabilities. Some of these systems are designed primarily to serve mobile users, while others will concentrate on fixed uses. In general, these systems can be divided into two types: geosynchronous and low-Earth orbiting (LEO).

Geosynchronous Satellites

Geosynchronous satellites orbit the Earth 22,300 miles directly above the equator. At this height and location, satellites move at the same speed the Earth is rotating. Thus, the satellite appears to be stationary in the sky. This is what enables geosynchronous satellite communications to work—they are always able to communicate with the satellite receivers on the ground.

Today, an increasing number of satellite data transmission systems use very small aperture terminals (VSATs). VSATs, introduced in the early 1980s, are small satellite dishes (approximately 1.8 meters in diameter) that are connected in a network through a central hub, which broadcasts information to the VSATs in the network and can connect individual VSATs directly (see figure 4-5). VSATs are capable of two-way voice, data, and video communication, but are usually used to send data to and from far-flung company locations. Networks of VSATs are commonly used to connect car dealerships, gas stations, and grocery stores, for example. Such a system enables a company headquarters to keep daily track of inventory and speed up shipments and deliveries. An increasing number of VSATs are being used to deliver video (live and recorded) training materials

FIGURE 4-5: Generic VSAT Network

SOURCE: Office of Technology Assessment, 1995.

to remote sites, and to connect company LANs at different sites.[59]

Using geosynchronous satellites, several companies are planning new services that will deliver data to businesses and consumers. Hughes Communications has announced plans to construct a bandwidth-on-demand system, Spaceway, that would provide a range of communication services to end-users (see chapter 5). Hughes Network Systems plans to launch a service that would allow users to download large files, software, or images from the Internet. The system is expected to operate at speeds up to 400 kbps, use a 24-inch satellite receive dish, and cost $16 per month plus a $1,495 setup fee.[60]

In addition to these primarily fixed-location systems, satellites also promise to make mobile data more ubiquitous. Inmarsat currently offers service to small satellite terminals that can be packed in a suitcase, enabling them to be carried to any location. Such services are designed to

---

[59]Over 6,000 VSATs are now being used to connect LANs. Dennis Conti, "LANs & VSATs," *Satellite Communications*, August 1994.

[60]"Hughes Network To Offer Data Retrieval Via Satellite," *The Internet Letter*, vol. 2, No. 4, Jan. 1, 1995.

Hand-held terminals will enable consumers and business users to send and receive short messages around the world and to determine their location anywhere on Earth within 100 meters.

support remote research locations and provide communications in times of disaster or emergency. Commercial satellite mobile data services, however, are very limited; only a few companies offer services, and these are primarily aimed at fleet management operations—messaging to truck drivers. In the future, however, new satellite systems promise to provide a full range of mobile data applications.

Low-Earth Orbiting Satellites (LEOs)

In addition to the geosynchronous satellite systems, several companies plan to use satellites orbiting the Earth at lower altitudes to deliver data services. There are three types of LEO services. The "big" LEOs (discussed in chapter 3) will focus primarily on voice services, but will also offer data services with capabilities similar to those offered by terrestrial PCS and cellular companies. Such systems will greatly exceed the coverage offered by terrestrial systems such as cellular or Ram/Ardis.

A second type of LEO satellite system has been proposed that would provide a wide range of services—bandwidth on demand—including voice and video telephony, interactive multimedia, as well as high-speed data communications. The Teledesic system, for example, will focus on providing high-bandwidth interactive applications to fixed locations in the United States, and to both mobile and fixed users abroad (see chapter 5 for more discussion of these multipurpose systems).

A third group of companies is developing so-called "little" LEO satellite systems that will provide ubiquitous (and eventually global) data-only messaging, tracking, and monitoring services to individuals and businesses.[61] The first application for a LEO system was filed at the FCC in 1988, and currently eight companies have proposed to launch little LEO systems.[62] To date, only one of these, Orbcomm, has received an FCC license to launch and begin offering service. It launched the first two of 48 satellites in March 1995. VITA now expects to launch in June 1995, with service beginning by the end of the year.

---

[61]The term "little" refers to the fact that all little LEO systems will use frequencies below 1 GHz, and that the services will be non-voice. The satellites used for little LEO systems are also physically smaller than those used for "big" LEO operations (see below). The FCC refers to this class of satellite services as *non-voice, non-geostationary* (NVNG).

[62]The eight are CTA Commercial Systems, Inc., E-Sat, Final Analysis Communication Services, Inc., GE American Communications, Inc., LEO One, Orbital Communications Corp. (Orbcomm), Starsys, and Volunteers in Technical Assistance (VITA), "Four New Applicants Join Leo One in Proposing 'little LEO' Systems for Second Processing Round," *Telecommunications Reports*, Nov. 28, 1994.

<hr>

BOX 4-5: Global Positioning System (GPS)

Initiated as a test program by the Department of Defense in the early 1970s, the Global Positioning System (GPS) has provided position location service for military and civilian applications since 1992. The system uses 24 satellites that orbit the Earth at an altitude of 10,900 nautical miles. Portable or vehicle-mounted GPS devices receive signals from the satellites and calculate the user's position to within 100 yards for civilian purposes and even closer for the military.[1] GPS operates 24 hours a day, can serve an unlimited number of users, and operates in all weather conditions.[2] The system may eventually replace many ground-based navigation systems, such as the current U.S. air traffic control system, helping to expand the capacity and improve the safety of the aviation system in the United States and the world.[3] Civilian GPS products are already used by boaters and trucking companies.

In the future, GPS services will form an integral part of many intelligent transportation system services, such as map and navigation programs for cars and portable computers. Many of the proposed satellite communication systems, including some of the Low-Earth Orbiting (LEO) satellite proposals and American Mobile Satellite Corp., plan to integrate the GPS location services into their service offerings.

SOURCE: Office of Technology Assessment, 1995.

<hr>

[1]For security reasons, the Defense Department scrambles the civilian GPS signal to limit its accuracy to approximately 100 yards.

[2]Keith D. McDonald, "Course 101: Fundamentals of GPS," presented at the Loews L'enfant Plaza Hotel, sponsored by Navtech Seminars, Inc., Washington, DC, July 11, 1994.

[3]U.S. Congress, Office of Technology Assessment, *Federal Research and Technology for Aviation,* OTA-ETI-610 (Washington, DC: U.S. Government Printing Office, September, 1994).

The little LEOs companies plan to offer ubiquitous, two-way messaging and data services, for either fixed or mobile users, potentially on a global basis.[63] Initially, service providers plan to target the transportation industry and remote monitoring applications (oil or gas pipelines or wells, for instance). In the longer term, proponents also perceive a market for emergency and personal communications; law enforcement, such as stolen vehicle location; environmental monitoring; fleet and cargo management for marine shipping companies and trucking companies; and other similar services. Most little LEOs will also couple data offerings with position location service, using the Global Positioning System (see box 4-5). To serve diverse customer needs, little LEO providers are designing a variety of consumer equipment expected to cost between $100 and $400. Message delivery is expected to cost about $0.25 per message.[64]

Although technical differences exist between the proposed little LEO systems, it is possible to describe a generic little LEO system (see figure 4-6).[65] Each system will consist of between 25 and 50 satellites in low-Earth orbit, about 1,000 kilometers above the Earth's surface. Each system

[63]VITA, for example, plans to offer e-mail and short file transfers between remote sites. Orbcomm, however, while offering services in a number of countries, will not transmit between countries. In addition, each little LEO system will have to obtain a license to provide service in *every* country in which it plans to operate. Negotiating these contracts could slow the deployment of worldwide services.

[64]"Orbcomm Gets First 'little LEO' License for Satellite Data Service," *Telecommunications Reports International,* Oct. 28, 1994.

[65]The exception is the VITA system, which will use only two satellites in fixed orbits serving about 1,000 stationary ground regional gateways. The system will be managed by a single control center near Washington, DC.

The U.S. Global Positioning System uses a network of satellites that allows users (in aircraft, on ships, in vehicles, or equipped with hand-held devices) to determine their location almost anywhere on Earth.

will also consist of at least one terrestrial control center, and sometimes secondary and tertiary "gateways" that will serve as the relay and control point between the customer units, the satellites, and public and private communications networks, including the Internet.[66]

Depending on complexity, the systems are expected to cost between \$100 and \$200 million each, with the exception of the VITA system, which will be significantly less expensive since it will use only two satellites.[67]

ISSUES AND IMPLICATIONS

Mobile data applications are quite promising and the industry has much potential. However, despite predictions for explosive growth, use of mobile

data applications and services remains relatively low. In many ways, the use of wireless technologies to serve mobile/portable computing needs is a microcosm of the larger world of wireless communications. A wide range of technologies is being developed and deployed to meet the perceived needs of wireless data users—traditional paging, satellite, and cellular, as well as PCS, LEOs, and public and private wireless data networks. But applications vary in their technical characteristics (speed, throughput, etc.), ease of use, and capabilities, and it is still unclear which models of wireless data will be successful and when. For equipment vendors, this makes it difficult to decide what systems and services to include with their hardware, and for users, it may be difficult to determine which product/service(s) best meet their needs.[68] Several issues will have to be addressed before mobile data reaches the levels many analysts have predicted.

■ Technical Issues

Fundamentally, using radio waves to send information is more difficult than using a wire because the environment—the atmosphere—is much harsher. Noise, interference, and attenuation are much harder to anticipate and overcome in the open air than in the more protected environment of an insulated wire. To overcome these problems, radio engineers are working on a variety of improvements to radio technology, including better data compression, higher capacity transmission and spectrum-sharing methods, improved error correction, and greater resistance to natural and man-made interference.

Despite the great strides made in the use of radio for sending information, the wireless data industry must still overcome several technical obstacles before wireless data applications become more widespread: 1) the speed and capacity

---

[66]All little LEO systems will use spectrum below 1 GHz.

[67]U.S. Congress, Office of Technology Assessment, *The 1992 World Administrative Radio Conference: Technology and Policy Implications,* OTA-TCT-549 (Washington, DC: U.S. Government Printing Office, May 1993).

[68]Seybold, op. cit., footnote 13.

FIGURE 4-6: Generic Little LEOS System

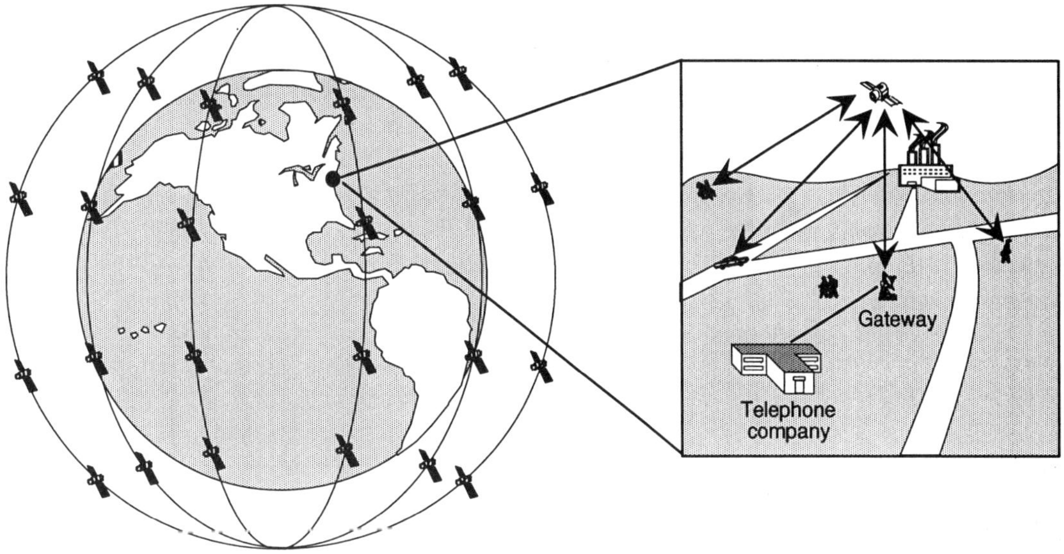

SOURCE: Office of Technology Assessment, 1995.

of the radio link, 2) its susceptibility to environmental interference and signal loss, 3) interference from electronic devices and other radio services, and 4) interoperability problems with other wireless systems and wireline systems. Because of these factors, current wireless data communication technologies generally cannot offer the same level of performance (measured by speed, reliability, and/or capacity) as wireline technology, although some individual wireless systems may offer comparable service.

Spectrum Limitations

Many of the problems confronting the wireless data industry come down to a limited number of radio frequencies on which to operate systems. Limited spectrum constrains the numbers of users who can use or offer a service, limits the speed at which information can be sent, and often creates interference problems between users when they have to share the same frequencies—further limiting performance and capacity. In recent years, the FCC has allocated more spectrum to wireless data services, and more is being considered.

Speed is limited

The most serious drawback to wireless data applications today is the limited speeds at which they operate. Wireless LANs generally operate at between 1 and 2 Mbps, compared with 10 Mbps on most wired LANs. Most commercial two-way wireless data services (packet radio, circuit-switched cellular) now achieve effective throughputs of 300 to 4,800 bps, compared to wireline modem speeds up to 28,800 bps. Commercial providers are working to upgrade speeds to 19,200 bps, and CDPD will operate at similar speeds, but these technologies are not yet widely deployed, and actual throughput is likely to be lower.

In part, slow speeds are a function of technology, and in part they are due to the limited bandwidth that is currently available for wireless data applications. Radio waves are limited in the amount of information they can carry—often measured in the number of digital bits per Hertz transmitted. Current systems can transmit about 1 or 2 bits/Hertz, but researchers are working to expand this to 6 or even 10. Digital compression and transmission technologies will help increase the carrying capacity of radio waves, but physical limitations may continue to limit high speeds. As a result, adequate spectrum allocations are critical—more spectrum is required to transmit more information. Recent frequency allocations made by the FCC in the 2390 to 2400 MHz band and proposed in bands above 40 GHz have the potential for offering much higher bandwidths and speeds, perhaps allowing some wireless applications to match wireline performance. Speed problems are also a matter of perception; customers used to high speeds on wired LANs are likely to be frustrated with the slow speeds available over wireless.

Interference

Another consequence of limited spectrum is the increasing likelihood of interference between different kinds of radio services sharing a band of frequencies, as well as between different systems providing the same service. For example, manufacturers of medical telemetry devices—such as electrocardiogram monitors—have asked the FCC to allow them to expand their operations and increase their power on vacant VHF and UHF television channels in order to overcome severe congestion and interference from other radio users. Sometimes interference is only a minor annoyance—static or voices on your cordless phone, for example—but at other times, interference can be severe enough to prevent transmissions from being received at all. For data communication systems, which are sensitive to interference and which depend on reliable transmissions to communicate information accurately, interference in the wireless environment can be a significant issue.

Interference problems are experienced by many wireless data systems, but they are currently acute in the unlicensed bands, which are home to a wide variety of systems and radio devices. Data systems in the 902 to 928 MHz band must share the spectrum with a number of other users—most of whom have priority. In the past several years, a number of companies have flocked to the unlicensed bands to develop various data transmission technologies. The band has been a boon to designers of new services, just as the FCC intended.

Unfortunately, the low status of the unlicensed data users in the band has become problematic. Several companies providing Automatic Vehicle Monitoring (AVM) services sought FCC authority to expand their operations and give them exclusive use of a major portion of the spectrum. The unlicensed data community fought this proposal, believing that it would essentially put them out of business. The FCC recently set regulations for AVM service and granted some new spectrum, but also established protections for unlicensed users.[69] A number of parties have filed petitions with the FCC to reconsider parts of its ruling, and it is still unclear how the issue will finally be resolved.

In addition to interference between different radio services sharing the same range of frequencies, there is concern that portable computing or other devices may interfere with or be affected by interference with the radio links. Electrical devices, like computers, often leak spurious radio energy produced as a by-product of their normal operation. This interference can affect nearby radio devices, including a radio modem/transmitter connected to (or inside of) the computer itself, causing serious performance problems for the radio device. Adequate shielding or redesigning the computer's internal layout can solve the problem.

Radio signals can also interfere with computers and other electronic devices. Because several ranges of frequencies are used for wireless data applications, computers must be designed to limit

[69]Federal Communications Commission, *Amendment of Part 90 of the Commission's Rules to Adopt Regulations for Automatic Vehicle Monitoring Systems*, Report and Order, PR Docket 93-61, released Feb. 6, 1995.

the effects of unwanted radio energy and shielded from devices using these frequencies.[70] There is, as yet, no group or process in place to determine if a portable computer is properly protected ("radio ready"). Chapter 12 discusses interference issues in greater detail.

Lack of Interoperability

One of the most serious problems facing wireless data users is the lack of interoperability—both between different wireless systems and between wireless systems and different wireline networks. Connecting users on different wireless networks and integrating an individual company's use of multiple wireless networks has been extremely difficult. Different companies have developed a variety of proprietary, incompatible wireless data technologies, such as those for the Ram and Ardis networks and CDPD, standards are almost nonexistent, and applications work differently on different systems (see below). Different wireless LAN manufacturers, for example, make equipment that is not compatible. And while the personal computer (PC) cards discussed in box 4-3 are standardized, there are reportedly so many different implementations of the standard that true interchangeability is not possible—a user cannot necessarily switch cards between different machines.[71] Overall, the use of radio-based technologies to support mobile data needs is often slow and tricky—users must be willing to endure complicated connections and poor quality to gain the advantage of mobility and portability.

The main problem with multiple technologies is that they complicate the development of applications software—word processing, electronic mail, and spread sheets, for example. Software developers do not want to incur the additional cost of writing a different version of their program for each type of wireless data system, especially when, as is the case today, the markets are small.

As a result, there are few off-the-shelf applications designed for use with wireless data systems. Most wireless data applications have been one-of-a-kind, written for a particular job by companies that can afford to do their software development in-house. However, even these companies are concerned about the lack of standards because they will have to rewrite their software to change providers.

In addition to the problems of incompatible wireless systems, one of the most serious concerns facing wireless data users is the transfer of information—interoperability—between wireless and wireline networks. Standard interfaces do not yet exist for sending data between cellular and Integrated Services Digital Network and other public switched telephone network (PSTN) services, for example. Speed is an important part of the problem. Wireless networks, because of spectrum limitations and the current state-of-the-art technology, cannot operate at the speeds now common in wireline applications.

A similar incompatibility problem exists with applications software designed to work on wireline and wireless networks. Most existing computer/data applications were written based on the parameters and characteristics of wire-based systems, and developers have years of experience in writing software that uses wireline protocols. This software, however, often does not work well when used over a wireless network. Software developers are now modifying some of their products to work in a wireless environment; this would reduce the cost of developing and adopting wireless data services. However, this process is difficult for developers, who have to learn specialized protocols in order to develop wireless data applications. It is also unclear how extensive or difficult it is to redesign such programs, and how many applications will have to be retrofitted to work well in a wireless setting.

---

[70]Data services are, or will be, offered on cellular frequencies, SMR systems, paging systems, PCS (licensed and unlicensed) frequencies, several satellite frequency bands, and general unlicensed (e.g., 902 to 928 MHz) frequencies. See Seybold, op. cit., footnote 13.

[71]Strom, op. cit., footnote 2.

Another solution is to develop a common interface for the software that could work in either a wireline or wireless environment. This would allow users to move between wireless and wireline networks more easily. For example, the same software could be used to access office computers from home via wireline and from the road via wireless.

Yet another solution to the problem of incompatible systems and standards is technological. More than a dozen companies are developing "middleware," an extra layer of software that translates information from the general application into the specialized wireless data protocol.[72] Middleware saves developers from having to learn the details of wireless data protocols: they write their applications to work with the middleware, which then handles the details of sending the data over any of a number of wireless networks. Middleware can also mask the differences in wireless data systems because it is usually able to translate into several different wireless data protocols. For example, middleware allows a user on the Ram network to communicate with a user on the Ardis network. Once an application has been written to work with the middleware, the user could switch to a different wireless data provider without having to make extensive modifications. In many cases, middleware mimics the behavior of a wireline network, allowing the large number of applications written for the wireline environment to be used over a wireless network.[73] Even middleware suffers from interoperability problems, however;

> Applications written to one vendor's middleware package don't necessarily work with middleware from a different vendor. The main reason for this interoperability gap is that most makers of wireless middleware products now use proprietary application program interfaces (APIs) to connect to network applications and services.[74]

It is possible that not all of today's wireless data services will survive in the marketplace. Software developers may write applications for some services, but not for others. Users would then tend to choose the wireless data service for which there is the widest choice of applications, enlarging that service's market share further and encouraging developers to write more software. Over time, the market may converge on only one or two of the systems available today. This is similar to personal computer operating systems, where a single operating system—DOS—came to dominate the market.

The wireless data industry is at an early stage in its development, and users and developers are only beginning to sort out the options. As various segments of the industry mature, better technology and increasing standardization is expected to alleviate many of the interconnection and interoperability problems that are now common. The speed with which this transition will take place, however, is still uncertain—most analysts believe it will take at least 3 to 5 years.

■ Demand Issues and Applications Development

Mobile computing is a reality and will become a more dominant part of computing later in the decade. Vendors are investing billions of dollars into the creation of new types of devices, new communications links and new software applications. There is a real danger that all this technology will be developed and made available without the existence of any real demand. Vendors must understand the segmentation of the mobile market to build the right products.

---

[72]For additional discussion of middleware, see Johna Till Johnson, "Middleware Makes Wireless WAN Magic," *Data Communications,* op. cit. footnote 8.

[73]One example is the *Winsock* interface. See Mobeen Khan and John Kilpatrick, "MOBITEX and Mobile Data Standards," *IEEE Communications,* vol. 33, No. 3, March 1995, p. 96.

[74]Johna Till Johnson, "The Wireless API Standards Watch," *Data Communications,* op. cit., footnote 8, p. 72.

Users have to understand the benefits, as well as the pitfalls, of mobile computing to get excited about using it.[75]

A substantial problem that has not yet been solved is how to move mobile data services more into the business and consumer mainstream. To develop applications for today's wireless data users: first, needs are identified; then, technology is produced or adapted to fit needs; and finally, pilot tests are conducted. Due to the nascent stage of technology and applications, customization is usually the first step. As a result, horizontal markets for mobile data applications may be difficult to develop because of the specific nature of the tasks vertical solutions are designed to serve.

Once concepts, products, and services have been validated across a number of business applications, a broadening of software can be expected. This is, in fact, what companies like Ardis and Ram are attempting to do—move from vertical to horizontal markets. Over the next few years, more general wireless data products and services are expected to come on the market. Developers are already writing more software and applications for the mobile environment,[76] and the expected explosion of mobile data users has prompted a flurry of alliances between software developers and wireless data companies. Microsoft and Mtel, for example, have teamed up to offer services on Mtel's Nationwide Wireless Network (NWN). GTE and IBM recently announced an agreement to allow GTE cellular customers to access IBM's data network. Analysts point to the availability of good applications as the key to the future growth of the market.[77]

As a result, the market for wireless data services is becoming increasingly crowded, but many analysts question whether the market can support all the different levels and kinds of competition. Traditional paging companies face competition from new PCS providers, as well as potential competition from little LEO systems. The original two-way data service providers, Ram and Ardis, will face increased competition from CDPD, narrowband PCS, and perhaps a range of satellite services. Some of these systems will provide competing services for some applications, but may also offer different combinations and levels of service. Some analysts believe that the systems currently serving vertical markets are unlikely to be able to broaden their customer base significantly. Ardis and Ram, for example, may be confined to vertical markets, while cellular data services will become the technology of choice for most business/mobile professional users due to the integration of cellular systems with the public telephone network.[78]

A final part of the problem of broadening the use of wireless data involves users themselves. Many businesses and consumers are less aware of the uses and benefits of mobile data than they are of a cellular phone or even a wireless LAN. As a result, demand has been unfocused, and applications developers have not had a clear direction to pursue. "If you think in terms of mobile data . . . it's far less obvious what the benefits of using mobile data are. It's a matter of education and awareness."[79]

[75]Dulaney, op. cit., footnote 1.

[76]Susan D. Carlson and Craig J. Mathias, "Big Guns Target Mobile Middleware," *Business Communications Review*, November 1994.

[77]Some analysts point to the development of mobile data in the United Kingdom as an example of the importance of developing good applications the market wants. There, five licenses were made available, and four were actually developed. Of these, the most successful, Cognito, has only 4,000 subscribers (compared to 3 million cellular users). Pat Blake, "Wireless Data: The Silent Revolution," *Telephony*, Dec. 5, 1994.

[78]Ibid.

[79]Ibid., p. 32.

In particular, the needs of residential consumers for such business-oriented services are likely to remain unclear for several more years. Most of the applications discussed in this chapter are designed to meet business needs. The benefits for individuals in their personal lives remain highly speculative. "Educating" mass market consumers about the benefits of new wireless data technologies has begun (e.g., Motorola pager television advertisements), but will continue to be one of the industry's more difficult challenges. With the proliferation of portable computers and PDAs, this awareness is expected to grow, user needs should become clearer, and the use of wireless data services should grow.

Prices

One of the key issues of demand for wireless data solutions is cost. The price of wireless data equipment is still high. Radio modems can cost up to $800. Economies of scale and mass market economics have not yet driven the price of equipment down to a level that is affordable to most companies or consumers. This relatively high up-front cost, in addition to activation fees, per month charges, and usage fees, may prevent some users from signing up—especially residential consumers. As economies of scale are realized, equipment prices are expected to drop.

In addition to high initial equipment costs, the ongoing costs of service are also an issue. Some mobile data service providers offer flat-rate payment plans that allow users unlimited use for a set fee. Others will charge a combination of flat rate plus additional charges for use over a set limit. In the future, businesses will likely demand flat-rate pricing based on large volumes of traffic. Individuals and small businesses, however, are more likely to want per-call charges because they will not want to pay for anything they do not use. And like cellular and PCS services, the question of who pays any air time charges for calls to the user—the user or the caller—will continue to be studied.

Coverage

Another important issue for users is coverage— "where can I use it?" Users want ubiquitous coverage within the area in which they travel. This geographical range varies by user. Some businesses, such as real estate companies, need primarily local/metropolitan coverage. Salespeople may need a larger coverage area—statewide or even multi-state regional coverage. Traveling executives may need an even wider coverage area—national or even global in scope. Different technologies can provide different levels of coverage. Paging networks are generally local/regional in scope, but, using satellite technology to connect local transmitters, some systems can offer nationwide or global coverage. Cellular circuit-switched or CDPD applications are also technically local, but, with roaming capabilities and their connection to the PSTN, can also achieve national or even international reach.

Defining "coverage" is not necessarily straightforward. Ram and Ardis, for example, are often referred to as "national" services; however, while they cover many metropolitan areas, they do not cover the whole country. In addition, a user's specific location within a coverage area may determine whether or not service is available. Users tell stories of having to switch hotel rooms from the north to the south side of a building in order to use their service.[80] For some business users, these may be minor inconveniences, but many will not tolerate such performance.

Security Concerns

Some companies are afraid that moving data over the airwaves, especially sensitive data about clients or products, might make them vulnerable to potential eavesdroppers who could be listening in.

---

[80]While terrestrial data services are not technically line-of-sight, position within a building does matter. Often, users will congregate near a window on a specific side of a buildin—gwhere the coverage is best.

Users are also concerned about the possibility that saboteurs could somehow use the systems to destroy computer files. New spread spectrum systems are relatively secure because of the way they transmit information, but users are still wary. Encryption is thus an important issue for wireless data users. Many large corporate and government users will not send data without encrypting it, and most wireless LAN providers offer some type of encryption software. A more complete discussion of the security issues associated with cellular and PCS data applications is found in chapter 10.

Broadcast and High-Bandwidth Services 5

Wireless communication systems will play an increasingly important role in the delivery of a wide range of high-bandwidth entertainment, information, and communication services. Radio-based technologies have been used for decades to transmit one- and two-way communications in support of a wide variety of applications. Radio and television broadcasting, for example, have long been a staple of the nation's communication infrastructure, supplying information and entertainment to millions of Americans for over 50 years. Since the early 1970s other wireless systems—microwave networks and satellites, for example—have been providing high-capacity links primarily for large corporate, industrial, and government users (the only users with bandwidth requirements large enough, or who could aggregate enough traffic to need a high-capacity system). Today, as the demand increases for high-speed data, multimedia, and video communications, wireless systems are increasingly being designed to provide high-bandwidth capabilities directly to individual users and businesses. This chapter examines the role of new and existing wireless technologies in delivering broadcast programming, video, and other high-bandwidth services as part of the evolving National Information Infrastructure (NII).

FINDINGS

- High-bandwidth radio technologies will play a somewhat paradoxical role in the NII. **At the local level, wireless systems will *compete* with established wireline and other wireless service providers.** From a national policy perspective, however, **wireless technologies will *complement* wire-based systems in extending video-based NII services to more**

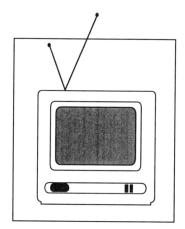

American citizens and businesses, and could be important in extending universal service to underserved populations.

As a competitor, high-bandwidth wireless systems are expected to bring substantial benefits to consumers and businesses, including lower prices and more diverse services. Direct broadcast satellite (DBS) services, for example, and several new terrestrial wireless systems will compete with cable companies and broadcasters in the market for video programming. Satellite-based digital audio broadcasting (DAB) will compete with local broadcasters for radio listeners in cars. Terrestrial and satellite-based "bandwidth on demand" systems will compete with local telephone and cable companies to provide "last mile" NII connections to businesses and consumers who need high-bandwidth communication services capable of handling video communications, image transfer, high-speed data, and multimedia applications.

As a complement to wire-based systems, wireless systems have great potential for extending NII resources to rural or underserved populations. In particular, satellite-based systems may bring the full range of NII services and applications to more users because of their ubiquitous nationwide coverage. This single-source coverage also assures consistent services across different local areas for users with national communication needs —multiple services, whether wireline or wireless, will not

have to be "stitched together." The architecture and cost structures of wireless technologies— terrestrial and satellite-based—may allow them to deliver NII services to some areas faster, and perhaps less expensively, than traditional wireline systems, especially in areas that are remote or undergoing new construction. High-bandwidth technologies may even be used by traditional wireline carriers to deliver services—at least one local telephone company has invested in a wireless video provider, and cable companies are actively involved in the DBS industry.

■ Although it is too early to assess the general effect of competition on price because the systems are too new, many analysts and policymakers believe that competition will drive prices down or at least hold them steady.[1] **Because some of these technologies, markets, and industries are still in their infancy, it is difficult to determine how effective competition in new markets will be, which technologies will survive, and which companies will prosper. Similarly, claims about the benefits new wireless technologies can bring to the national economy must be regarded cautiously.[2]**

Each system has advantages and benefits that will be attractive to consumers and businesses, but that will also splinter markets and frustrate analysis and policymaking. As technology advances and demand sharpens, systems will become increasingly differen-

---

[1] Some anecdotal and statistical evidence does exist, for example, that a second cable company in a given franchise area will reduce cable rates. See Federal Communications Commission, *Annual Assessment of the Status of Competition in the Market for the Delivery of Video Programming*, CS Docket 94-48, released Sept. 28, 1994, paragraphs 57-60 and 203. At least one MMDS provider claims similar reductions in cable rates as a result of its entry into the local market. Letter from Todd Rowley, Peoples' Choice TV to Andrew Kreig, Wireless Cable Association International, Jan. 16, 1995.

[2] The Federal Communications Commission noted this problem explicitly in an ongoing proceeding:

"...it must be noted that the proposals before us are largely that. There is little evidence in the record regarding the likely public interest benefits of the various proposals, including increased access to high-quality, affordable, and innovative services, and stimulation of economic growth through increased competition for existing services and introduction of new services that may be expected to stimulate demand and create jobs." Federal Communications Commission, *Rulemaking To Amend Part 1 and Part 21 of the Commission's Rules to Redesignate the 27.5-29.5 GHz Frequency Band and To Establish Rules and Policies for Local Multipoint Distribution Service*, CC Docket 92-297, released Feb. 11, 1994, at para. 23.

tiated—not only in the products and services they offer, but in what they can actually deliver. The unique capabilities and disadvantages of wireless technologies, combined with changing consumer demand, will lead to markets that overlap for some services, but diverge for others provided by the same systems. Consumers will benefit from a wider range of services and competition among many different types of providers—both wireline and wireless. Assessments of the overall market will lose meaning as many smaller submarkets form. In addition, **the uncertainties of technology advances, consumer and business demand, and regulatory treatment make it difficult to judge their overall effects on the wireline portions of the NII.**

- **As technology advances and competition develops, the implementation of universal service (whatever its definition) and other public interest obligations becomes more problematic for both wireless and wireline carriers.**[3] Historically, universal service has been associated with the provision of basic telephone service (see chapter 9). Today, the development of new technologies, coupled with changing societal needs, is forcing the concept of universal service to evolve as well. In the future, universal service is likely to include a wide range of advanced communication and information applications, such as voice, data, and video services. Exactly what the new universal service will encompass is unclear, but because wireless providers are expected to be significant competitors in various markets, how these issues are resolved will directly affect their operations and economics.

An evolving definition of universal service will pose serious challenges for policymakers regarding wireless services. First, if universal service comes to include access to high-bandwidth information and entertainment services—such as those offered by the wireless providers discussed below—new segments of the wireless industry will be subject to new regulations. Additionally, if universal service mandates two-way, broadband access to NII resources, the majority of wireless providers—those who cannot technically offer such services—could be put at a regulatory disadvantage. Mandating such a level of service for all telecommunications providers fails to account for legitimate technology differences and could penalize companies that made rational technology and business decisions in the past.

A system of universal service based on designation of essential carriers—such as that envisioned in recent legislation—or a tiered system of universal service obligations based on technology and services delivered might represent a more flexible, and hence long-term, approach to setting universal service obligations and rights.[4] Such an approach would be consistent with current congressional initiatives for deregulation and belief in the market as the most efficient and effective means of delivering services to consumers. However, until decisions are made about what constitutes universal service, and what mechanism will be used to move its subsidies, evaluating the effects on providers of all sorts would be guesswork at best. Even when these fundamental decisions are made, more data will be needed on wireless system costs, wireline upgrade costs, and the extent of the universal service "problem" before these questions can be answered.

Second, identifying the companies that will bear the cost of providing new levels of universal service, and those that will receive financial

---

[3] For more discussion of these issues, see Leland L. Johnson, *Toward Competition in Cable Television* (Cambridge, MA: The MIT Press, 1994).

[4] U.S. Congress, Senate, S. 652, *The Telecommunications Competition and Deregulation Act of 1995* (Washington, DC: U.S. Government Printing Office, June 15, 1995).

help in meeting these obligations have already become significant issues. Providers who have traditionally borne public service obligations will be increasingly subject to competition from newer providers who use different technologies and do not carry similar burdens. For example, broadcasters—in return for their free use of the public airwaves—have been subject to public service requirements, while Multichannel Multipoint Distribution System (MMDS) and DBS are not.[5] Cable television systems have been subjected to many types of franchising requirements in return for their use of public rights-of-way; MMDS and DBS are not because they do not use public rights-of-way per se.

From a competitive standpoint, such inequities may skew the ability of different firms to compete, although the extent of such inequities is unclear. For example, "[w]ere the wireless systems taxed and the proceeds used to benefit their wireline competitor in its high-cost area also served by the wireless systems, competition from these wireless systems might be weakened."[6] It may be possible to adopt a consistent set of regulations to guide competition. However, if attempts to reduce technical and regulatory inequities are too broad, they may not work because the inherent capabilities of the technologies are often quite different. Satellites, for example, inherently have national reach, but does that mean they should be subject to franchise fees in every local jurisdiction in the country? At least one analyst has proposed that extending license auctions to new video service providers might be one method for recovering value from the public use of spectrum—eliminating the need for franchise fees and public service obligations—while still

allowing different technology systems to compete.[7]

- **In the emerging NII, the role and function of television broadcasters will have to evolve to fit new competitive models.** Broadcasters have played an important role in American life for 50 years. They were long the sole providers of video programming, and have had exclusive access to what has become a very sought-after portion of the radio frequency spectrum. Despite increasing competition from cable television and other smaller programming providers, television broadcasting has remained relatively strong. However, an uncertain regulatory future and new forms of competition from program distributors with far greater capacities have made the outlook for the industry increasingly unclear.

Even with a conversion to digital technology and the capability to broadcast multiple channels of video and perhaps other (data) services, broadcasters' ability to compete with interactive cable television, telephone company services, DBS, and other wireless broadcasters is unknown. Broadcasters have several advantages in the emerging competitive environment—including programming resources, prime spectrum, local community ties, advertiser-supported free (to consumers) programming, and a broad base of political support. However, they also suffer some significant disadvantages, including a lack of channel capacity and an unfocused vision of what their new role is likely to be. In considering the future of broadcasters, a range of issues must be considered by both the industry and Congress that are beyond the scope of this report. These include national and local ownership rules, allowing

---

[5] DBS providers were included in a 5 to 7 percent channel capacity public interest set-aside included in the 1992 Cable Act, but that requirement is not being enforced pending court review. The FCC does have a rulemaking examining whether and how DBS should be subjected to programming obligations. Federal Communications Commission, *Implementation of Section 25 of the Cable Television Consumer Protections and Competition Act of 1992, Direct Broadcast Satellite Public Service Obligations*, MM Docket 93-25, 8 FCC Rcd. 1589, para 1 (1993).

[6] Ibid., p. 168.

[7] Johnson, op. cit., footnote 3.

broadcasters to provide nonbroadcast services, and what the impacts would be on viewers if broadcasters stopped broadcasting free over-the-air programming altogether.

BACKGROUND

The technologies and systems discussed below share a number of important characteristics that will shape their contributions to the NII. First and foremost, the advent of digital technologies lies at the heart of many of the changes now taking place in radio communications. Each of the technologies discussed in this chapter is either in the process of converting to digital technology or is being designed from the outset to work digitally. This switch will fundamentally affect the services companies can offer and at what cost.

Second, many of the systems discussed below were originally designed to be one-way. Although two-way wireless systems are used—satellite networks, for example—and some wireless systems are supplemented by return communications supplied by the telephone network, most use of radio waves for high-bandwidth communications remains concentrated in a one-way broadcast or point-to-multipoint format. It is only recently that companies have begun to develop interactive, broadband wireless networks for the consumer and business markets.

Finally, many of these systems were designed to serve users at fixed sites. The ability to broadcast radio waves over a wide area has proven to be a remarkably efficient way to reach many people quickly, easily, and at relatively low cost. In the future, the low cost and ease of deployment of broadcast technologies will enable them to compete with wire-based alternatives in many mar-

kets, especially one-way entertainment programming.

RADIO BROADCASTING

Radio broadcasting is one of most familiar wireless services. Commercial radio broadcasting began in 1921, and within 10 years, more than 50 percent of all American households had a radio receiver. In 20 years that figure climbed to 90 percent, and today, radio broadcasts blanket almost the entire nation and radio receivers are almost everywhere. The average American home has 5.6 radios, and it is almost impossible to buy a car without a radio—there are nearly 200 million radios in American cars and trucks.[8] People listen, on average, to a little more than three hours of radio per day, mostly while commuting or at work. However, although there are more than 11,000 radio stations operating in the United States today—almost evenly divided between AM and FM—many of these are concentrated in and around metropolitan areas, and the most rural areas of the country may have access to only one or two stations.

Radio broadcasters use a single high-powered transmitter, operating in either the AM or FM frequency band, and a tall antenna to beam programming—including music, local news and information, education, talk radio programs (mostly on AM stations), and emergency information—to listeners in a radius of approximately 25 miles.[9] Because of this relatively limited range, radio broadcasting traditionally has been closely linked to the communities in its broadcasting area. National radio networks also use satellites to share programming. For example, the 25 Native American radio stations use a satellite link provided by

---

[8] Radio Advertising Bureau, *Radio Marketing Guide and Fact Book for Advertisers 1993-1994*, Dallas, TX, 1994.

[9] Repeaters/translators are used to extend the broadcast signal and serve outlying areas. AM stations are capable of beaming programming over far longer distances at night. The differences between AM and FM radio are significant (see app. A). Amplitude modulation (AM) uses relatively little spectrum—each station needs only 10 kHz—but the signal is easily disrupted by noise and interference (the signal is lost under bridges, for example). Due to poor quality, many listeners have shifted over to FM radio, making it the dominant radio format. Frequency modulation (FM) is more resistant to noise and signal loss, but each station needs a wider range of frequencies (200 kHz) to operate. Although both formats are capable of carrying stereo signals, most FM stations broadcast in stereo and most AM stations do not, and the majority of existing radios are not compatible with AM stereo.

the National Public Radio satellite system to receive programming through the American Indian Radio on Satellite (AIROS) project. Broadcasters are now trying to broaden their services to include low-speed data transmission that could provide local travel information, as well as supplementary information for advertising and audio programming (see chapter 4). In the future, radio broadcasters will switch to digital technology, and satellites may increasingly be used to deliver radio programming over wider areas.

■ Digital Audio Broadcasting (DAB)

The next generation of radio broadcasting will use digital transmission technologies. While no such services are operating yet, broadcasters and startup companies are developing systems that will replace traditional AM and FM modulation techniques with digital signals that will allow them to broadcast compact disc (CD), or near CD-quality, programming that is more resistant to noise and interference. DAB may also enable new types of information services to be delivered. Consumers will have to replace their existing ana-

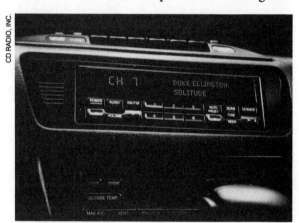

Satellite radio receivers similar to this prototype will have three bands: AM, FM, and satellite.

log radios with new digital ones to receive the better sound and new information services.

Two types of DAB systems are being developed in the United States. Existing AM and FM radio broadcasters are planning to implement DAB technology using existing radio channels. The new digital signals will be sent simultaneously alongside the analog signals. Meanwhile, a small number of startup companies is developing satellite-based DAB systems that will use new frequencies recently allocated for this purpose.

This divided approach has slowed the development of DAB in the United States, as the two sides have battled bitterly before the Federal Communications Commission (FCC). The result is that in the United States—unlike in many other countries where integrated systems are being planned—digital radio services will likely be delivered by two different kinds of systems: existing broadcasters, who will have to upgrade their facilities, and satellite-based providers, who are building their systems from scratch.[10] The two systems will not be directly compatible, although future radio receivers probably will be able to receive both terrestrial and satellite-delivered DAB as well as existing AM/FM broadcasts. The FCC is still in the process of developing the rules for future DAB services.

Satellite DAB

The idea of broadcasting radio programming directly from satellites dates back at least 45 years.[11] In the 1980s, a small number of companies around the world proposed satellite-based (formally known as Broadcast-Satellite Service-Sound, or BSS-Sound) systems that would use frequencies in the L-band (roughly 1.4-1.6 GHz) to transmit their programming. Because these types of systems would use frequencies other than the tradi-

---

[10] Some other countries are planning to use new internationally allocated frequencies in the L- or S-bands to deliver DAB services using both terrestrial and satellite transmitters working in a single system.

[11] The concept of using satellites to transmit programming was first described by Arthur C. Clarke in 1945. Arthur C. Clarke, "Extra-Terrestrial Relays," *Wireless World*, October 1945. More recently, satellite broadcasting was considered at international conferences dating back to 1979.

tional AM/FM broadcasting bands, they are often referred to as ""out-of-band" or "new band" systems.[12]

The first U.S. out-of-band system was proposed by Satellite CD Radio, now CD Radio, in 1990, and in December 1992, five other companies submitted applications to the FCC to offer satellite radio services.[13] In January 1995, almost exactly three years since the frequencies were allocated internationally, the FCC formally allocated radio frequencies for satellite DAB in the United States.[14] Now the FCC must develop licensing and operating rules to govern the provision of satellite DAB services. The FCC anticipates that this process will last until the end of 1995, and that licenses will be granted shortly thereafter. Once applications are granted and licenses issued, proponents expect it will take about three years to construct and launch the satellites, making service available in roughly 1998-99. CD Radio is currently testing its system using two NASA satellites, and predicts a startup date of 1998.[15]

Services

Proponents of satellite DAB are planning a variety of programming targeted to audiophiles, users with specific musical tastes, and groups with differing ethnic and cultural backgrounds. These small audiences may not be able to support a local radio station, but when aggregated across the country, make a national service possible. This "narrowcasting" concept is analogous to the programming philosophy of cable television. Satellite DAB may be especially popular in rural areas that lack access to the wide range of programming available in most metropolitan areas. The inherently national nature of the satellite technology, however, means that no locally originated programming—news, weather, or sports—can be transmitted. In addition, for technical reasons discussed below, satellite DAB is being developed primarily to serve radios in vehicles, although other markets are being considered. As currently planned, the CD Radio system would broadcast 30 commercial-free music channels to subscribers who would pay a $5 to $10 monthly fee. Other companies plan to offer some channels on a subscription basis, and others as advertiser-supported programming.

In addition to audio programming, the transmission of data services directly to users is also being explored. Proponents envision broadcasting data services to support educational needs, paging operations, and navigation and traffic management systems for the nation's cars and highways. Up to 20 channels may be broadcast to support these services.

---

[12] Although out-of-band systems can technically be satellite or terrestrial, development of out-of-band systems has focused almost exclusively on satellite technologies in the United States. Other countries, including Mexico and Canada, are experimenting with out-of-band solutions using both terrestrial and satellite delivery.

[13] In addition to Satellite CD Radio, American Mobile Radio Corp., Digital Satellite Broadcasting Corp., Loral Aerospace Holdings, Inc., Primosphere Limited Partnership, and Sky-Highway Radio Corp. petitioned the FCC in 1992 to offer satellite DAB. Since then, Loral and Sky-Highway have merged with Satellite CD Radio, leaving a total of four applicants. Carol Horowitz, "DAB: Coming to a Car Near You?," *Satellite Communications*, October 1994, pp. 38-40.

[14] The frequencies allocated were 2310-2360 MHz. This action was consistent with the position taken by the United States at the 1992 World Administrative Radio Conference. The United States and India are the only two countries to use these frequencies. Other frequencies to be used include 1452-1492 MHz (in Europe, South America, Africa, and, importantly, Canada and Mexico) and 2535-2655 MHz (including Russia, China, and Japan, among others). This means that no common radio broadcasting system will exist across the world as the AM and FM systems do now.

[15] CD Radio has petitioned the FCC for a 319d waiver, which would allow them to begin construction at their own risk prior to receiving a license from the FCC. This would allow CD Radio to begin operating sooner after receiving their license.

Satellite dishes such as these will beam digital quality radio programming up to satellites that will then retransmit it across the country.

Technology

Satellite DAB systems are conceptually quite simple (figure 5-1). On the ground, large satellite dishes will beam programming up to one or two geosynchronous satellites that will then rebroadcast these signals nationwide. CD Radio, for example, plans to construct and deploy two satellites to be used to deliver its services. Other developers of satellite DAB systems plan to augment the satellites with terrestrial transmitters (so-called "gap fillers") that would improve reception in urban areas (e.g., between buildings and in tunnels). Satellite DAB systems will feature individually addressable radios that will require a signal from the system's operations center to be activated or deactivated. Receiving antennas are silver-dollar-sized discs built into a car's roof. Satellite DAB systems are likely to have difficulty serving radios in homes or offices because the frequencies involved will not penetrate buildings very well. Antennas could be mounted on roofs or windows, but additional wiring would be needed to connect to the radio.

Because satellite DAB will be a new service—an additional choice for consumers rather than a replacement for their existing radios—there are no real transition problems to new satellite DAB technology. For listeners, the important point is that existing analog radios will not be able to receive the new programming; consumers will have to buy new radios if they want digital sound. CD Radio has demonstrated a new receiver that receives the AM, FM, and satellite bands, but this receiver is not yet commercially available.

Terrestrial DAB

In response to local broadcasters' concerns about the transition to digital broadcasting technologies, competition from new satellite services, and the possible effects of these changes on smaller radio stations, several companies began developing digital technologies that would work "in-band" —using the same frequencies currently used by AM/FM stations. This approach would allow existing broadcasters to upgrade their facilities without bringing in new, unwanted competition.

Development of terrestrial DAB in the United States is now focused primarily on in-band, on-channel (IBOC) solutions that will allow a broadcaster to transmit its present analog signal simultaneously with a new digital signal without the two interfering (figure 5-2). No new spectrum is required. This development path indicates that terrestrial DAB is most likely to be treated as an extension or upgrade of existing radio services— better quality, some additional radio-related services and maybe data broadcasts—rather than as a new service like satellite DAB. IBOC will use existing broadcast facilities to a large extent, but will require new digital transmitters and radio receivers. The cost for a radio station to upgrade its facilities is somewhat unclear, but will depend on how advanced and up to date the station's existing equipment is. Estimates put the cost at approximately $50,000 to $150,000 per station; not prohibitive for large market stations, but potentially a

FIGURE 5-1: Satellite Digital Audio Broadcasting

SOURCE: Office of Technology Assessment, 1995.

problem for smaller ones.[16] Consumer radios are expected to be expensive initially, but fall into the $50 to $350 range—about the price of current high-end radios—once they are produced in quantity.

Like satellite DAB, the transition to terrestrial DAB should be relatively easy for consumers.

Those with older radios will continue to receive the existing analog signal, while newer radios will receive the new digital signal that is transmitted simultaneously. Past technical and institutional issues that divided the industry internally appear to have been largely resolved, and development of a terrestrial DAB standard is progressing.[17] While

[16] Bortz & Company, *Digital Audio Broadcasting: Phase I,* Mar. 4, 1993. Testimony of John R. Holmes, in Hearings before the Subcommittee on Telecommunications and Finance of the Committee on Energy and Commerce, House of Representatives, 102d Congress, Nov. 6, 1991, p. 9.

[17] The Electronic Industries Association (EIA) established a task group in August 1991 to develop a U.S. standard for terrestrial DAB. The group—composed of specific system proponents, manufacturers, and broadcasters—received 11 proposed standards, which were reduced to five by the end of 1992. Testing began in 1993, and EIA now expects to finish in mid-1995. The group will then forward its recommendation to the FCC for consideration as the final DAB rules are developed. Demonstrations of both AM and FM IBOC systems were held at the National Association of Broadcasters convention in April 1995.

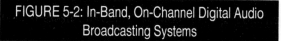

FIGURE 5-2: In-Band, On-Channel Digital Audio Broadcasting Systems

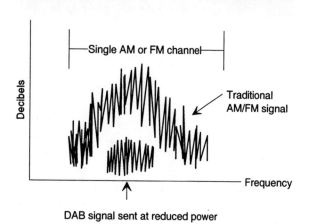

SOURCE: Office of Technology Assessment, 1995.

there may be some economic dislocation caused by the switch to digital broadcasting technologies, OTA believes disruption to the industry will be minimal.

■ Issues and Implications for the NII

The radio broadcasting industry is now at the beginning of a transition to digital technologies. It seems clear that two different DAB technologies will be deployed: satellite-delivered, out-of-band services and terrestrial systems using IBOC technology. Several regulatory and institutional issues remain unresolved, and competition from alternative programming providers is possible.

Demand and Competition

The primary issue now consuming the DAB industry is the battle between traditional broadcasters and satellite DAB proponents. This conflict has been bitterly fought for the past five years and

shows no signs of abating.[18] The conflict is based on different assessments of market demand—no one is really sure how consumers will react to these new services. Traditional broadcasters are concerned that satellite DAB will harm local broadcasters by taking significant audience share—and, hence, advertising dollars—from them and could cause some smaller (and more rural) stations to go out of business. Similar concerns have also been voiced by some FCC commissioners.[19]

Proponents of satellite DAB argue that the economic impacts of satellite systems will be minimal because the systems are expected to serve largely niche markets (audiophiles, special interest groups, and underserved customers). One report states that satellite DAB providers will achieve penetration rates of between 3 and 10 percent of the automobile market nationwide, while others put the figure at between 5 and 15 percent for all radios.[20] Further, some proponents of satellite DAB contend that the health of traditional broadcast radio stations should not be a factor in the FCC's consideration of satellite DAB service.[21]

This is not technically a "one or the other" choice; consumers who subscribe to DAB services will continue to listen to their local stations—just as they switch between AM and FM now. What is unclear is the *extent* to which consumers will treat satellite DAB as a substitute for local programming—the time that they will spend listening to satellite rather than local services. It is this time, translated into market share, that local broadcasters are afraid of losing because of the potential corresponding losses in advertising revenue. Comments filed before the FCC indicate that national advertising makes up only a small portion of a station's total advertising revenue, but it

---

[18] The National Association of Broadcasters, for example, has promised a "tough fight" against satellite DAB as the licensing and operating rules are developed at the FCC. "FCC Takes First Major Step Toward Satellite DAB Service," *Audio Week*, vol. 7, No. 3, Jan. 16, 1995, p. 1.

[19] Comments of Commissioners Ness and Barrett, reported in ibid.

[20] First numbers are from InContext, Inc., *Satellite Radio*, August 1994; second numbers are from Bortz & Company, op. cit., footnote 16.

[21] "NAB Renews Attack on Satellite Digital Audio Radio," *Telecommunications Reports*, Jan. 9, 1995.

may be that the loss of even that small amount could force some marginal broadcasters out of business.[22]

In addition to satellite providers, traditional radio broadcasters also face competition from local cable operators, many of whom now offer digital music services using existing cable television facilities.[23] Digital Music Express (DMX) and Digital Cable Radio now offer digital audio services to cable systems nationwide, and DMX is also being delivered via satellite as part of Hughes DirecTV programming (see below). Each offers about 30 channels (to be expanded to about 120 channels) of commercial-free music programming on a subscription basis, but no local or informational programming. Programming packages range from about $11 per month to $75 per month for business users. Although rollout of the service to providers has been relatively rapid, consumer acceptance has been slow. Total penetration rates are now expected to peak at between 5 and 10 percent of cable-served homes. Some analysts believe this may indicate low demand in general for radio services listeners have to pay for.

No firm conclusions can be reached about demand and competition at this time. Doing a prospective analysis of the economic impacts of a new technology is always difficult, and DAB is complicated by current uncertainties in demand and product/service acceptance. Using past technology diffusion and interaction patterns to determine future acceptance and demand—as some industry studies do—is not sufficient for policy purposes. The tradeoff for consumers will be between free local programming with commercials and commercial-free programming that they must pay for. Take-up of terrestrial DAB services may exceed that of satellite services, if only because they are more familiar and can be positioned as an extension of an existing service. Mass mar-

ket data services may not do well in competition with many other data services (see chapter 4), but services narrowly tailored to radio listeners—auxiliary services like local travel information—may find acceptance.

Policy Considerations

The deployment of terrestrial and satellite DAB raises some difficult questions for policymakers at the FCC. In the short term, the FCC is wrestling with questions about operating rules. In the longer term, more fundamental questions need to be considered. The most difficult long-term issue facing policymakers is how satellite and terrestrial DAB will affect the local, terrestrial broadcast industry. How can the traditional strength of the U.S. local broadcasting industry be complemented by the new technologies of satellite delivery? How can new forms of competition in radio services be promoted, while acknowledging (but not necessarily protecting) the role and investments of local broadcasters? What might the future structure of the U.S. broadcasting industry look like?

Satellite broadcasting, because it injects new competition into the whole radio industry (not just local competition), could dramatically reshape the broadcast industry in this country. Satellite services could complement local programming, be limited to serving niche markets, or emerge as a substantial competitor to local broadcasters. In some countries—Canada, for example—terrestrial and satellite DAB may develop as complementary parts of one broadcasting system. In the United States, however, it now seems likely that the two industries will remain separate—the established broadcast industry controlling terrestrial DAB, and the new startups controlling satellite services.

Given this context, it appears that satellite and terrestrial DAB will compete on the local level—

---

[22] See, for example, various comments of CD Radio and the National Association of Broadcasters, before the Federal Communications Commission, *Amendment of the Commission's Rules with Regard to the Establishment and Regulation of New Digital Audio Radio Services*, Docket 90-357.

[23] The information in this section comes from Bortz & Co., op. cit., footnote 16.

radio listening in vehicles—while complementing each other at a broader level—extending coverage, meeting unserved needs. It should be possible to set the rules for satellite and terrestrial DAB such that both industries can thrive. A possible analogy may be the dual nature of cable television—local television stations and cable channels exist alongside "superstations" and nationwide cable channels that cater to specific interests. Nationwide DAB services may be able to supplement existing local services in the same way and would also fill in gaps in coverage of various programming formats; not every person in America can get the kind of radio station he or she wants, and not every market has 10, 15, 20 or more stations with a variety of formats available. For a listener in a remote location who would like to hear classical music, a satellite-delivered service may be the only option.

The broadcast industry's fears that nationwide satellite audio programming will force some radio stations out of business must be taken seriously. When satellite services start up, some smaller radio stations may not survive. On the other hand, satellite DAB proponents argue that development of satellite DAB technology will help the United States maintain its competitiveness in satellite and related broadcasting technologies.

For policymakers the issue is relatively simple, but difficult to solve: do the benefits of nationwide satellite radio services outweigh the loss of a number of smaller, likely rural, local radio stations? Relying on competitive forces is one way to approach the problem, but the social value of these stations may override the workings of the market.

If a local station cannot compete, should it be allowed to go out of business, or do the benefits of local information and entertainment call for some kind of protection? Could other local stations (where they exist) take up the slack? The industry should be prepared to present a good case for preserving small stations based not on past history—there can be little doubt of the historical importance of local radio stations—but on the prospects for future performance. Society in the 1990s and beyond is changing rapidly, and the nation's radio listeners are entitled to a radio system that best meets their needs. The public interest may need to be redefined to include not only local, but also national and international programming and services. Congress should be prepared to address the social value of local broadcasters, and whether that value may outweigh reliance on market-based outcomes alone.

VIDEO PROGRAMMING SERVICES

Video entertainment programming, which began with broadcast television in the 1940s, has become a pervasive part of American life. In this industry, broadcast and cable television are the dominant suppliers—broadcast television is available to roughly 96 percent of the American public, with cable television passing roughly the same percentage of households, 63 percent of which subscribe. Today, however, a number of wireless systems, as well as telephone companies, are poised to compete directly with cable and, to a lesser extent, broadcasters.[24] A full assessment of the competitive market for video programming services, including smaller local competitors such

---

[24] FCC definitions specifically exclude current broadcast television companies from this market because they do not provide multichannel service or use a fee-for-service model. The FCC notes, however, that "for at least some viewers, broadcast television service satisfies their demand for video programming." FCC, *Annual Assessment*, op. cit., footnote 1, para. 98. For purposes of this discussion, OTA takes the view that the aggregated channels provided by multiple local broadcasters essentially represent a multichannel service that does, in fact, compete with basic cable service. The future of multichannel individual broadcasters, as discussed above, also argues for including broadcasters in a future-oriented assessment of video programming services. In addition, by strict antitrust and economic definitions, competition with each of the wireless services discussed will be different because different technology systems offer slightly different packages of services. Because DBS, for example, cannot provide local broadcast programming, it is not a perfect substitute for local broadcast or cable service. DBS does, however, compete directly with the enhanced or premium services offered by cable companies. It is in this sense that competition (although not complete or perfect) is used throughout this section. This position is consistent with views taken by the Federal Communications Commission, *Reexamination of the Effective Competition Standard for the Regulation of Cable Television Basic Service Rates*, Report and Order and Second Notice of Proposed Rulemaking, 6 FCC Rcd 4552-53 (1991).

as low-power television and satellite master antenna television systems (SMATV), is beyond the
scope of this report, but several analysts and the
FCC have already examined these issues in greater detail.[25] Consequently, this section will focus
on the wireless entrants in the video programming
market, and assess the technical, economic, and
regulatory issues they will face in the coming
years.

∎ Broadcast Television

Broadcasting has been an important component of
the nation's communications infrastructure for decades—bringing entertainment and information
to millions of people, and having an undeniable
impact on the nation's culture. In a sense, television was the first broadband communications service. By using the airwaves, it was possible to
deliver hundreds of megahertz of video programming at a time when wired media could not.
Today, however, several different technologies—
including cable television, DBS, other wireless
systems, the local phone companies, and even video rentals—are putting competitive pressure on
broadcasters. Over the next decade, broadcasters
face the difficult task of managing the transition
to a new generation of digital technology.

Services

In one sense, *broadcasting* is a technology—the
use of the airwaves to distribute a high-powered
video signal over a metropolitan area. But broadcasters do not simply provide a conduit to the
home. Their real business is the selection of content for their channel and the sale of advertising
time. The more viewers that a station can attract
with its programming, the more advertisers will
be willing to pay. The content used to attract viewers includes news, sports, and entertainment.

A station's programming can come from three
main sources. Some of it, such as much of the station's news programming, is locally produced. If
the station is affiliated with a network, the network programming usually arrives at the station
over a satellite feed and is then rebroadcast. While
it would be possible to distribute programming to
stations over high-bandwidth fiber links, satellites
are more cost-effective, given the large number of
stations to which the programming is distributed
and the inherent point-to-multipoint nature of
satellite services. Finally, programming can be
distributed to the station by independent programmers, who provide programs either on tape or via
satellite.

The true value of broadcasting technology lies
in its ability to provide universal access to video.
Once the television station's tower is in place, almost everyone within the station's coverage area
can receive the signal. It costs the station nothing
to add additional viewers. By contrast, with wired
broadband media, each new subdivision or subscriber requires additional expense. Even after the
rapid build-out of cable systems over the past decades, over-the-air broadcasting is still the only
universally available source of video programming. Nearly all U.S. households can receive at
least one over-the-air broadcast television signal,
and nearly 95 percent can receive more than five
channels.[26]

The second hallmark of broadcast television is
that the service is "free," once the viewer has purchased a television. This is not strictly a consequence of the use of wireless technology. Wireless
technology makes it possible for every viewer in a
city to receive a video signal; advertiser-supported
programming makes the service free. This business model emerged in part because it was considered too difficult or expensive for each station to

---

[25] FCC, *Annual Assessment,* op. cit., footnote 1. Bruce L. Egan, "Economics of Wireless Communications Systems in the National Information Infrastructure (NII)," contractor report prepared for the Office of Technology Assessment, U.S. Congress, Washington, DC, November
1994. Johnson, op. cit., footnote 3, ch. 8.

[26] Federal Communications Commission, Office of Plans and Policy, "Broadcast Television in a Multichannel Marketplace," OPP Working
Paper Series 26, June 1991, p. 18.

try to recover fees for the service directly from the viewer.[27] Whatever the origins of the business model, however, policymakers in the United States have long attached considerable value to the availability of a video service that is both universally available and free.

Because broadcast television is so ubiquitous and is perceived to have considerable influence on U.S. cultural and political life, policymakers have periodically tried to influence programming content. Efforts to influence what broadcasters show have focused on violence, children's programming, and balance in news reporting. The FCC has the authority to impose standards on broadcasters because the spectrum that broadcasters use is considered to belong to the public. The Commission, acting on behalf of the public, requires broadcasters to meet programming standards as a condition of licensing. The FCC has not imposed similar conditions on programmers who distribute their content through cable because they do not use the public airwaves.

Licensing decisions also focus on the degree to which broadcasters tailor their programming to the community in which they operate, particularly through news and public-affairs programming. The natural coverage area of a broadcaster's signal matches a typical metropolitan area, and "localism" has long been cited as one of the hallmarks of the U.S. broadcasting system. But in practice, broadcasters distribute a mix of local and national programming. Many stations are affiliated with national networks, who pay their local affiliates a fee to broadcast network programming in ex-

change for the right to sell some of the affiliates' advertising time to national advertisers.

Technology

Current television technology

Over-the-air television broadcasting was first authorized more than 50 years ago. On July 1, 1941, the FCC allocated spectrum for channels 1 to 13 in the so-called Very High Frequency (VHF) band.[28] Subsequently, a much larger band of frequencies, for channels 14 to 83,[29] was allocated in the Ultra High Frequency (UHF) band.[30] Each of these channels is 6 MHz wide.

Broadcasters transmit their signal from a single antenna on a tower several hundred feet tall. The power output necessary for good reception throughout the city depends on the antenna height, the terrain, and the frequency at which the broadcaster operates. The signal can usually be received upwards of 50 miles from the tower, depending on the type of antenna employed by the user. In part of the coverage area, it may be necessary to use an outdoor antenna to get good reception, but in other areas simple "rabbit ears" are sufficient.

The basic format for transmitting television signals in the United States is referred to as NTSC (National Television Systems Committee), named after the group that developed the system.[31] It was chosen by the FCC as the U.S. national standard in 1941 and has proven remarkably durable. In 1953, color was added to NTSC in a compatible way—old black and white receivers could still receive the new signal. Later,

---

[27] Ibid., p. 4.

[28] Channel 1 was later reassigned.

[29] Channels 70 to 83 were later reassigned to cellular telephony and other land mobile radio services.

[30] The "very high frequency" and "ultra high frequency" terminology reflects broadcasting's long history. With advances in radio technology, television's frequencies, the highest of which is 806 MHz, are now considered to be at the lower end of the usable spectrum. By contrast, the new PCS services will operate at 2000 MHz (2 GHz), and many other services operate at still higher frequencies.

[31] Two other formats are used for television transmission around the world: Phase Alternation Line (PAL), which is used in Germany and the rest of Europe, and Systeme Electronique Couleur avec Memoire (SECAM), which is used in France, Africa, and Russia, among other countries. The three standards are not compatible.

in 1984, a stereo sound capability was added to the standard. In addition, "subcarriers" within the signal have been exploited for the transmission of closed-captioning information and other services. While the standard has remained much the same for more than 50 years, better camera, production, and receiver technologies have considerably improved the quality of the picture seen in most households

As good as NTSC has been, however, it is highly inefficient in its use of the spectrum. Many of the radio frequencies that are allocated to television cannot actually be used because there would be unacceptable interference between channels. In the UHF band, for example, only nine out of the 55 channels can be used in any given city.[32] On several occasions, the FCC has tried to encourage development of a receiver that would allow use of the unallocated channels, referred to as taboo channels, but their efforts have been unsuccessful.[33] Problems with interference also require that channels not be reused in adjacent cities less than 150 miles away. Even if the station's signal is not strong enough to be received in the next city, it may be strong enough to cause interference.

Advanced television systems and high-definition television
In the mid-1980s, technology advances made it possible to develop a new television format that would offer significant improvements over NTSC. Japanese companies had begun to demonstrate a new high-definition television (HDTV) system that offered better resolution, a wider screen, and better sound. However, the Japanese system was not compatible with NTSC, and required more spectrum than a conventional television channel to deliver the extra information re-

High-definition television receivers will offer film-quality images, digital sound, and a wider aspect ratio to enhance the home theatre experience.

quired for a sharper picture. Nonetheless, it was proposed for use in the United States, sparking a vigorous debate that was partly about industrial policy and partly about the future of over-the-air broadcasting.[34]

The FCC has played an active role in the development of HDTV technology. Fearing that the limited spectrum available in the broadcast band would make it impossible for them to compete with other media in the delivery of HDTV, broadcasters petitioned the FCC in 1987 to investigate the implications of HDTV. The FCC responded by opening a Notice of Inquiry,[35] and in November 1987 it established the Advisory Committee on Advanced Television Service (ACATS), which was charged with providing information to the Commission.

ACATS established a testing process to compare the candidate systems. Originally, the systems being proposed were based on analog technology; by 1990, however, new digital compression technologies allowed an HDTV signal to

---

[32] Federal Communications Commission, *Advanced Television Systems and Their Impact Upon the Existing Television Broadcast Service*, Notice of Inquiry, 2 FCC Rcd 5125, 5126, 5133 (1987).

[33] Ibid.

[34] "Super Television," *Business Week*, No. 3089, Jan. 30, 1989, pp. 56-63.

[35] Federal Communications Commission, *Advanced Television Systems*, op. cit., footnote 32.

be squeezed into a standard 6 MHz NTSC channel, and also allowed use of the unused taboo channels.[36] The number of HDTV system candidates dwindled as proponents dropped out or merged their efforts to develop digital systems. At the end of the ACATS testing process in 1992, there were few differences among the four remaining systems, and the proponents were encouraged to combine their efforts. A "Grand Alliance" was subsequently formed in May 1993, and testing of the Grand Alliance system is scheduled to conclude in late 1995. Once the tests are completed, ACATS will recommend the system to the FCC as a U.S. national standard.

To smooth the transition to HDTV, the FCC has proposed a transition plan that would move the industry and consumers to HDTV over the span of several years.[37] The transition will begin when the FCC picks a standard and assigns HDTV channels to each city in a new *Table of Allotments*. According to current proposals, all current high-power television stations will be eligible for a second channel to be used for HDTV (their original channel will be returned at the end of the transition period). Broadcasters will have three years to apply for an HDTV channel, and by the end of the sixth year are required to be broadcasting in HDTV. After nine years, broadcasters are expected to be simulcasting, showing the same programs on both their NTSC and their HDTV channels. The purpose of the simulcasting provision is to prevent NTSC viewers from being deprived of the opportunity to see the same programming as HDTV viewers.[38] The HDTV and NTSC channels are

currently not considered separate services. Before the ninth year, however, broadcasters will be permitted to show different programs on HDTV in an effort to experiment with the capabilities of the new medium or to use specialized programming to attract viewers to the new service. The FCC's preliminary decision is to require broadcasters to return their NTSC channel 15 years from the date that the transition to HDTV begins, but, as with the dates of all of these milestones, this will be reviewed at regular intervals during the transition process.

In part, the FCC schedule is designed to build momentum for HDTV. By specifying a date on which HDTV programming will begin, the Commission is hoping to encourage programmers and equipment manufacturers to invest in the development of the programs and receivers that will be needed for HDTV to be a success. The FCC is attempting to avoid a chicken-and-egg problem in which broadcasters do not begin HDTV broadcasts until sufficient receivers are available and manufacturers do not produce receivers until broadcasts begin. The FCC is using its jurisdiction over the broadcasters to position them as market leaders, hoping that receiver manufacturers, programmers, and other media will follow.[39]

In the past year, the debate over HDTV has shifted. Broadcasters have been quite reluctant to commit to HDTV in any meaningful way because they believe that viewers may not want it—or be willing to pay the thousands of dollars new HDTV sets are expected to cost. Instead, broadcasters have been pushing the more generic idea of digital

---

[36] By transmitting the video signal in digital rather than analog form, it is possible to do complex mathematical manipulations of the signal in order to reduce the bandwidth requirements. Federal Communications Commission, *Advanced Television Systems,* First Report and Order, 5 FCC Rcd 5627 (1990).

[37] Federal Communications Commission, *Advanced Television Systems,* Second Report and Order and Further Notice of Proposed Rulemaking, 7 FCC Rcd 3340.

[38] Federal Communications Commission, *Advanced Television Systems,* Memorandum Opinion and Order, Third Report and Order, and Third Further Notice of Proposed Rule Making, MM Docket No. 87-268, Sept. 17, 1992.

[39] "In addition, because over-the-air broadcasting reaches more than 98 percent of U.S. households, an ATV terrestrial broadcast system is the medium most likely to bring this technological advance to virtually all Americans. Consequently, it is the medium most likely to result in rapid penetration of ATV receivers and, hence, to contribute to higher sales volumes and eventually lower costs for these receivers." Federal Communications Commission, *Advanced Television Systems,* op. cit., footnote 37.

television (DTV) or advanced television (ATV). These concepts are designed to give broadcasters more flexibility to deliver different kinds of television services—depending on what viewers actually want and will pay for. For example, broadcasters could offer multiple channels of digital television at a level of quality that approximates the current NTSC system, or deliver one HDTV channel, and/or provide advanced information and data services. These issues are currently being discussed at the FCC and in Congress, where the terms and conditions of broadcasters' provision of data services is being debated.

Issues and Implications

Technology, standards, and spectrum

The main issue facing broadcasters is the transition to next-generation digital technology.[40] The FCC has not issued any rulings on HDTV since 1992, apparently waiting for ACATS to report its recommendation on the HDTV standard that has taken longer than expected to develop. Although the basic elements of a new digital television standard are in place, there are unresolved issues that will have to be addressed by the Commission. One issue is the question of interlace versus progressive scan. Traditionally, television receivers have used interlace scan, in which alternate lines are scanned in each frame, whereas computer monitors use progressive scan, in which all lines are scanned every frame. Because they believe that the distinctions between computers and televisions will blur, the computer industry has been pressuring ACATS to use progressive scan for HDTV. Currently, the Grand Alliance system offers both modes, but the FCC could impose policies that require broadcasters to transmit

progressive scan material, to encourage the sale of computer-friendly progressive scan displays.

A second set of technology issues involves efficient use of the broadcast spectrum. From a spectrum management standpoint, there are good reasons to develop policies that would result in the adoption of modern technologies as soon as possible. As long as broadcasters are permitted to continue using NTSC, the broadcast allocation will be underutilized. But new digital television technology, combined with the requirement that NTSC broadcasting cease at some point in the future, would make it possible to use the spectrum more efficiently. It is possible that at the end of the transition process, the entire VHF band would be freed for other uses, such as mobile or new in-building communications technologies.

Another spectrum/technology concern involves system architecture—whether to use the traditional model of a single tower broadcasting a high-powered signal, or several smaller transmitters broadcasting at lower power. This latter scheme is sometimes referred to as "distributed transmission" or "cellular television" because each tower broadcasts to only part of the overall coverage area. One advantage of these "single frequency networks" is that towers can be located wherever necessary to tailor coverage; for example, filling in coverage in a valley.[41] But the main advantage of this approach is that it leads to more efficient spectrum use because the same channel can be used in adjacent cities.

Finally, the cost of upgrading to digital transmission technologies is an important issue for broadcasters. Although costs will vary depending on how much digital equipment a station already has (digital film storage and tape playback machines, for example), costs could be high, especially for smaller stations that do not have the

---

[40] HDTV was, until perhaps two years ago, the preferred acronym. Now, in trying to move toward a more flexible use of the new technologies, broadcasters coined the digital television (DTV) term. DTV is conceived to be broader and more inclusive than HDTV, which is being portrayed as an overly narrow technology mandate.

[41] One single frequency network technology is COFDM (coded orthogonal frequency division multiplexing). Its consideration was mentioned in the FCC's last Report and Order, op. cit., footnote 38, but it is not currently part of the Grand Alliance system.

advertising revenues of stations in larger markets. Broadcasters will have to buy new antennas, towers, and production equipment. The cost of adding basic HDTV capability—allowing a station to "pass through" network programming and add local commercials, but not originate any local programming—has been estimated to be between $1.3 million and $2.2 million per station.[42] The cost could be significantly higher for the estimated two-thirds of all stations that would need to build a new tower for HDTV broadcasting.[43] Stations will also incur higher costs to buy the production and studio equipment needed to originate programming in an HDTV format. However, the ability to pass through network programming will meet the FCC requirements outlined in its transition plan.

Demand

In recent years, broadcasters have begun to question whether there is enough demand for HDTV to warrant the expensive technology upgrades that would be required to provide it. It is unclear how many viewers will be willing to pay the (initially) high cost of HDTV receivers to receive better pictures. The advantages of HDTV are most apparent on large screen displays, which are inherently more expensive. Because their service is not by subscription, broadcasters will be unable charge viewers extra for a premium HDTV service, as would a cable company. Nor will they capture any of the revenues from the sale of HDTV receivers. In the 1950s and 1960s, NBC used the transition to color in part to spur the sales of color receivers produced by its parent, RCA.

Faced with what they perceive to be high costs and low demand, many broadcasters are actively resisting the mandated transition to HDTV. Instead, they argue, they should be allowed to use the spectrum more flexibly to offer multiple digital channels (instead of just one HDTV channel) or even other services, such as data transmission. Such uses, industry representatives point out, could increase spectrum efficiency, enhance diversity, and provide a way to offset the cost of deploying any new technology the FCC requires. The debate over what the FCC should require now occupies center stage in the digital television/HDTV debate. There is concern that broadcasters are being forced by the FCC in a direction that consumers will not want to go—HDTV.[44]

The viability of HDTV is, in part, a separate issue from the question of whether the FCC should encourage broadcasters to adopt digital broadcast technology. If HDTV is not considered to be viable, one option is "multicasting," the use of the digital channel to broadcast multiple standard-definition channels (SDTV). The same technology that squeezes a high-definition signal into a single channel can also be used to transmit four or more standard-definition signals. Viewers could continue to use their existing television sets, but would need a set-top box to translate the digital signal into the NTSC format understood by their television. This box would be much less expensive than an HDTV receiver, most of whose cost is in the display, not the decoder. The additional channels could provide broadcasters with additional revenue sources (through subscriptions, perhaps) and provide an incentive to move to more

---

[42] National Association of Broadcasters, *NAB Guide to HDTV Implementation Costs* (Washington DC: NAB, 1993), p. 39.

[43] Ibid., p. A-7.

[44] "What also comes through in the industry's comments, however, is trepidation—and understandably so. After all, the Commission is *mandating* the development of this new technology in only one sector of the video marketplace: broadcast television. Other segments of the industry—program producers, film studios, cable programmers, DBS providers— can elect to watch from the box seats as the broadcasters enter the Colosseum. While shouldering only a fraction of the risk, they will have the luxury of awaiting the answers to the fundamental questions that broadcasters, and the Commission, must grapple with today: Will consumers rally around high-definition? Will compellingly crisp pictures and sound make HDTV indispensable to America's 90 million television households?" Statement of Commissioner Ervin S. Duggan, federal Communications Commission, *Advanced Television Systems,* op. cit., footnote 38.

efficient technology. This is, in fact, the strategy now being pursued by cable, wireless cable, and satellite companies (see below), that are converting to video distribution systems that use digital transmission technologies and set-top decoders to deliver services to current analog televisions. However, manufacturers who have invested in the development of HDTV receivers and production equipment are opposed to standard-definition multicasting. In addition, this strategy would perpetuate the use of NTSC's interlaced display technology, which is opposed by the computer industry.

In addition to multicasting more video programming, broadcasters are considering many other services that could be delivered over a high-bandwidth digital channel. These include data delivery or paging. As a wireless medium, broadcasters can quickly deliver services to locations that do not have wireline facilities and to mobile users. But because these services are not seen as being part of the broadcasters' traditional service, the ability to use spectrum in this way is seen by some as a windfall. The issue of "flexible use" of broadcast spectrum was debated in the last Congress, and in the current Congress, proposed legislation would give broadcasters the freedom to offer "ancillary and supplementary" data services, subject to certain restrictions. The meaning of "ancillary and supplementary services," however, will have to be defined by the FCC. Broadcasters would have to pay a fee for spectrum used for these services.

Competition and the role of over-the-air broadcasting

Broadcasters' main business—programming a channel and selling advertising—is no longer completely tied to broadcast technology as its sole means of distribution. While over-the-air broad-

casting made the television business possible, today more than 60 percent of households now receive broadcasters' programming over cable and some rural viewers receive programming directly from satellites.[45] While new cable programming competes with broadcasters for advertising dollars, cable technology is also an essential conduit for broadcasters to reach viewers. For this reason, the terms under which cable systems carry broadcast signals have been the subject of intense policy debates and negotiations between networks and cable providers.[46]

To some extent, the fate of broadcasters as programmers (creating and selling programming) may be separate from their role as program distributors. Whether or not over-the-air broadcast technology will continue to be a significant mode of distributing entertainment programming depends on a variety of factors. While wireless technology was a good way to deliver television service quickly to all of the people in a metropolitan area, there is a limit to the amount of available spectrum. By contrast, the cable and telephone companies are rapidly upgrading their distribution plant to deliver an even wider range of programming; over the past decade, there has been significant growth in the number of viewers preferring to receive programming using cable or other "multichannel" services such as DBS or wireless cable. In addition, many of these companies are proposing new interactive services that may attract even more subscribers. Even with digital compression and multicasting, it is unlikely that broadcasters will be able to match the number of channels or range of services these other providers will offer, unless more spectrum is made available to individual stations—an unlikely prospect.

Some have suggested that if other distribution media were to provide programmers with satisfac-

---

[45] Viewers can only receive network programming via satellite if they cannot get a broadcast signal or have not recently been a cable subscriber. For those viewers who qualify, packages of network programming are available to C-band system owners from NetLink and PrimeTime 24. DirecTV/USSB owners are subject to the same qualifications.

[46] See, for example, Federal Communications Commission, *Amendment of Part 76 of the Commission's Rules Concerning Carriage of Television Broadcast Signals by Cable Television Systems*, Report and Order, MM Docket No. 85-349, Nov. 28, 1986.

tory access to the viewing audience, it is conceivable that broadcasters could choose to stop distributing programming over the air altogether. If this were to occur, there would be difficult questions about the fate of remaining viewers who still relied on free over-the-air broadcasting. Currently, however, control over their own distribution medium provides broadcasters with significant advantages. They can sell advertisers access to the 40 percent of households that do not have cable, as well as to a significant number of second televisions in cable households that are not connected to cable. In addition, their status as broadcasters entitles them to carriage on cable systems by "must carry" regulations. Other programmers have to compete to be included as one of a cable system's channels.

■ Alternative Video Service Providers

The market for video entertainment programming is becoming increasingly crowded and competitive. Broadcasters face competition not only from cable television providers, but also from a small—but growing—number of companies that use radio-based technologies to provide similar services. Recently launched DBS services bring hundreds of channels of premium and pay-per-view programming to subscribers, and terrestrial wireless systems promise similar, if fewer, services. Telephone companies are preparing to enter the video distribution market by upgrading their own wire-based networks, but also through the use of wireless.

The emergence of these new wireless distribution technologies is undercutting the traditional preeminence of the television networks and local broadcast stations, and could provide substantial competition for cable television as well. Wireless companies provide, or plan to provide, program-

ming packages similar to those offered by cable television, and each of the alternatives has brought competition for viewers and advertising dollars. Some analysts expect new wireless services to be the main source of competition to cable television and broadcasters in the market for alternative video programming—not the local telephone companies that have been planning and fighting for the right to offer video programming for years.[47]

Multichannel Multipoint Distribution Service (MMDS)

MMDS providers, commonly known as "wireless cable," offer entertainment programming services in competition with traditional cable television providers. To date, the industry has grown very slowly in the United States, amassing only 750,000 users—served by 175 systems—across the country.[48] In recent years, however, growth has picked up noticeably, and individual companies have been successful in some local markets. Industry representatives predict that by the end of 1995, the number of subscribers will more than double to 1.8 million viewers served by 200 systems, and by the year 2000 analysts expect wireless cable systems to be serving between 3.2 million and 4 million subscribers and earning between $1.5 billion and $2 billion in annual revenue.[49] Other countries are installing wireless cable systems instead of wired systems because of its lower costs and faster installation times.

MMDS providers use low-power microwave signals broadcast from a central tower to deliver their services. No local franchise is required. Programming packages typically include movie channels like HBO, premium programming (Disney channel), some local broadcast stations (and national "superstations"), and pay-per-view. Wireless cable providers, however, do not pro-

---

[47] Johnson, op. cit., footnote 3.

[48] Much of this paragraph is based on materials provided by the Wireless Cable Association International.

[49] Andrew Kreig, Wireless Cable Association International, personal communication, March 20, 1995; Louise Lee, "Wireless Cable-Television Sector Is on Acquisition Binge," *The Wall Street Journal*, June 8, 1994; Tom Kerver, "The Wild World of Wireless Video," *Cablevision*, May 23, 1994.

duce their own programming, such as local news or sports. To receive the MMDS signal, subscribers must purchase about $200 worth of equipment, including a rooftop antenna, signal converter, and a set-top box, and pay a monthly service fee roughly between $17 (basic package) and $25 (basic plus one premium channel).[50] The major advantage of MMDS over cable and DBS is the low initial construction costs—$1 million to $2 million for the tower and transmitting equipment—no expensive satellites to build and launch, and no expensive cable to lay. This lower cost structure is what allows MMDS providers to charge less for their services (although usually for fewer channels).

MMDS systems operate at 2.6 GHz, limiting them to line-of-sight delivery, and use analog transmission to deliver video to consumers.[51] The number of channels used (and offered to consumers) by individual MMDS providers varies. FCC rules allow MMDS companies to use up to 33 channels, but only 10 of these channels are dedicated to MMDS. Twenty of these channels are allocated to the Instructional Television Fixed Services (ITFS), and another three are allocated to the Private Operational Fixed Service. ITFS license-holders will often lease some or all of their capacity to a local MMDS provider, or the channels can be shared by time of day. Complex rules govern sharing between the three services, resulting in a situation where not all 33 channels are available to MMDS providers in all markets.[52] This is likely to hamper the ability of MMDS providers to compete effectively in some areas.

"Wireless cable" systems will provide consumers with another choice in the increasingly competitive multi-channel video distribution market.

Over the last several years, the MMDS industry has grown considerably and is now preparing for serious competition with other video service providers—cable, DBS, and Local Multipoint Distribution System (LMDS) (see below). Rapid consolidation has taken place as companies seek to develop the economies of scale and cost advantages that will bolster the industry's competitive position.[53] Until three years ago, MMDS companies were often denied access to programming—or charged exorbitant rates—by many video programmers who were owned by or locked into contracts with cable television companies. In 1992, Congress passed the Cable Act, which pro-

---

[50] John Ramsey, "MMDS: The Advent of Latin American Pay TV," *Satellite Communications,* p. 17, August 1993; Kreig, op. cit., footnote 49.

[51] Specific frequencies are 2500-2655 MHz and 2655-2690 MHz. Line-of-sight restrictions, including blockages by trees and buildings, may be overcome by technological advances that will allow the signals to be "bent," but to date they have limited MMDS to relatively flat topography. MMDS systems' range is about 30 miles.

[52] Bennett Z. Kobb, *Spectrum Guide: Radio Frequency Allocations in the United States, 30 MHz-330 GHz* (Falls Church, VA: New Signals Press, 1994), pp. 149-151.

[53] Lee, op. cit., footnote 49.

hibited video programmers from discriminating against program distributors like MMDS and DBS.[54] The Cable Act opened up access to programming that had been held back for many years, and allowed the wireless companies to compete more effectively and evenly on product and price.

In addition, the MMDS industry is now developing digital compression schemes that are expected to increase the number and variety of channel offerings, perhaps allowing providers to offer as many as 200 channels. A digital upgrade could also enable MMDS providers to offer interactive programming. Also, the ITFS service has channels specifically identified as "return" or "response" channels, allowing voice and data communications to be sent back to the broadcaster.

As a result, wireless cable has become a more attractive technology choice for both consumers and suppliers. Pacific Telesis recently announced plans to acquire a wireless cable company in Southern California, and Bell Atlantic/NYNEX will team up to invest in another MMDS provider.[55] These companies see wireless cable as a way to deliver advanced digital video services to their customers until they can upgrade their existing telephone systems to carry video signals. This allows them to enter the video programming distribution market significantly faster than waiting for new fiber optic systems to be installed. This strategy is a preemptive response to cable company provision of telephone services later in the decade.

Satellite Television Services

Satellites have been an integral part of the communications infrastructure since the first communications satellite, Hughes' Early Bird, was launched in 1965. Early satellites transmitted telephone calls across the Atlantic Ocean, and were soon used to distribute television programming to network affiliates across the country. Today, satellites deliver video programming directly to over 5 million people.

C-band and Ku-band satellites

C-band satellites have been carrying television programming for more than 20 years. These satellite systems were primarily designed to distribute programming from television networks to their local broadcast affiliates, and premium cable channels (HBO, Discovery, and Disney) and television "superstations" to cable television systems across the country. However, in the early 1980s consumers began putting up their own dishes—so-called backyard dishes—to receive the programming directly.[56] Today, satellite television services provide video, data, and music services, mostly to people in rural areas where broadcast and/or cable do not reach. By 1994, there were about 4.5 million backyard satellite dishes in use in the United States, roughly 3 million of which are in areas with access to cable television.[57]

C-band systems account for the bulk of consumer satellite TV systems.[58] Consumers use 7-to 10-foot-diameter dishes, costing from $2,000 to $3,000 installed, to receive analog video signals from geostationary satellites in orbit 22,300 miles above the Earth. C-band dish users can receive approximately 150 free, unscrambled signals and roughly another 100 scrambled channels, such as HBO, can be ordered through various program packagers for a monthly subscription fee. The

---

[54] The Cable Television Consumer Protection and Competition Act of 1992, Public Law No. 102-385, 106 Stat. 1460 and codified at 47 U.S.C. section 151.

[55] "PacTel To Buy Tiny Wireless-Cable Firm For $120 Million To Speed Video Project," *The Wall Street Journal*, Apr. 18, 1995, p. A4.

[56] These satellite receiving dishes are also referred to as "home satellite dishes" and "television receive-only dishes. At first, the programming transmitted over satellites was unscrambled and free to anyone with a receiving dish. Soon, however, programmers began scrambling their services and charging for use.

[57] Johnson, op. cit., footnote 3, pp. 115, 151.

[58] Most cable programming services still use C-band for program delivery.

number and types of programming packages available vary widely, but for about $25 a month, a subscriber can receive approximately 25 basic cable channels and eight movie channels, in addition to the 150 free channels. These systems also use subcarrier frequencies to offer multiple channels of audio, such as music and talk radio stations. C-band services also provide data services for an additional fee. By attaching a data terminal to their home equipment, customers can receive a host of information services, such as financial information, stock updates, and specialty services.[59]

Ku-band satellite services use higher frequencies that allow smaller dishes, and are used mostly by businesses, broadcast and cable companies, the government, and others to supply private communication networks. These networks often use very small aperture terminals (VSATs) to link far-flung company sites (see chapter 4). Ku-band satellites also provide commercial radio and television distribution, teleconferencing, private data networks (such as remote credit card verification), high-speed image transmission, distance learning, international and domestic long-distance telephone transmission, and other services. In addition, Ku-band satellites have helped establish telephone service for remote and/or less developed countries.

Direct broadcast satellite (DBS)

DBS systems represent the next evolution of satellite-delivered television.[60] DBS was originally conceived to serve households not passed by cable, but as that number shrank from 18 million in 1984 to approximately 4 million in 1992, services were targeted more directly at existing cable markets.[61]

High-power DBS satellites allow receiving dishes, seen here on the corner of the garage roof, to be quite small.

The FCC authorized DBS service in 1982, and established rules for the service that regulate it not as a broadcasting or common carrier service, but according to its own rules. Despite support from some large companies, all early attempts to establish a successful DBS venture failed. The satellites for the new service were very expensive to build and launch, premium programming was difficult for some providers to obtain, and consumer demand was low—the systems could only transmit a half dozen channels.

In the past four years, however, two new DBS systems have begun offering packages of video programming, as well as pay-per-view events, directly to consumers' homes. These new systems use high-power and digital technology to provide a wide selection of programs and CD-quality sound, using smaller dishes than traditional large-

---

[59] Harry Thibedeau, Satellite Broadcasting and Communications Association, personal communication, Jan. 20, 1995.

[60] Direct Broadcast Satellite (DBS) technically refers to a specific type of high-powered satellite operating in the 12.2-12.7 GHz (Ku) band. This was the way that most analysts and policymakers thought video programming would be delivered directly to consumers when the service was established in 1981, and the name has gained widespread acceptance. Primestar, discussed below, is not technically a DBS system, because it uses a lower powered Ku-band satellite that operates according to the FCC's Fixed Satellite Service rules. For purposes of clarity, Primestar will be discussed in this section because it provides the same services historically ascribed to DBS.

[61] Johnson, op. cit., footnote 3.

dish satellite TV. Although the systems should appeal most to users who cannot receive cable television or have chosen not to subscribe, early indications are that the market for such services may be much broader. Initial sales of DBS services have exceeded expectations, with nearly 750,000 subscribers signing up in the first year of operation. Some DBS proponents have interpreted these figures to indicate consumer discontent with cable television providers. Various types of direct-to-home satellite services are being developed around the world.[62]

Conceptually, DBS systems are quite simple (figure 6-3). Programmers send their material to a central facility similar to a cable system's headend, where the programming is compressed and sent up to orbiting geosynchronous satellite(s). The signals are then broadcast over the United States for reception by the user's receiving dish. From the dish, a cable feeds the programming to the set-top receiver, which decodes the compressed programming and records billing information for pay-per-view (PPV) events. One system remotely polls the subscriber units each month (via phone-line connection) to collect the billing information.

Despite the advantages offered by DBS—including national coverage, high-quality sound, and wide selection—the systems suffer some competitive disadvantages as well. Perhaps the biggest is that the receiving dish must have a clear line-of-sight to the satellite in the southern sky with no obstructions such as tall trees, mountains, or buildings. Analysts estimate that 50 percent of all U.S. households, including apartment buildings, have this capability, meaning that the other 50 percent cannot receive DBS programming at all.[63] The other significant disadvantage, which some consumers are apparently still unaware of, is that the systems cannot carry local programs, and most DBS customers cannot get network programming (ABC, CBS, NBC, FOX, and PBS) at all. The Satellite Home Viewers Act of 1994 allows subscribers to receive network programming only if the consumer cannot receive it off the air, and if they have not subscribed to cable in the last 30 days.[64] Finally, the systems are not expected to be able to offer true video-on-demand services (in which the user can control "Stop," "Review," and "Search" functions) in the near future, although they do offer near video-on-demand in which movies begin every 15 minutes or so. The nature of the broadcast satellite beam combined with the large number of subscribers makes it currently infeasible to dedicate a single channel to an individual subscriber.[65]

Two systems offer direct-to-home services today—Primestar, owned by a consortia of cable companies and GE American Communications, Inc.; and Hughes' Communications Galaxy DirecTV/United States Satellite Broadcasting

[62] For an overview of these activities, see Michael S. Alpert and Marcia L. De Sonne, *DBS: The Time is Now* (Washington, DC: National Association of Broadcasters, 1994).

[63] Satellite Broadcasting and Communications Association, presentation to OTA staff, Apr. 7, 1994. The number of single-family homes affected is likely to be significantly lower.

[64] Satellite Home Viewer Act of 1994, Public Law 103-369, Oct. 18, 1994. Dawn Stover, "Little Dish TV," *Popular Science*, January 1995. One company, Local DBS, Inc., has proposed to use spot beams to relay local programming to viewers. See Alpert, op. cit., footnote 62.

[65] Johnson, op. cit., footnote 3.

FIGURE 5-3: Direct Broadcast Satellite

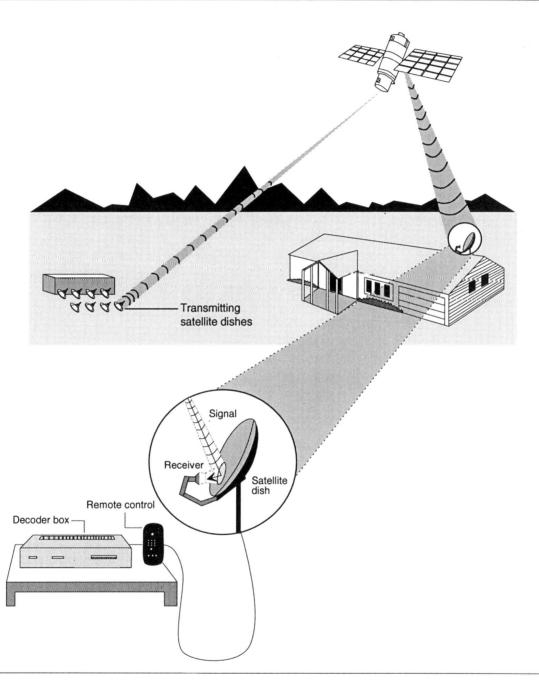

SOURCE: Office of Technology Assessment, 1995.

(USSB), which uses RCA's Digital Satellite System (DSS).[66] Other companies have received licenses for DBS, but are not yet operating.[67] Although the two services differ in many respects, each is digital and uses significantly smaller sized dishes than C-band systems.[68]

Primestar initiated service in 1991 as an analog system, but in 1994 converted to digital to expand its channel offering and improve quality. The Primestar system uses a commercial Ku-band satellite, and operates under the FCC's Fixed Satellite Service (FSS) rules. This classification restricts the Primestar system to medium-power broadcast, which requires the use of a receiving dish of either 36 or 40 inches. The dishes cost about $900, but most Primestar subscribers lease the equipment for a small monthly fee. Depending on distributor, subscribers pay between $21 and $54 a month for 77 channels, plus an installation fee of about $200. Users receive a number of pay-per-view (PPV) channels, which cost about $4 per movie or event.[69] The Primestar system currently serves about 400,000 customers.

DirecTV/USSB began offering service in October 1994. By March 1995, it had signed up 500,000 customers, and expects 3 million by 1996 and 10 million by 2000.[70] The system uses two satellites in geosynchronous orbit, compared with Primestar's one,[71] and broadcasts at higher power, resulting in a smaller (18 inches) receiving dish.[72] DirecTV controls the majority of the capacity on the Hughes satellites (27 of 32 transponders), and therefore offers more channels and more diversity than USSB. The full DirecTV package includes 150 channels of traditional cable programming, as well as sports packages, and many PPV options. The RCA dish[73] sells for $699 for the basic model and $899 for the model that allows two TVs to be hooked up. However, if consumers want the option of watching different channels on the two TVs simultaneously, they need to pay an additional $649 for another receiver. Professional installation costs $150 to $200, while a do-it-yourself installation kit is $70. Programming packages range from $17.95 to $34.95, plus PPV charges for USSB, and from $21.95 to $29.95 for DirecTV.[74] PPV movies are $3. Users who subscribe to both services can pay upwards of $65 per month plus any pay-per-view charges.

---

[66] DirecTV and USSB are actually two separate programming services, but use the same Hughes Communications satellite. The two companies offer services that complement, rather than compete with, each other (with some overlap). Users need only one set of equipment to receive both services, and many subscribe to both.

[67] Other potential DBS providers include: Echostar, Direct Broadcast Satellite Corp. (25 percent owned by Echostar), Advanced Communications Corp. (Tempo holds their license), Continental Satellite Corp., Dominion Video Satellite, and Tempo. Echostar is the furthest along—satellites are built and programming alliances are in place. Alphastar is planning to offer service by the end of 1995 using an AT&T fixed service satellite. Primestar was planning to transition to true DBS through Tempo's control of Advanced Communications licenses, but Advanced was turned down by the FCC for a license extension, putting Primestar's DBS plans in jeopardy.

[68] Stover, op. cit., footnote 64.

[69] Ibid.

[70] Eric Schine, "Digital TV: Advantage, Hughes," *Business Week*, Mar. 13, 1995.

[71] Due to differences in orbital spacing between these two classifications of satellites, BSS satellites are less susceptible to interference from adjacent satellites. This difference, along with their higher power, allows DSS systems to use smaller dishes.

[72] The Primestar system broadcasts at 45 watts, whereas the two DSS satellites broadcast at 120 watts.

[73] RCA has an exclusive license to manufacture the equipment for 18 months after the launch date or until one million units are sold, whichever comes first. After this point, Sony will enter the market with its dishes.

[74] In addition to the standard packages, DirecTV offers a $5.95-per-month package consisting of only one channel, but it allows subscribers to select the full complement of specialty sports packages and pay-per-view options. Some of the specialty sports packages offered by DirecTV are the Golf package, for $6.95 a month, the NFL season package for $119.95, and the NHL season package for $69.95.

Issues and Implications

Competition

The historic context for video programming services, and for the emerging NII specifically, is clearly based on competition. The video programming market is in its infancy, but already shows signs of becoming quite competitive.

> Any investigation of competition and public policy in such a dynamic arena [video programming] is handicapped by uncertainties about future technological advances and social needs. The only certainty is that surprises are in store. Before the end of the decade, we must anticipate achievements and disappointments going far beyond anything foreseeable in this monograph... Fortunately, these developments do not critically depend on the widespread deployment of any one technology or on the success of particular firms. The possibilities are so numerous, in terms of alternative technologies and the roles of diverse firms, that the public will benefit almost regardless of which path is taken through the maze. The challenge for public policy is to facilitate and to guide this dynamic process in ways that maximize these benefits.[75]

Congress demonstrated its commitment to competition in the 1992 Cable Act, where it expressed its preference for competition over rate regulation and its belief that the promotion of competition through new distribution technologies was critical.[76] The FCC has now taken over the congressional mandate to encourage competition in video services.[77] In September 1994, the FCC concluded that, despite substantial growth in alternative delivery systems, competition in multichannel video programming still did not exist for most Americans. Competing cable systems are still few in number, local telephone companies are only operating experimental video-delivery systems, and wireless competitors still do not have enough subscribers to make the market truly competitive.[78] The FCC further concluded that lowered entry barriers—to let more competitors enter the market—were likely to lead to significant benefits for consumers. Even if competitors do not actually enter the market, the threat of competition may provoke incumbents to improve services and cut costs.[79]

It now seems likely that, as the video programming market matures and technology continues to advance, services and providers will become increasingly differentiated. In part, this will be due to the different capacities and characteristics of the systems noted above. Provision of video programming packages may continue as the "core" market, but ancillary markets will form as well. Because the technology systems discussed above are not perfect substitutes for each other—some national, some local; some more interactive than others—they are likely to compete in the core market, but not necessarily in the splinter markets. The result will be that consumers will have a wider array of choices—that are more likely to match their needs more closely—than in the previous era of broadcast television's "one size fits all." It will be difficult to generalize policy nationally when competition will vary from location to location. Much more research will be needed to determine the nature and effectiveness of competition in these highly diversified markets.

The transition to future services, such as HDTV and interactive applications, will be a substantial

---

[75] Johnson, op. cit., footnote 3, pp. 187, 179.

[76] The Cable Television Consumer Protection and Competition Act of 1992, Public Law No. 102-385.

[77] The Federal Communications Commission undertook several actions in response to congressional mandate in the act. See Federal Communications Commission, *Implementation of Sections 12 & 19 of the 1992 Cable Act—Development of Competition and Diversity in Video Programming Distribution and Carriage*, First Report and Order, MM Docket 92-265, adopted Apr. 1, 1993; FCC, *Annual Assessment*, op. cit., footnote 1.

[78] FCC, *Annual Assessment*, op. cit., footnote 1, at para 15.

[79] Glenn A. Woroch, "The Evolving Structure of the U.S. Wireless Communications Industry," contractor report prepared for the Office of Technology Assessment, U.S. Congress, Washington, DC, December 1994. Johnson, op. cit., footnote 3.

issue for alternative video providers. Satellite service providers have said they are capable of and will provide HDTV if demand warrants it. Like other video providers, however, they are not rushing to HDTV. From a consumer's standpoint, HDTV may be viewed not as revolutionary, but as an upgrade—like color television. Because demand for HDTV is so uncertain, some analysts have called on the FCC to rethink its policies toward it.[80]

Interactive services are also likely to pose a competitive challenge for the video providers discussed in this section. Today's companies are primarily one-way providers of entertainment programming. Cable companies, however, are rapidly positioning themselves as information service providers as well. For example, several offer Internet access—something one-way services cannot do. It seems likely that the market for multichannel video programming will splinter as different technology systems exploit their technology and regulatory status, but it is still unclear which of these providers might begin to offer interactive services and when.

Technical constraints

Today, the primary technical challenge facing video service providers is the conversion to digital technology or the upgrading of digital capabilities to improve capacity and service. Because most systems are up and running (or are expected to be soon), technical concerns are not expected to substantially slow or stop the development of new services. Providers' use of different technologies, however, entail limitations or restrictions on what the systems can do and the services they can offer. Such differences are the basis of the competitive diversity of the industry.

As the industry matures, technical and regulatory differences will become more important. Programming limitations—due to lack of capacity or regulation—may hamper some providers' competitive positions. Satellite TV providers, for example, cannot deliver local programming because of technical limitations, are severely limited by regulations in the number of customers they can deliver network programming to, and will not likely be able to offer true video-on-demand due to capacity constraints. MMDS providers will likely continue to have fewer channels compared with their cable and satellite rivals. Individual broadcasters, too, will only be able to offer a limited number of video channels—even the aggregate of all local television stations' digital channels will be unable to match the hundreds of channels offered by cable and DBS.

Another technical constraint for MMDS, LMDS, and most satellite providers is the limitations of line-of-sight transmission. The number of people that can actually be served by wireless systems may be considerably less than first thought due to these physical constraints. Technology advances and better engineering are expected to alleviate some, but not all, of the limitations of line-of-sight systems.

Many alternative video programming providers are also affected by restrictions that have been placed on receiving antennas and dishes. Despite FCC regulations preempting local zoning ordinances and rules, many localities and homeowners associations continue to enact local regulations in violation of FCC rules.[81] Chapter 8 discusses these issues in more detail.

Integration and concentration issues

The economics of the wireless video programming industry will not be fully discussed here. Rather, this section will identify some of the issues that may affect the industry as it matures. Policymakers are concerned about the extent to which the competition and the diversity it implies can be sustained over the long term. Because it is still a young industry—many services are not operating yet—it is difficult to determine what it will

---

[80] Johnson, op. cit., footnote 3.

[81] FCC, *Annual Assessment*, op. cit., footnote 1, para. 76.

look like in five or 10 years. Costs, revenues, and future plans generally are closely guarded secrets. As a result, even getting a baseline of data to work from is difficult.

During the next five to 10 years, it is likely that the industry will continue to grow, adding new entrants as new companies emerge. Beyond about five years, it also seems likely that consolidations and mergers among some industry players will increase. Consolidations already have been seen in the MMDS market, and analysts expect other industries to follow suit as they mature.[82] Mergers also are likely between various wireline and wireless carriers, if regulations permit, and wireline carriers are investing in MMDS providers.

Such combinations, however, may have both short- and long-term negative effects. In the short term, horizontal integration between directly competing firms, such as in the DBS industry, could reduce the level of competition in individual markets—whether or not this is harmful would be determined case-by-case. Because most markets do not have multiple providers of the same service—currently each area tends to have one cable service, one MMDS (if that), and several local broadcasters—the more important potential problem is mergers between indirectly competing firms, or firms that provide not the same service, but a close substitute. For example, cable and DBS, DBS and MMDS or LMDS, and telephone companies with any of these. Because of these concerns, cross-ownership restrictions currently exist between cable and MMDS (and SMATV) providers. However, no such restriction exists between cable and DBS, and the local telephone companies are reportedly interested in LMDS technology.

In the long term, the ultimate outcome and extent of this trend are unclear, as are the final impacts. It is conceivable that, if cross-ownership became widespread across the various segments of the video programming industry, both the di-

versity and quality of services could decline, and overall prices could rise. Policies that are procompetitive now—to allow wide latitude in mergers and acquisitions—could turn out to be anticompetitive in the long run. Again, the immature state of the industry makes analysis highly speculative. Firms will merge or not based on the economics of individual situations that have not yet developed.

Interconnection issues

The extent to which the wireless video service providers discussed in this section will interconnect or interoperate with other parts of the NII will only be determined over time—absent government intervention to require specific levels and kinds of interoperability. The systems now function primarily as the final delivery (one-way) link to consumers and businesses. In this regard, their connections to the NII may be quite limited. The NII would serve as a resource base—or a backbone—for supplying the information or entertainment that is then sent on to customers. It seems likely that the cable/telephone networks will serve as an important way for video service providers to get programming in addition to the satellite delivery systems that already exist. Very little information if any is likely to travel back through the NII core from the users of these systems.

If these services become two-way or interactive, however, their integration with other networks is likely to be greater. One-way broadcasting systems, for example, may be relatively isolated from other communications systems now, but may link up with interactive programming provided by the Interactive Video Data Service (IVDS). DirecTV/USSB is also primarily one-way, but gathers billing data over phone lines. In the future, real-time interactive services may also be provided through such combinations. The next generation of DBS could add an element of interactivity by allowing users to download large amounts of information—mov-

---

[82] In the DBS industry, for example, one analyst believes that after four to five companies enter the market over the next three to five years, they will begin to consolidate. Michael Alpert, Alpert and Associates, personal communication, Mar. 23, 1995.

ies, for example—that they could then use as they wish. Such developments depend on continuing advances in memory technology (movies require large amounts of storage) and declining costs.

It is also conceivable that wireless programmers could eventually concentrate more on programming, and become less involved in the distribution side of the business. In the future, what are now wireless companies ironically may come to depend on the wire-based NII backbone to deliver some or all of their programming. In addition to their broadcast operations, for example, broadcasters could move their products over many competing delivery systems including cable, MMDS, and the public switched telephone network (PSTN).

EMERGING HIGH-BANDWIDTH SERVICES

In addition to the services described above, a new class of entertainment/information service provider is emerging—one that is capable of delivering a wider range of high-bandwidth, even interactive, services. Only one of these services is operational, and all, in fact, are still vying for spectrum before the FCC. They represent a mix of local and national services targeted at both businesses and individuals.

■ Local Multipoint Distribution Service (LMDS)

LMDS, also known as cellular television, is being developed primarily as another alternative to cable television, MMDS, and DBS services. In the future, LMDS technology may be able to deliver telephony and interactive data services as well. Proponents believe that the high-bandwidth capabilities of the system, combined with its interactive potential, make it a natural extension of the NII. Currently, only one provider, CellularVision of New York, is offering commercial video programming service, serving about 200 customers, but 12 other companies have received experimental licenses.[83]

LMDS proponents plan to use frequencies in the 27.5-29.5 GHz band (line-of-sight is required) and low-power transmitters in a cellular-like arrangement to deliver up to 50 channels of analog one-way video programming (figure 6-4).[84] For about $30 a month, customers can receive local broadcast stations, as well as popular enhanced programming, such as ESPN, movie channels, and pay-per-view channels.[85] A central tower uses an omnidirectional antenna to transmit programming to each individual cell site—between three and 12 miles in diameter—which then retransmits it to subscribers' homes. Thus, to cover a major metropolitan area of 1,000 to 2,000 square miles would take between 20 and 40 transmitter sites.[86] This configuration allows the provider to tailor the coverage areas of each transmitter to provide the best possible service. At the subscriber's home, a small antenna (there are several designs, including small dish antennas and 6.5-inch-square flat panels) on a windowsill or roof connects to the user's television.

---

[83] In January 1991, the FCC granted Suite 12 Group (now CellularVision of New York) a license to provide LMDS service in the New York City metropolitan area. Service began in June 1992. Since that time, the FCC has received over 971 applications to build similar systems across the country.

[84] Each channel is very wide—20 MHz. Using a special transmission technique (opposite polarization of signals), proponents and the FCC believe that the number of channels can be doubled—each original channel matched by a new one. These new channels could be used to carry more video programming or interactive services.

[85] B.J. Catlin, ed., "Wireless Cable TV FAQ," unpublished paper, Colorado State University, Department of Computer Science, May 3, 1994 (rev.).

[86] Egan, op. cit. footnote 25, p. 37.

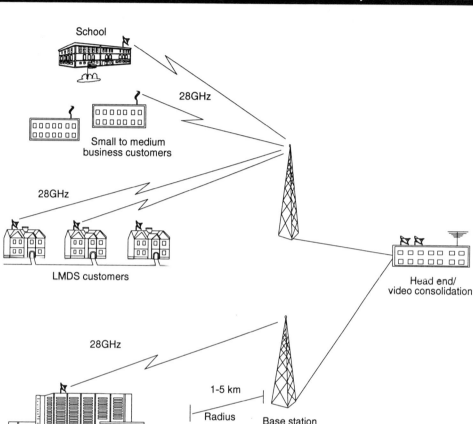

FIGURE 5-4: Local Multipoint Distribution Service System

SOURCE: Office of Technology Assesssment, 1995.

In the future, proponents plan an even wider range of services, including more video channels, telephone services, and various interactive services.[87] System capacity could be at least doubled by using digital compression technology and different transmission schemes. This extra capacity would then be used for new channels or services. By combining the wide (20 MHz) LMDS channels with interactive capabilities (LMDS systems can offer interactivity by inserting return-path communication channels between the video channels),[88] LMDS proponents envision delivering applications such as video-on-demand, video-conferencing, telephone service, and various data services, including computer networking to homes and businesses.[89] These applications are

---

[87] Except where noted, the services and applications discussed in this paragraph are from the Federal Communications Commission, op. cit., footnote 2.

[88] Return channels will use opposite polarity signals to avoid interfering with the video programming.

[89] Texas Instruments presentation to OTA staff, Nov. 9, 1994.

expected to serve distance education, telemedicine, and a number of business communication needs. Providers of rural telephone and broadcasting services have also expressed an interest in LMDS as a way of serving remote customers. Proponents claim that the systems will be able to accommodate future digital communications advancements, including HDTV.

LMDS offers a number of potential advantages over competing video delivery systems, primarily stemming from its point-to-multipoint cellular architecture. First, construction costs are lower compared with satellite systems, and savings can be passed on to consumers in the form of lower monthly bills.[90] Costs can also be spread out over time as the system increases its service area (this is different from the PCS/cellular model where mobility requirements mean that broad coverage areas are much more important at system startup). Second, the cellular-like architecture allows the system to be built quickly and implemented in areas with the highest potential demand—sites can be added as needed. Finally, the cellular design makes the system very spectrum efficient because frequencies are reused in each cell. This reuse also increases the capacity of the systems, which is greater than MMDS but less than DBS. DBS, however, cannot match the interactive services provided by LMDS.

The rules and regulations that will govern LMDS are currently being determined at the FCC. The 28 GHz band being used by LMDS is currently allocated to FSS, and the satellite community would like to use the spectrum for a number of applications. To resolve the conflict, the FCC began

a proceeding to consider redesignating the band for shared use by LMDS and satellite providers. As part of that process, the FCC convened a Negotiated Rulemaking Committee (NRMC), consisting of representatives from all interested parties, to develop consensus on the technical rules for sharing the 28 GHz band. The NRMC, after weeks of difficult debate, was unable to agree on a sharing plan, and some participants believe that sharing is impossible. The FCC will now make its own decision based on the information provided through the NRMC and the normal rulemaking process. A decision is expected sometime in late 1995.

■ New Satellite Systems[91]

Spaceway

Hughes' Spaceway system plans to offer high-speed, high-quality data, video-telephony, and voice services to fixed sites, including homes with personal computers, telecommuters, and businesses. Hughes predicts applications—including medical image transfer; connecting to online services, such as America On-Line and Prodigy; as well as personal video-telephony—will drive demand for their service. Spaceway will also be capable of providing basic voice service to remote regions on a global basis.

The Spaceway system ultimately will employ constellations of satellites in each of six orbital locations.[92] Its design will utilize intersatellite links to provide global communications, much as Iridium plans. Spaceway will use Ka-band frequencies to deliver these services, and Hughes

---

[90] Alpert and De Sonne, op. cit., footnote 65. This is true, of course, only on a local basis; to achieve comparable national coverage to a satellite system, costs would be substantially higher.

[91] Loral Corp. announced in May 1995 that it would be providing services similar to the systems described. The CyberStar system would use a single satellite to provide high-speed data communications to support video conferencing, computer networking, distance learning, and other applications. The system is estimated to cost $500 million, and company officials plan to begin service to all 50 states in 1998. Jeff Cole, "Loral Plans a Data Service Using Satellites," *The Wall Street Journal*, May 3, 1995, p. B5.

[92] Hughes plans to launch its Spaceway system in two phases. The initial phase will consist of nine satellites, two in each of the four orbital planes and one interconnection satellite between North America and Asia-Pacific. Hughes anticipates operation of the first phase by the year 2000. The second phase will introduce two additional satellites in the four orbital planes and keep just one interconnectional satellite between Asia-Pacific and North America.

will offer interconnection to the PSTN through terrestrial operations control centers. Users will have to purchase their own sending and receiving equipment.

Teledesic

Teledesic's proposed system of low-Earth-orbit (LEO) satellites is singled out because it differs from the other "big" and "little" LEO systems in both scale and the services it hopes to deliver. In its original FCC application of March 21, 1994, Teledesic applied to provide fixed satellite services in the United States and abroad; in late 1994 the company amended the application to include mobile services provided outside the United States.[93] According to FCC regulations, Teledesic cannot provide mobile services in the United States in the bands it is currently seeking.

Teledesic plans to offer telephone, high-speed data, and video services in the United States to fixed users, and these same services to both fixed and mobile users outside of the United States. The company also expects to offer full interconnection to the PSTN with access to the various online services, such as Compuserve. Teledesic plans to market its network to other service providers in the United States, acting as a wholesaler of services rather than selling directly to the end-users.

The Teledesic system design calls for a constellation of 840 satellites in low earth orbit, roughly 621 miles above the earth. Satellites will use the internationally allocated Ka FSS band. The network will feature intersatellite links using fast packet switching technology, a ground component composed of end-user terminals, and gate-way terminals serving groups of users. Teledesic plans to offer a variety of end-user terminals to accommodate various user needs, with the upper end allowing bit rates of 1.2 Gbps.[94]

Issues and Implications

The primary issue facing the industry is the allocation of spectrum for the various service providers. Rules regarding what frequency bands the systems will use and how much bandwidth they will get are yet to be determined, and the FCC has delayed any decisions on operating rules until the spectrum issues are resolved.[95] LMDS proponents are fighting to gain full access to spectrum in the 28 GHz band, while various U.S. satellite service providers also want to use the band.[96] The National Aeronautics and Space Administration (NASA), for example, is currently using this band for its Advanced Communications Technology Satellite (ACTS) experiments, which it launched in 1993 at a reported cost of $1 billion. Other satellite providers, including Hughes and Teledesic, are developing satellite systems that would also use the band. Finally, several of the LEO satellite systems are supposed to use this band to provide mobile satellite services (see chapter 3). The technical, service, and other regulatory uncertainties that flow from this unknown outcome have seriously slowed LMDS development.

The FCC has indicated that it intends to allocate spectrum to all these potential services as part of its overall mission to encourage the development of competitive systems that will bring new services to the public as quickly as possible. It now appears, based on the conclusions of the

---

[93] Teledesic Corp., *Amendment of Application of Teledesic Corporation for Authority to Construct, Launch, and Operate a Low Earth Orbit Satellite System in the Domestic and International Fixed Satellite Service*, before the Federal Communications Commission, File No. 22-DSS-P/LA-94, Dec. 30, 1994.

[94] Teledesic Corp., *Application of Teledesic Corporation for a Low Earth Orbit Satellite System in the Domestic and International Fixed Satellite Service*, before the Federal Communications Commission, Washington, DC, Mar. 21, 1994.

[95] All other issues pertaining to establishment of LMDS will await development of frequency coordination and sharing criteria for space and terrestrial services and technical parameters for the service. Federal Communications Commission, op. cit., footnote 2.

[96] The band is currently allocated only to point-to-point services, but LMDS services have been operating in the band on a waiver of existing rules. For a summary discussion of the various satellite proposals, see Federal Communications Commission, op. cit., footnote 2.

NRMC, that sharing is not a viable option, given today's technology. LMDS systems would interfere with satellite systems and vice versa. Given these factors, the FCC has essentially three options: 1) divide the spectrum between the various systems; 2) move LMDS operators to another band; or 3) move satellite operations to another band. Existing use of the band by NASA's ACTS, and the varied uses of the band already proposed by satellite interests, appears to make the third option the least likely. Dividing the spectrum between the proponents probably could be done technically, but all future services likely would suffer from spectrum shortages and capacity constraints.[97] Systems may have to be reengineered. In addition, since the LMDS spectrum is to be auctioned, and the value of the licenses is closely tied to the amount to be offered, companies cannot plan auction strategies until such concerns are worked out.

Because only one operator is currently using the band—although there is more extensive experimental use—moving LMDS operations to another band also seems to be a viable option, and the 40 GHz band, which is now part of a reallocation proceeding at the FCC, is one possibility.[98] Other countries are developing systems similar to LMDS in these bands, although Latin American countries reportedly are experimenting in the 28 GHz band. If either group of users is forced to relocate to other frequencies, systems will have to be reengineered, increasing costs and time to market.[99]

Although these new systems have some way to go before they begin full-scale operation, they represent the best efforts to date to replicate the full range of services proposed for the NII. If such services eventually begin operation, they have the potential to meet the bandwidth requirements of many, if not most, users, and to extend the reach of high-bandwidth services to all areas of the country, regardless of location. The technical and regulatory hurdles that must be overcome, however, are substantial.

---

[97] This conclusion is premised on today's technology. Future developments in compression technology and spectrum-sharing methods could make band segmentation and spectrum-sharing possible.

[98] Federal Communications Commission, *Amendment of Parts 2 and 15 of the Commission's Rules To Permit Use of Radio Frequencies Above 40 GHz for New Radio Applications*, ET Docket 94-124, released Nov. 8, 1994.

[99] Although the band is currently lightly used for satellite services, the time and costs of relocating satellite operators is unknown, because many of the systems involved are still under development and costs are closely guarded. The record on whether such a move is feasible or practical for LMDS is similarly unclear, but generally seems to indicate that such a move is possible, if potentially costly. "Commenters Like FCC Proposals To Open Above-40 GHz Bands..." *Telecommunications Reports*, vol. 61, No. 5, Feb. 6, 1995, pp.19-21.

Part C:
Issues
and
Implications

As wireless technologies and systems are deployed, a host of technical, legal, and social issues will need to be addressed. Some will be amenable to marketplace solutions; others will not and will require a policy response. The policymaker's task is complicated because the implications of ubiquitous wireless information services are poorly understood due to uncertainties in technology, user demand, and regulation. The greatest unknown in the rollout of the National Information Infrastructure (NII) and wireless services is what type of implications the NII generally, and wireless technologies specifically, will have for people and businesses. In addition to the technical problems associated with the wide–scale use of radio–

based communications, there are also likely to be a host of administrative and social problems associated with wireless that must be addressed. Chapters 6 through 12 survey the various issues and implications associated with the widespread use of wireless technologies.

- Standards and Interoperability
- Regulation of Interconnection
- Zoning Regulations and Antenna Siting
- Wireless Technologies and Universal Service
- Privacy, Security, and Fraud
- Health Issues
- Electromagnetic Interference and Wireless Devices

Standards and Interoperability | 6

T oday's telecommunications and information infrastructure consists of many independently operated networks and systems, including the telephone network, cellular systems, cable television systems, broadcast radio and television networks, and various satellite and data communications systems. Some of these can connect and exchange information, while others cannot. The National Information Infrastructure (NII) initiative was designed to bring together these various networks—and a variety of new services—into a seamless network of networks that would allow users to send information across systems easily and efficiently.[1] In order for this to happen, different networks must be interconnected and interoperable. Standardized interfaces and connections will be critical in bringing this about and allowing the NII to develop. This chapter describes the technological requirements for building a seamless and integrated infrastructure that includes both wireless and wireline networks.

FINDINGS
- **A proliferation of wireless voice technologies and standards is leading to a patchwork of potentially incompatible systems that may make it more difficult for some mobile telephone users to "roam" outside their home system, or to easily switch service providers.** Until the early 1980s, the Federal Communications Commission (FCC) played an active

[1] U.S. Department of Commerce, Information Infrastructure Task Force, "The National Information Infrastructure: Agenda for Action," Sept. 15, 1993, p. 7.

role in standards-setting, specifying the technologies that licensees were required to use. For example, all cellular licensees were required to use a technology called Advanced Mobile Phone Service (AMPS). During the past decade, however, the FCC has largely withdrawn from standards-setting for wireless communications. Today, the FCC usually leaves it to industry to decide whether there will be a standard and which technology will be chosen as the standard. The FCC is following this approach for Personal Communications Services (PCS) and digital cellular air interface standards.

Various industry groups tried to settle on a single standard for PCS and digital cellular services, but were unable to reach agreement. Individual carriers are now choosing the technology standard/system they will deploy from among several contenders. Many cellular carriers have announced their technology choice, but most PCS carriers have not. Among the carriers that have announced which technology they will use, there is no consensus; two different cellular technologies will be deployed, and it appears that at least three different PCS technologies will be used. As a result, there is a danger that incompatible systems will make it more difficult or impossible for some users to make and receive calls as they travel from city to city.

The final impact on customers of the deployment of multiple standards is not yet clear. To some extent, carriers are coordinating their technology choices with carriers in other regions. Carriers are also acquiring additional licenses to enlarge their service areas, allowing them to provide expanded roaming without the need to coordinate technology choices with other carriers. At least three carriers plan to provide near-nationwide service to their customers. Consumers and businesses will have to shop carefully for the next generation of mobile services.

- **Technical challenges and incompatibilities may slow the integration of wireless systems**

into the NII, but pose no insurmountable obstacles. Wireless carriers have a clear incentive to ensure that their networks are interoperable with wireline networks because their customers want to be able to call users of the landline network, access the Internet, and download information from online services. If wireless users were unable to communicate with the much larger number of wireline users, wireless networks would not survive in the marketplace. However, there are technical challenges that must be overcome. Most of today's networking protocols were developed for wireline networks and do not work well in the more challenging radio environment. Because it is often necessary to use specialized protocols in wireless networks, interoperability cannot be achieved unless the wireless carrier makes provision for a translation between wireless and wireline protocols to occur at the interface.

▌ Options

In order to encourage the more orderly integration of wireless technologies into the NII, Congress has several broad options. One is for Congress to encourage the FCC to play a more active role in ensuring that cellular and PCS carriers do not deploy multiple technologies. However, the FCC's current approach allows considerable flexibility in the service offerings of carriers and spurs a continuing competition among technologies. It is consistent with the trend toward deregulation and competition that individual carriers be allowed to choose the technology that they believe will give them a competitive edge. Moreover, it would be difficult for the FCC to reverse course at this time. Manufacturers have invested in developing their systems and service providers have begun making their technology choices.

Congress may still wish to hold hearings and monitor the process closely. The technology selection process for digital cellular and PCS can be viewed as an experiment that will show whether interoperability can be achieved in the decentralized and competitive telecommunications industry of the future. Moreover, the federal government, as a user, may want to ensure that seamless nationwide services are available to support its activities. Through their procurement decisions, federal agencies may be able to encourage carriers to coordinate their technology choices and create a seamless network.

THE WIRELESS STANDARDS-SETTING PROCESS

In wireless networks, as in all networks, there are many pieces that must work together to ensure seamless communications. From the user's perspective, the most important connection is the radio link between the service provider's transmitter and the user. The user's equipment must be able to understand the radio signals transmitted by the service provider's network, and vice versa. For example, televisions must be able to decode the signals broadcast from television stations and cellular phones must be able to send signals through the air in a format that the cellular network understands.

It is sometimes sufficient that user equipment work with only one service provider's network. For example, wireless data users can obtain nationwide coverage from a single carrier—they may have no need for a modem that works with several carriers' networks. For other services, however, users may want to be able to access different networks with the same device. For example, cellular users can use cellular systems all over the country because their phone is interoperable with the visited carrier's network. Television sets can receive signals from different stations as well as from cable and satellite services.

One way to guarantee that user equipment will operate with several service providers' systems is to develop an industry standard—a common technology that all service providers agree to deploy. In the past, because of the value to consumers of interoperability, the FCC played a major role in ensuring that wireless network operators deployed a standard technology for the radio link between the network and the user.[2] However, a new model has emerged in which government leaves it to "the market" to decide whether a standard technology is required and what it will be.

■ The FCC Standards Process

Until the early 1980s, it was generally accepted that FCC involvement in wireless standards-setting was in the interest of the public and the industry.[3] The alternative—the deployment of different technologies by different service providers—was considered too chaotic, and there was a fear that technology development would be slowed if consumers were uncertain about which of many competing technologies to buy. Setting a standard was also thought to create the certainty that the industry needed before it would make the potentially large investment in manufacturing and deploying a new technology.[4] FCC-selected standard technologies are still used in many segments of the wireless communications industry, including radio, broadcast television, and cellular telephony.

In setting a standard, manufacturers would propose different technologies for adoption, and the FCC would compare them—often by means of a competition. The FCC would then select the "best" technology and designate it as the standard that had to be used by all service providers. Much of the actual work involved in testing and comparing the candidate systems was done by committees established by the FCC, but the ultimate decision was made by the FCC itself.

---

[2] U.S. Congress, Office of Technology Assessment, *Global Standards: Building Blocks for the Future,* OTA-TCT-512 (Washington, DC: U.S. Government Printing Office, March 1992).

[3] Mark J. Braun, *AM Stereo and the FCC* (Norwood, NJ: Ablex, 1994), p. 10.

[4] In its proceeding on high definition television (HDTV), the FCC observed that "establishing a standard may overcome audiences' and broadcasters' reluctance to invest in ATV technology by increasing the amount of programming available to audiences and ensuring that receivers will be compatible with broadcast signals." Federal Communications Commission, *Tentative Decision and Further Notice of Inquiry, Advanced Television Systems and Their Impact on the Existing Television Broadcast Service,* 3 FCC Rcd 6535 (1988).

An important benefit of FCC standards-setting was that the chosen technology had to be licensed on equitable terms to other manufacturers, allowing competition in the manufacture of equipment to develop. Furthermore, the standard created a large national market, attracted competitors, and created manufacturing economies of scale. This competition also led to significant innovations in equipment and services. For example, competition among the many manufacturers building to today's AMPS standard has led to cellular phones that are dramatically smaller and less expensive than those available when cellular service began.

Beginning in the early 1980s, however, the FCC has withdrawn from most standards-setting activities. The Commission will not, for example, select a standard for the next generation of cellular telephones or for PCS. This change in direction is part of the trend towards deregulation in the 1980s. One component of telecommunications deregulation is giving service providers the freedom to select the technology that they believe will attract the most customers. According to proponents of this approach, consumers benefit from having a range of technology choices and also benefit from service providers' flexibility to introduce new technologies as they become available.

But the FCC withdrawal from standards-setting is also the result of practical considerations. In many cases, it was difficult for the Commission to determine which of the contenders had developed the "best" technology. The process was often long and contentious because the contending technologies were often quite similar in their performance, making it difficult to assemble a rationale for the choice that was sufficiently solid to preempt lengthy litigation by the losing proponents. With AM stereo, the first technology for which the FCC left standards-setting to the market, the Commission had first tried unsuccessfully to set the standard itself.[5]

The notable exception to the FCC's new policy of leaving technology choices to the market is High Definition Television (HDTV), for which the FCC followed the old model of establishing an advisory committee and organizing a competition between proponent technologies. There are several reasons why the FCC may have decided to play a more active role with HDTV. First, there was great political pressure to develop a national champion technology that could compete with systems developed in Japan and Europe.[6] Second, there was no interest on the part of broadcasters in deploying anything other than a standard technology. Third, there were severe constraints on the freedom that designers could be allowed, given the need to squeeze the HDTV signal into unused channels. Fourth, pressure from Congress to avoid multiple standards may have played a role in preventing the FCC from leaving the choice to the market.[7] The HDTV standards process is described in more detail in chapter 5 and in box 6-1.

■ The Marketplace Approach

If the government does not set a standard, then the private sector decides whether there will be a standard and which technology will be chosen. The telecommunications industry often uses standards committees to determine a common technology. Committee-developed standards have many of the same advantages as a government-selected standard. For example, network operators all deploy the same technology, reducing confusion for consumers. In addition, as with government-selected standards, a committee-developed standard is not proprietary. All manufacturers are free to build to the specification contained in publicly available standards documents. Companies par-

---

[5] Braun, op. cit., footnote 3.

[6] See, for example, William D. Marbach et al., "Super Television," *BusinessWeek*, No. 3089, Jan. 30, 1989, pp. 56- 63.

[7] Braun, op. cit., footnote 3.

BOX 6-1: Interoperability of Video Services

One issue that has attracted considerable attention is the interoperability of video services. There is growing recognition that video is no longer synonymous with broadcast television, but is an important component of many industries and can be delivered by a variety of media, both wired and wireless. Interoperability, in this context, means the ability to use the same video equipment and standards for as many of these applications and media as possible.[1] This lowers the cost of equipment and makes it possible for users to receive information from a variety of sources.

Government plays a special role in ensuring video interoperability because the FCC is leading the selection of a High Definition Television (HDTV) standard. While the FCC will only mandate a standard for broadcasters, the Commission has recognized that the selection of an HDTV standard will have significant implications for other industries. Through the committee structure that was established by the FCC, computer, cable, and other industries have attempted to push the broadcasters toward a technology that takes their needs into account. In fact, the HDTV system under development is compatible with the international MPEG-2 framework,[2] which has been adopted by the new DBS services, the LECs for their new video dial-tone networks, and many players in the cable industry.

A remaining issue is whether the broadcast industry should be required to broadcast programming in interlace mode or progressive mode. Current televisions display in interlace mode, in which alternate lines of the screen are scanned in each frame. Progressive mode, in which each line is scanned every frame, is considered by many to be more suitable for display on computer monitors. The computer industry has campaigned for the inclusion of this capability in the terrestrial broadcast system for HDTV. While it now appears that the standard will permit progressive-scan broadcasts, the FCC still has to determine whether broadcasters will be required to use this capability (see chapter 5).

---

[1] For a discussion of video and the NII, see Technology Policy Working Group, Committee on Applications and Technology, Information Infrastructure Task Force, "Advanced Digital Video and the National Information Infrastructure," Feb. 15, 1995.

[2] MPEG is the Motion Picture Experts Group, an international standards committee that is developing standards for video compression.

SOURCE: Office of Technology Assessment, 1995.

---

ticipating in the activities of a standards committee usually have to agree to license, on reasonable and nondiscriminatory terms, any of their technology that is included in the standard.

However, the participants in industry standards committees do not always agree on which technology should be the standard. Manufacturers work to promote the technologies that they have developed, and campaign against those that other companies have developed. There is no mechanism for ensuring that agreement will be reached quickly or at all, and the process of developing and agreeing to a standard can often take many years. Moreover, the existence of a committee-developed standard does not guarantee that it will be the

only technology that is deployed. In contrast to an FCC-selected standard, a committee-developed standard is voluntary. Manufacturers may choose to sell, and service providers may choose to deploy, a different, proprietary technology. Finally, it is possible that different standards committees will produce contending standards.

If standards committees fail and multiple technologies are manufactured, the market still has an opportunity to create a de facto standard. Service providers and others who are responsible for choosing from among the contending technologies may eventually converge on a single technology. This is what happened with videocassette recorders, as the VHS technology gradually

---

BOX 6-2: Proprietary Interfaces and Lock-in

In the cellular systems deployed in the United States, the interface between cellular switches and base stations is proprietary. Switches only work with base stations built by the same manufacturer. If network operators choose to change suppliers for one component of their network, either the switch or base stations, they have to rebuild the whole system. This tying was of concern to the Department of Justice (DOJ) when it evaluated AT&T's acquisition of McCaw. Because cellular companies that used AT&T equipment were to a certain extent locked in, the DOJ felt there was a risk that AT&T could hurt a competing carrier by delaying development or delivery of equipment or software, and imposed several safeguards. [1]

By contrast, in the European cellular standard, GSM, the switch to base station interface is not proprietary—base stations and switches from different manufacturers can work together. In fact, the use of *open* interfaces is a basic principle of GSM. The GSM standards committee unbundled all important network functions and defined open interfaces between them. Because of the number of interfaces involved, the GSM specification is over 5,000 pages long.

---

[1] U.S. Department of Justice, "Proposed Final Judgment and Competitive Impact Statement; *United States of America* v. *AT&T Corp. and McCaw Cellular Communications, Inc.,*" notice, *Federal Register* 59(165):44158, Aug. 26, 1994 at 44168, 44172.

SOURCE: Office of Technology Assessment, 1995.

---

took over the market.[8] Although in the early stages of the marketplace process, limited interoperability and customer confusion may slow the convergence to a single standard; because interoperability is so valuable to users, situations in which multiple incompatible technologies are marketed are often transient.

In addition, it is possible that the de facto standard will be a proprietary technology, limiting competition among manufacturers and keeping equipment prices high. Proprietary interface standards enable manufacturers to lock in future sales in an adjacent market: if an interface is proprietary, equipment can often connect only to other equipment made by the same manufacturer. For example, the subscriber equipment that works with the new high-powered DBS service is currently only available from one manufacturer[9] and cellular base stations usually work only with

switches made by the same manufacturer (see box 6-2).

Europe and Japan have not followed the new U.S. model of standards-setting. They also rely on standards committees, but their governments do not permit the deployment of multiple technologies. This creates an incentive for committees to come to agreement. In Europe, strong centralized standards-setting is viewed as essential to knitting together disparate national networks. In the first generation of analog cellular service, different technologies were deployed in different parts of Europe, and some technologies were deployed in only one country. It was impossible for a user to roam outside their home country and difficult to achieve economies of scale in the manufacture of cellular phones. To avoid a recurrence of this problem, the European Union launched a coordinated effort to develop a European standard for

---

[8] For an economic analysis of this phenomenon, see Michael L. Katz and Carl Shapiro, "Technology Adoption in the Presence of Network Externalities," *Journal of Political Economy,* vol. 94, No. 4, 1994, pp. 822-841.

[9] After the first million units are sold, however, a second company will begin selling equipment.

next-generation digital cellular. This system, the Global System for Mobile Communications (GSM), is now being deployed across Europe and in many other countries.

CELLULAR AND PERSONAL COMMUNICATIONS SERVICE STANDARDS

The development of digital cellular and PCS technologies is a prime example of how the marketplace tries to set standards. It shows the tension between giving competing service providers the freedom to choose their own technologies and the desire for nationwide interoperability. The advantage of the marketplace approach is that it allows carriers considerable flexibility in choosing the services they offer. Moreover, by fostering a competition among technologies, the less rigid U.S. standards-development process may ultimately lead to a better technology choice than the European approach, which is now locked in to a single technology, GSM. But there is a real danger that different technologies will be deployed in different cities, limiting the possibilities for seamless nationwide roaming. Users may find that they are unable to use their phones when away from their home city, contrary to the vision of "anytime, anywhere" mobile telephone service.

The problem is, in fact, a combination of "no standards" and the FCC's decision to divide the nation into many license areas. In developing the cellular licensing plan, for example, the FCC created 734 cellular license areas—with two licensees per area. Although some cellular carriers now operate across several areas, the wireless industry remains fragmented. With so many companies, establishing seamless nationwide service requires that many carriers across the nation deploy the same technology. When cellular service began in the early 1980s, the FCC solved this coordination problem by requiring all carriers to use the AMPS standard. For the next generation of digital cellular, however, the FCC did not specify a standard, preferring to let industry committees settle the issue. They could not, and two standards—TDMA and CDMA—will be deployed (see below).

In the PCS industry the situation is much the same. The licensing plan for PCS established two licenses in each of 51 Major Trading Areas (MTAs) and four licenses in each of 493 Basic Trading Areas (BTAs)—MTAs and BTAs overlap, meaning that each local area could have up to six PCS carriers. PCS industry committees also could not agree on a standard, and several technologies are being developed. In both digital cellular and PCS, individual companies will have to decide which technology is best for them. Because each carrier has different business priorities, different companies are likely to initially select different standards, making the coordination problem potentially quite formidable.

■ Multiple Air Interface Standards

Today's cellular phones use AMPS for the *air interface*—the radio link between the phone and the base station. Two incompatible digital air interface technologies have been proposed as a replacement for AMPS, one based on Time Division Multiple Access (TDMA) and the other based on Code Division Multiple Access (CDMA) (see box 3-3). In the late 1980s and early 1990s, the cellular industry attempted to choose between the two technologies but was unable to reach a consensus (see box 6-3). As a result, some carriers are deploying the TDMA system, while others will deploy the CDMA system.

PCS operators have also been unable to agree on a standard. A standards committee established to determine which air interface technology would be used in the PCS band only managed to reduce the number of contenders from 16 to seven (see box 6-4). Two of the proposed PCS technologies are based on the cellular CDMA and TDMA systems, but modified to work at the higher PCS-band frequencies. A third PCS technology is based on the European GSM cellular system, but modified to work at the U.S. PCS frequencies and renamed DCS-1900. The four other technologies were developed specifically for the new PCS band.

While the digital cellular and PCS standards committees were unable to reach agreement, they

---

BOX 6-3: Cellular Standards

The development of digital cellular standards is the responsibility of a committee of the Telecommunications Industry Association (TIA) called TR45. In the late 1980s, it appeared that the industry would be able to agree on a single digital cellular system, based on a technology called Time Division Multiple Access (TDMA).[1] But in 1990, Qualcomm, a company based in San Diego, CA, proposed that a second technology, called Code Division Multiple Access (CDMA), be used instead.[2] This proposal was supported by some cellular carriers, and, in 1992, the cellular industry's trade association, the Cellular Telecommunications Industry Association, abandoned the idea of selecting a single technology as a U.S. standard and asked that TR45 establish a new subcommittee to work on a CDMA system.[3]

TR45 has developed two U.S. "standards," the TDMA-based system, referred to as IS-54, and the CDMA-based system, referred to as IS-95. These are standards in the sense that TR45 has written publicly available specifications that any manufacturer can use to build a conforming system. However, neither IS-54 nor IS-95 is a national standard in the way that the current analog cellular system, the Advanced Mobile Phone Service (AMPS), is a standard: a single specification that all manufacturers and cellular service providers have agreed to adhere to.

---

[1] Steven Titch, "The Digital Dilemma," *Telephony,* Oct. 14, 1991, pp. 33-36.
[2] Steven Titch and Charles F. Mason, "Digital Cellular: What Now?" *Telephony,* Feb. 10, 1992, pp. 30-36.
[3] Charles F. Mason, "CTIA Approves CDMA Standards Setting," *Telephony,* June 15, 1992, p. 3.

SOURCE: Office of Technology Assessment, 1995.

---

will publish specifications for each of the proposed systems. Manufacturers will be able to use these specifications to build any of the proposed systems, although they may have to obtain licenses to any patented technology that the systems incorporate. It does not appear that manufacturers will try to sell proprietary equipment that is not based on one of the published air interface specifications. Carriers would be unlikely to choose a proprietary air interface technology because they would not have as wide a choice of manufacturers and the future development of their technology would be in the hands of a single company.

In part, the wireless industry was unable to agree on a single technology for either the cellular or PCS bands because it was difficult to assess the strengths and weaknesses of the newly developed systems before large-scale deployment. System proponents argued at length about the relative performance and technical feasibility of the proposed technologies. But these arguments were based largely on theoretical calculations, simulations, and small-scale tests. None of the proposed systems had been tested with real world traffic at the time that the standards committees were deliberating. There was no conclusive way to evaluate the claims made by system proponents.

Another significant cause of the industry's failure to agree on a single technology was the competitive nature of the wireless equipment industry. Standards-setting requires compromise; however, manufacturers who had invested in the development of prototype systems and owned intellectual property rights to the technologies they had developed tried to prevent rival technologies from being chosen as a national standard. Although cellular and PCS service providers played a less active role in the standards committees, they also differed in their perception of the features that they thought their customers would value and in their evaluation of the contending technologies.

Because the standards committees were unable to reach consensus, some analysts have suggested that the FCC should have acted as an arbiter and selected a standard. However, it is doubtful that an

---

BOX 6-4: Personal Communications Services (PCS) Standards

The standards controversies in the 2 gigahertz PCS band are even more complex than those in the cellular band. At first, two different committees, a new Telecommunications Industry Association (TIA) committee known as TR46, and T1P1, sponsored by the Alliance of Telecommunications Industry Solutions (ATIS), were working on PCS standards. ATIS historically has worked on wireline standards for the public switched telephone network (PSTN), not wireless standards. Its involvement in the development of PCS standards reflects the fact that PCS was initially viewed as a *low tier* service that would be integrated to a greater extent with the PSTN than had been the case for cellular. In 1992, the two committees recognized the overlap in their work and formed a joint committee, the "Joint Technical Committee on Wireless Access" (JTC).

A total of 16 technologies were proposed to the JTC for consideration as a U.S. PCS standard. The committee was only able to reduce the number of contenders to seven; subcommittees are writing standards for each of these technologies.[1] One of the main reasons that there are so many more contenders in the 2 GHz band is that there are different conceptions of what this band is to be used for. Originally, the PCS band was thought to be for a new kind of wireless technology that would be different from cellular. Compared to cellular, PCS was supposed to be simpler, use smaller cells and lower power handsets, and be aimed more at pedestrian than vehicular use. However, many carriers have since come to believe that the PCS band will be used in much the same way as the cellular band. The diversity of views has made it even more difficult to agree on a single standard.[2]

---

[1] Charles I. Cook, "Development of Air Interface Standards for PCS," *IEEE Personal Communications,* vol. 1, No. 4, Fourth Quarter 1994, p. 30.

[2] "The ideal goal of the [committee] would be to arrive at a single air interface that meets the needs of everyone. However, the wide diversity of potential service providers has caused this to become an unrealistic goal." Ibid., p. 31.

SOURCE: Office of Technology Assessment, 1995.

FCC-led competition between the proposed systems could have resolved the issue sooner, if at all. The same technological uncertainties and competitive factors that made it impossible for the industry standards committees to select a single system would also have made it difficult for the FCC.

It is now too late for the FCC to take any action that could force agreement on a single digital cellular or PCS standard. Manufacturers have begun to build equipment, and service providers have begun to make their technology choices. If government is going to be involved in standards-setting, it cannot easily step in at the last minute; instead, it must act early in the process to establish the expectation that a single technology will be chosen. In Europe, the development of GSM followed from a clear objective to create a single standard that would tie the formerly incompatible national cellular networks together into a continent-wide system. Furthermore, the GSM project began at an early stage in the development of digital cellular, before manufacturers had a vested interest in any particular approach.

■ Mobility Management Systems

In addition to the problem of incompatible air interfaces, a second standards problem—incompatible *mobility management* technology—may be a greater challenge. Cellular and PCS networks use mobility management technology to connect systems and exchange information about roamers. For example, a cellular system can send messages to a roamer's home system, informing it of its customer's current location so that any incoming calls can be forwarded. The switches and other network equipment that use a particular air interface also come with a particular mobility management technology—when carriers choose their air interface technology, they are also choosing a mobility management technology.

Fortunately, all of the cellular air interface technologies and most of the PCS-band air interface technologies are usually sold with switches that use the same mobility management technology, known as IS-41. Users could roam between IS-41-based systems as long as they had multimode phones to overcome any air interface incompatibilities. However, the European DCS-1900 system is sold with a mobility management system that is not compatible with IS-41. Therefore, users could not roam between DCS-1900 systems and IS-41-based systems, even though it is possible to build a multimode phone that incorporates both the DCS-1900 air interface and a second air interface. This may dissuade some carriers from choosing DCS-1900, although some manufacturers are trying to make it possible for the two mobility management systems to work together.

■ Carrier Technology Choices and Interoperability

Because the industry has failed to agree on an air interface standard, carriers have been evaluating the contending systems and trying to determine which technology to deploy. There are significant risks associated with their technology choice because the construction of a digital cellular or PCS network requires the investment of millions of dollars and the wrong choice could leave a carrier at a competitive disadvantage. Among the factors of concern to carriers are coverage, capacity, and voice quality. The most important consideration is the per-user cost of building and operating the network, because this factor most directly affects a carrier's ability to compete with its rivals.

Carriers are also concerned with the technological maturity of the contending systems. For example, some cellular carriers have chosen TDMA because it is commercially available and they have an immediate need for the greater system capacity afforded by digital technology. Other carriers will wait for CDMA, which is still being tested. The maturity of the technology is given special weight by the new PCS entrants because delays caused by unforeseen problems with a new technology would give cellular carriers even more of a head start in the market. One of the selling points of the DCS-1900 system is that its GSM and DCS-1800 cousins have been in commercial service in Europe for several years. American Personal Communications, one of the "pioneer's preference" winners, has selected DCS-1900 for this reason.

Because of uncertainties about the contending systems' capabilities and because of differences in their business plans, different carriers are choosing different technologies. Most cellular carriers have announced their technology choices; Bell Atlantic Mobile, NYNEX Mobile, and AirTouch plan to deploy CDMA, while AT&T (formerly McCaw) and Southwestern Bell Mobile Systems are deploying TDMA. Among the carriers with PCS licenses, most have not yet announced their technology choices. However, it appears that two technologies, the U.S. CDMA system and the European DCS-1900 system, are attracting the most interest.

Because there is no clear favorite among the technologies at this time, there is a risk that a patchwork of technologies will be deployed, making it difficult for users to roam in all cities. The impact of multiple standards on roaming depends not on how many technologies are deployed, but the pattern in which they are deployed. Some major players in the wireless industry intend to build networks with near-nationwide coverage through acquisitions of other carriers, mergers, and alliances (see chapter 3). Other carriers are working to coordinate their technology choice with carriers in neighboring regions. These companies or alliances could then guarantee seamless roaming by deploying a single technology throughout their license areas. In addition, the technology choices of these major players will influence the choices of smaller carriers and thereby determine which of the contending technologies will survive in the marketplace.

Technological Solutions to Interoperability

To a certain extent, there may be technological solutions to the multiple-standard problem.[10] It may be possible to use *multimode* phones that work with more than one type of air interface. However, a multimode phone built with today's technology requires additional circuitry that increases the cost and weight of the phone. In the future, it may be possible to minimize this penalty by implementing most of the phone's functions in software.[11] This approach is the focus of research sponsered by the Advanced Research Projects Agency,[12] but the required signal processing technologies are still several years away from commercialization.

Dual-mode phones will indirectly allow interoperability between cellular companies that deploy different digital technologies. These phones will not be TDMA/CDMA phones; instead, they will incorporate AMPS and one of the digital technologies. The AMPS capability is being included with all digital phones mainly because it allows users to make calls in areas where digital has not yet been deployed—all cellular carriers will continue to support AMPS until most of their customers own digital phones. However, users who roam into an area that does not employ the digital technology the user has will be able to fall back on AMPS to complete their calls. Falling back to analog incurs a significant performance penalty; when operating in analog mode, phones deplete their batteries at least twice as quickly. In addition, the continued use of AMPS to support roamers could slow the transition to more efficient all-digital networks.

Because there is no existing common technology in the PCS band, PCS carriers would have to either use phones that incorporate multiple PCS technologies or *dual-band* phones that incorporate both a PCS air interface and an analog or digital cellular air interface. These dual- or multimode phones would be more expensive to design and build than a single-mode phone, and would take longer to develop. The added cost would depend in part on the degree of similarity between the air interface technologies combined in the phone. It would also depend on manufacturing volumes; the price of a multimode phone would only be reasonable if it could be sold in large quantities. Manufacturers are trying to determine which air interface combinations the market will demand, if any.

Coordinated Technology Choices

Although multimode technology may provide a partial solution to the multiple-standard problem, several carriers are taking more direct action to ensure that roaming is possible. They recognize that nationwide roaming is of value to users and that they will have a competitive advantage if they can offer nationwide roaming. They are working to coordinate their technology choices with carriers in other regions. In several cases, a group of carriers has established an alliance whose members agree to deploy a common technology.[13] For example, US West New Vector, AirTouch, Bell Atlantic Mobile, and NYNEX Mobile have formed an alliance that is committed to CDMA.

Carriers are also working to expand the area that they are licensed to serve, reducing the need to

---

10 "On the other hand, the next generation of mobile radio may well be 'computers with an RF front end' with the capability of performing many signal processing functions. Perhaps different format translations and emulations will be performed by the mobile unit itself so that it can operate in different modes. Perhaps the mobile unit will be able to be updated to perform new capabilities in the same way that computers today are updated with new software, expansion boards, and the like." Federal Communications Commission, *Notice of Inquiry, Advanced Technologies for the Public Radio Services*, FCC Gen. Docket No. 88-441, Dec. 11, 1989.

11 Joe Mitola, "Software Radios," *IEEE Communications,* vol. 33, No. 5, May 1995, p. 24.

12 Robert J. Bonometti, "Integration of Space and Terrestrial PCS in the Information Infrastructure," *Proceedings of the 1994 Third Annual International Conference on Universal Personal Communications* (Piscataway, NJ: IEEE, 1994), p. 455.

13 Gutam Naik, "Alliance Planned for National Wireless System," *The Wall Street Journal,* Nov. 7, 1994, p. A3.

coordinate with other carriers. One strategy is to acquire other carriers; there is a clear trend toward consolidation in the wireless industry. Another strategy for building nationwide coverage was afforded by the FCC's design of the PCS auctions. The licenses in all regions are being auctioned simultaneously, allowing a carrier to bid for contiguous license areas. In theory, it would be possible to assemble a nationwide system by winning all of the available licenses. While this did not occur in the first round of auctions, several companies assembled licenses covering very large areas. For example, one consortium won licenses with a total of 182.5 million potential customers.[14]

Some of the biggest winners in the first round of PCS auctions were cellular companies who will use their new PCS spectrum to fill in the gaps between their cellular properties.[15] In order to knit their cellular and PCS licenses together into a nationwide service, these companies' customers will have to use dual-band phones that work in both the 800 MHz cellular band and the 2 GHz PCS band. It is possible that these will be *dual-band, dual-mode* phones that would use a different air interface technology depending on whether they were operating in the PCS or cellular band. But phones that used the same air interface technology in both bands would be simpler and less expensive. Two of the proposed PCS technologies are simply *upbanded* versions of the cellular CDMA and TDMA systems, facilitating this dual-band strategy. To some extent, the technologies deployed in the PCS band will be determined by the technologies deployed in the cellular band. For example, the alliance of US West New Vector, AirTouch, Bell Atlantic Mobile, and NYNEX Mobile plans to use CDMA in both its cellular and PCS properties.

Alliances and consolidation represent the industry's attempt to overcome the FCC's decision to divide the wireless service map into a large number of license areas. Almost every other country grants licenses on a nationwide basis to begin with, guaranteeing nationwide roaming. When there are nationwide networks, the deployment of multiple technologies would only be of concern to users if they decided to switch carriers, in which case they might have to buy a new phone. The lack of a national standard would not limit roaming. While the FCC has withdrawn from standards setting, it should be recognized that its decisions about the structure of the wireless industry critically affect the pattern in which technologies are deployed.

Narrowing the Choices

The technology choices of the larger PCS carriers and alliances will begin the process of reducing the number of contending PCS technologies from seven to, most likely, two or three. The larger carriers will be looking for partners in the regions where they do not have roaming agreements. As a result, many mid-sized and smaller operators will follow the lead of the larger carriers and alliances. For example, if a high percentage of a small operator's customers were roamers from a large city, it would likely follow the lead of the larger operator. The technologies that receive only limited initial support may not survive long in the marketplace. Manufacturers would be less likely to build to these standards, and the price of the phones would not benefit from economies of scale.

Over time, the number of incompatible air interface technologies in the market is likely to be further reduced. Although it is costly to do so, carriers may switch technologies as more is learned about the performance of the competing systems or about the choices of competitors and alliance partners. Carriers may choose to deploy a more mature technology today, knowing that in a few years they will exchange it for a better technology. For example, some carriers believe that CDMA may prove to be a better technology in the long

---

[14] "Broadband PCS Auction Nets $7.7 Billion," *Telecommunications Reports,* vol. 61, No. 11, Mar. 20, 1995, p. 3.

[15] Ibid.

run, but that TDMA is the best technology for solving immediate capacity problems. Some manufacturers support this strategy by designing their products so that much of the equipment purchased for a TDMA rollout can later be used for CDMA.

■ Effect of PCS and Cellular Standards on Trade

One side effect of the U.S. approach to standards-setting is that it has left the United States without a national champion technology to sell in other countries. The worldwide market for cellular telephone equipment is large, especially when the possibilities for wireless local loop applications are considered. Because the battles over standards in the United States have slowed the commercialization of U.S. digital cellular, more and more countries are adopting GSM. GSM has a significant head start, with 1.8 million phones in service worldwide in mid-1994 compared to 100,000 U.S. digital phones.[16] It has been adopted by 78 network operators in 59 countries.[17] Outside of the European Union, GSM has been selected by carriers in China, Australia, New Zealand, Russia, and Hong Kong, for example.[18]

The openness of the U.S. technology selection process creates other imbalances. Because Europe and Japan have specified the technology that all licensees must use, these markets are closed to the U.S.-developed technologies. For example, even if the U.S. CDMA system does turn out to offer significant advantages, service providers in Europe would not be able to adopt it in place of GSM. At the same time, however, the technology-neutral U.S. licensing process allows PCS carriers to adopt the European DCS-1900 technology. The real effect on U.S. manufacturers is unclear, however. The largest suppliers of GSM equipment are all European companies,[19] but U.S. companies build GSM and DCS-1900 equipment and are selling it around the world.

INTEROPERABILITY OF WIRELESS AND WIRELINE NETWORKS

The first section of this chapter discussed the radio link standards that enable interoperability between a user's phone or other wireless device and a service provider's network. But it is equally important that different networks be interoperable with each other, allowing their users to exchange information with users of other networks. The future NII is often envisioned as a network of networks—a diverse collection of networks that are independently operated but still interoperable. Therefore, it is necessary that the wide variety of wireless networks currently being deployed—PCS, cellular, wireless data networks, and others—be interoperable with wireline networks as well as with each other.

Although there are technical challenges that need to be overcome to ensure wireless-wireline interoperability, it is unlikely that the infrastructure will be segmented into separate wireless and wireline worlds. There are clear incentives for the operators of wireless networks to ensure that there is interoperability between wireless and wireline networks. Wireless carriers know that their customers want to be able to talk to wireline users of the public switched network, exchange e-mail with users of the Internet, and retrieve information from their companies' computer networks. Wireless networks would not survive in the marketplace if their users were limited to isolated islands, unable to communicate with the far larger number of wireline-connected users.

---

[16] Gail Edmondson, "Wireless Terriers," *BusinessWeek,* May 23, 1994.

[17] Mark Newman, "GSM Takes On the World," *CommunicationsWeek International,* Issue 133, Oct. 24, 1994, p. 1.

[18] "GSM Gold Mine," table in *CommunicationsWeek International,* Issue 132, Oct. 10, 1994, p. 26.

[19] Ibid. A table lists the four largest suppliers of GSM equipment as Ericsson, Siemens, Nokia, and Alcatel. Motorola and AT&T appear on the list, but sales volumes are considerably smaller. For example, according to the table, AT&T has sold four GSM switches, Ericsson 33, Siemens 30, Nokia 15, and Alcatel 14.

Wireless-wireline interoperability also allows for communication between disparate wireless networks. Because most wireless networks act as an extension to a larger wireline network, the wireline network can serve as a common core through which incompatible wireless networks exchange traffic. For voice or fax traffic, this common core would be the public switched telephone network; for data, it might be the Internet. For example, the fact that both CDMA and TDMA cellular networks are designed to interoperate with the public switched telephone network (PSTN) will also allow them to interoperate with each other. The wireline standards can act as a common language, allowing users of incompatible wireless networks to communicate.

■ Translation of Protocols

Despite the incentives for wireline-wireless interoperability, it is not always easily or inexpensively achieved. It would be easier to achieve if wireless and wireline systems could use the same protocols—the rules and formats that govern how communication occurs. But many wireline protocols do not work well over wireless links, because wireless links are noisier, have less bandwidth, and may have a long transmission delay. Therefore, it is often necessary to use specialized wireless protocols.[20] Because these protocols are incompatible with their wireline counterparts, interoperability requires that there be some type of translation or "gateway" at the interface between wireless and wireline networks.

For example, interconnection of digital cellular networks to the public switched network requires that the voice signals be translated from the wireless to the wireline format—wireless networks have to use a much lower bit rate because of the limited bandwidth available. Cellular carriers also need to install "modem pools" at their switches to translate between ordinary wireline modem standards and special modem protocols that work better over a noisy wireless link. Operators of wireless packet data networks need to translate the specialized protocols that they use into the protocols used in the Internet or in corporate data networks. E-mail may have to be translated from a wireline format into the format used by paging networks, permitting instantaneous delivery of e-mail from wireline users to alphanumeric pagers or laptop computers equipped with paging receivers.

Because different types of services require separate translation schemes, it is often the case that services that have the most commercial value are supported first. For example, the new digital cellular services will support the interoperability of voice services from the beginning because voice is considered to be the core service. But interoperability of fax and data services will not be supported until the appropriate *interworking* equipment is installed. More specialized services, such as secure voice services, which have only a limited market, may have to wait even longer. Where these services are essential to the mission of a government agency, the agency will have to get involved with industry groups and standards committees to ensure that the services are available.

Most of the cost of ensuring interoperability falls on wireless network operators because wireless networks are newer and have fewer users. For the most part, wireline protocols have been developed without regard to the needs of wireless. Satellite operators, in particular, have complained that wireline protocols were developed and standardized based on assumptions about short transmission delays that do not hold true for satellite services.[21] Many of the technical issues of integrating wireless access with Asynchronous Trans-

---

[20] John A. Kilpatrick and Mobeen Khan, "MOBITEX and Mobile Data Standards," *IEEE Communications*, vol. 33, No. 3, March 1995, p. 96.

[21] It takes about half a second for a signal transmitted to a geosynchronous satellite to reach its destination.

fer Mode (ATM) networks, which are expected to play a key role in the future wireline infrastructure, still have to be addressed.[22] In the future, however, the increasing interest in wireless may mean that network designers will use a more integrated approach that takes both wireless and wireline into account. Government can reinforce this direction by supporting testbeds and demonstration projects that include both wireless and wireline components.

∎ Wireline Networks and Mobility

Another challenge to integrating wireless and wireline networks is that existing wireline networks, such as the PSTN and the Internet, do not recognize that users can be mobile. They associate a telephone number, for example, with a fixed location. As a result, wireless operators have had to develop their own specialized call routing procedures. For example, the cellular industry's IS-41 mobility management system, used to forward calls to a user's cellular phone as they travel, operates separately from the wireline network's call-routing mechanism.

The lack of integration between wireless and wireline call routing mechanisms causes inefficiencies.[23] With IS-41, for example, calls are first delivered to the user's home system and then forwarded to the city where the user is currently located. In fact, the called user could be in the next room, but the call would still be routed all the way to the user's home city and then back again, requiring two long distance calls and turning an inexpensive call into a very expensive call. More efficient call routing would send the call directly to the user's current location. For this to be possible, however, the LEC or long distance carrier at the originating end of the call would have to have to be able to recognize that the number belonged to a mobile user, look up the user's current location in a database, and then route the call appropriately.

As more and more users become mobile, wireline networks will have to begin to recognize the concept of mobility. The first step toward incorporating mobility concepts into the landline network is now being taken with the assignment of special "500" numbers. If this nongeographic prefix is used in place of an area code (e.g., (500) 123-4567), it indicates to wireline switches that the user could be mobile. Wireline carriers are currently using "500" numbers for an advanced call-forwarding service. Customers use a touch-tone phone to update a database that records the phone number to which calls should be forwarded. However, with current technology, it is not possible for a wireless network to automatically update this location database as a customer moves from city to city. True integration will require that the wireless industry's mobility management technology work with the wireline industry "Intelligent Network" call routing technology, which is only now becoming possible.[24] It will also require business arrangements that permit wireline and wireless carriers to have access to each other's location databases (see chapter 7).

---

[22] "News from JSAC," *IEEE Communications,* vol. 33, No. 5, May 1995, p. 12.

[23] See discussion in National Regulatory Research Institute (NRRI), *Competition and Interconnection: The Case of Personal Communications Services,* July 1994, Columbus, Ohio, pp. 20-24.

[24] Brenda E. Edwards and Paul B. Passero, "Testing PCS in Pittsburgh," *Bellcore Exchange,* September 1993, p. 14.

Regulation of Interconnection | 7

T he nation's telecommunications industry consists of many independently operated networks. In order to create a seamless infrastructure, these networks must interconnect. The Federal Communications Commission (FCC) has long required local exchange carriers (LECs—the local telephone companies) to interconnect with cellular carriers, making it possible for cellular and wireline users to call each other. But as new wireless carriers—Personal Communications Services (PCS), Enhanced Specialized Mobile Radio (ESMR), and mobile satellite—enter the market, and as the wireless industry evolves from a niche player into a central component of the infrastructure, the interconnection rules will also have to evolve.

FINDINGS

- **Ensuring wireless carriers fair and affordable interconnection to the public switched telephone network (PSTN) will be critical in determining what role they will play in the National Information Infrastructure (NII).** Wireless carriers pay interconnection charges for every minute of traffic they send to the LEC, and often these charges are above the cost the LEC incurs in providing interconnection. Interconnection charges are an important component of wireless carriers' cost structure. As new digital technology reduces the per-user cost of operating a wireless network, interconnection charges will assume even greater significance. Elevated interconnection charges would increase the price and reduce demand for both mobile and fixed wireless services. Interconnection charges priced too far above cost could keep mobile communication prices artificially high and stunt its potential growth. The level of interconnection charges could even determine

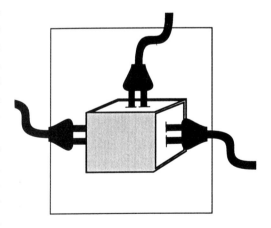

whether wireless carriers will be able to effectively compete in the local telephone service market, where bills have to remain affordable even if customers use their phones for hundreds of minutes per month.

Rethinking interconnection charges, however, is a complex problem. Under current law, the states have primary jurisdiction over interconnection charges and the process by which they are determined. State regulators have kept the price of wireless interconnection above cost in order to provide the LEC with additional revenues that support its universal service obligations. Before wireless interconnection charges can be reduced, policymakers would have to determine that universal service would not be affected if the contribution from wireless carriers were reduced. Alternatively, they would have to find a new mechanism to further universal service goals that did not disadvantage wireless carriers or other new competitors to the LECs.

- **To ensure that wireless systems can achieve their full potential as a mass-market service, regulators and policymakers may need to play a more active role in determining the cost of wireless carriers' interconnection to the LEC.** Congress has the option to establish guidelines for the states to follow in setting interconnection charges. Both S. 652 and H.R. 1555, the telecommunications bills currently being debated in Congress, provide a mechanism for carriers, including wireless, to ask state regulators to intervene in interconnection disputes. Congress could also expand the FCC's jurisdiction over mobile radio services by giving it more power to determine interconnection charges.

Part of the problem in ensuring fair and affordable rates is the way in which interconnection charges are set. In most states, the cost of interconnection is based on contracts negotiated between the wireless carrier and the LEC. In negotiating these contracts, the LEC has considerable bargaining power because it has a near-monopoly in the provision of wireline telephone service. In addition, wireless systems depend critically on the LEC to complete the vast majority of calls made to and from wireless phones—wireless-to-wireless calls on the same system account for less than 2 percent of all wireless traffic.[1]

The FCC does not permit LECs to discriminate among wireless carriers in the price of interconnection or other terms of interconnection agreements. No wireless carrier should be disadvantaged because it is paying higher interconnection rates than its competitors. However, the new entrants in the wireless marketplace, especially smaller PCS carriers, fear that the established cellular carriers are more familiar with the process of negotiating interconnection agreements and will be able to obtain better terms, despite the requirement that the LECs not discriminate unreasonably.

One barrier to determining whether there has been discrimination is that not all states require that interconnection agreements be made public. It is difficult to enforce the nondiscrimination requirement without knowing the terms under which competing carriers are obtaining interconnection. **Regulators may have to require that interconnection agreements be made available for public inspection.** A public filing requirement would improve the bargaining position of new entrants by giving them access to the agreements that cellular carriers have been able to negotiate. Both S. 652 and H.R. 1555 would require that interconnection agreements between the LECs and other carriers, including wireless, be filed with state regulators and made public.

---

[1] 80 percent of all mobile calls are wireless to land line, 18 percent are land line to wireless, and 2 percent are wireless to wireless. The 2 percent, however includes wireless to wireless calls on the same system as well as to other cellular systems. Tim Rich, CTIA, personal communication, June 5, 1995.

Chapter 7   Regulation of Interconnection I 187

- A key issue is whether wireless carriers should be required to provide their customers with *equal access* to long-distance services—allowing customers to choose a preferred long-distance carrier as they do now with their wireline telephone. Different rules govern wireless carriers' provision of long-distance service, depending on whether or not they are subject to equal access requirements. As a result, **some wireless carriers may be at a competitive disadvantage not only in providing long-distance services, but also in providing a wider variety of services and pricing plans.** Currently, only the wireless affiliates of AT&T and the Regional Bell Operating Companies (RBOCs) are subject to equal access rules. All other wireless carriers do not have to give their customers a choice of long-distance carrier, and are permitted to sell a bundled package of local and long-distance service. However, the FCC has recently launched a proceeding to determine if all wireless carriers should be subject to equal access rules.

  The entry of new competitors into the wireless market calls into question the need for equal access rules. These rules were first developed in the wireline context because the LEC could use its local monopoly to also dominate the long-distance market. The cellular affiliates of the RBOCs and AT&T are subject to equal access rules in part because competition in the cellular industry was also limited, with only two carriers in each market. With the entry of ESMR and PCS carriers, however, the market power of any one wireless carrier will be substantially reduced. Both S. 652 and H.R.

1555 would allow wireless carriers to provide a weaker form of equal access than the wireline LECs.

LEC INTERCONNECTION OBLIGATIONS

In order to guarantee that wireless users are linked to the PSTN, the FCC mandates that LECs interconnect with all wireless carriers (see box 7-1). Until recently, regulators were concerned primarily with ensuring that the right of interconnection was well defined and enforced. However, as wireless carriers become a more integral part of the NII and develop into potential competitors to the LECs, the cost of this interconnection is becoming a more central issue.

∎ Regulation of Interconnection

The FCC began to develop the rules that govern wireless interconnection in the proceeding that created cellular telephone service.[2] These regulations were later clarified and strengthened in a series of rulings in the 1980s.[3] In 1993, Congress created the Commercial Mobile Radio Service (CMRS) regulatory classification, which brought most Specialized Mobile Radio (SMR), PCS, and mobile satellite carriers under the same regulatory umbrella as cellular.[4] All CMRS service providers are entitled to interconnect with the LEC on the same terms as cellular carriers.[5]

The FCC's policy on wireless interconnection has two main components. First, LECs must provide interconnection when it is requested by a wireless carrier.[6] Interconnection is critical because users of wireless services want to be able to call anyone on the PSTN; they do not want to be restricted to calling only other wireless users. A

---

[2] Federal Communications Commission, *An Inquiry Into the Use of the Bands 825-845 Mhz and 870-890 Mhz for Cellular Communications Systems*, Report and Order (Cellular Report and Order), 86 FCC 2d 469, 496 (1981).

[3] Federal Communications Commission, *The Need to Promote Competition and Efficient Use of Spectrum for Radio Common Carrier Services*, Memorandum Opinion and Order, 59 RR 2d 1275 (1986); Declaratory Ruling, 63 RR 2d 7 (1987); Memorandum Opinion Order on Reconsideration, 66 RR 2d 105 (1989).

[4] Omnibus Budget Reconciliation Act of 1993. Public Law 103-66.

[5] Federal Communications Commission, *Implementation of Sections 3(n) and 332 of the Communications Act*, Second Report and Order, GN Docket No. 93-252 (1994), pp. 87-88.

[6] Federal Communications Commission, *Cellular Report and Order*, op. cit., footnote 2.

BOX 7-1: Interconnection to the Local Exchange Carrier

Interconnection requires a connection between the cellular carrier's switch and a nearby local exchange carrier (LEC) switch. This connection, which can be a microwave link or a high-speed digital line leased from the LEC, allows the cellular carrier to complete calls to the LEC's customers and connect calls originated in the wireline telephone network to its customers. Over time, a standard set of interconnection arrangements has evolved, designated as Type 1, Type 2A, or Type 2B, depending on the sophistication of the cellular switch and the type of LEC switch involved. These configurations are well known and described in reference documents published by Bellcore, the LECs' technical organization.

Similar interconnection arrangements will be used to connect other types of wireless services to the public switched telephone network (PSTN), including Personal Communications Service, Enhanced Specialized Mobile Radio, and satellite. Satellite networks are interconnected to the PSTN at earth stations known as gateways. User traffic is beamed down from the satellite to the earth station and routed through the satellite network's switch to the PSTN. While a cellular network may have several switches that are interconnected to the PSTN, there may only be a single earth station that handles all of the traffic from the satellite.

Interconnection also requires that the LEC provide wireless carriers with blocks of telephone numbers that they can assign to their customers. Wireless carriers are part of the PSTN's numbering plan, and, in each area code, the LEC is the *code administrator*, responsible for assigning numbers. Cellular numbers have the same 10 digit format as landline numbers, and, in most cases, they have the same area code as a landline number in the same region. When cellular numbers are assigned, the LEC programs its switches to recognize that calls to these numbers are to be routed to the wireless carrier.

SOURCE: Office of Technology Assessment, 1995

---

connection to the PSTN is necessary for wireless carriers to attract customers and survive in the marketplace. If the LECs, who have a near-monopoly in the provision of wireline telephone service, were able to withhold interconnection, wireless carriers would have no other way of connecting calls to wireline users and would likely go out of business.

The second part of the FCC's policy on wireless interconnection requires the LECs to provide independent wireless carriers with interconnection of the same quality and cost as they provide to their own wireless affiliates.[7] In order to police this requirement, the FCC requires structural separation of most LECs' wireline and cellular operations.[8] While the FCC recognized that there were potential economies of scope in greater integration of the LECs' wireless and wireline operations, it also believed that integration could give the LECs' wireless affiliates an unfair competitive advantage. As a result, the LECs have to build their cellular networks independently of the wireline network, as would any other carrier. LEC and independent cellular carriers have similar interconnection requirements, making it easier to determine if the LEC is discriminating against the competing cellular carrier.

The Cost of Interconnection

Wireless carriers are required to pay the LECs for interconnection.[9] The interconnection charges are

---

[7] Ibid.

[8] Ibid., p. 495.

[9] Charles H. Kennedy, *An Introduction to U.S. Telecommunications Law* (Norwood, MA: Artech House, 1994), pp. 44-46.

intended in part to cover the costs the LEC incurs in handling its part of the call. The most important charge is a per-minute fee paid by the wireless carrier for every call completed by the LEC. Typically, this charge is about three or four cents per minute, but it can be over 10 cents per minute, depending on the state, the duration of the call, and the distance of the call. In addition to the per-minute charge, the wireless carrier usually pays the LEC for a leased line between its switch and the LEC's switch. To minimize the cost of this leased line, some wireless carriers locate their switch across the street from a LEC central office or at another nearby location.

Currently, the states have primary jurisdiction over the cost of interconnection.[10] The FCC can only step in if the cost of interconnection is so high as to make wireless service prohibitively expensive.[11] As a result, the interconnection charges vary from state to state. In addition, the means by which states exercise their jurisdiction over interconnection charges differ.[12] In some states, such as New York and Florida, interconnection charges are specified by a tariff, a schedule of rates approved by state regulators. In most states, however, there is no formal tariff; instead, wireless companies and LECs negotiate an agreement with little or no involvement by state regulators. Some states require that these negotiated agreements be filed with state regulators, while others do not. Some states then make the agreement public, while others do not.

Regardless of whether interconnection charges are tariffed or negotiated, state regulators have generally allowed the LECs to impose interconnection charges that are above the cost they incur in handling their part of the call. Moreover, the compensation arrangements are usually one-way: wireless carriers compensate LECs for completing their calls, while the reverse is not true. Above-cost interconnection charges and unbalanced compensation arrangements reflect the fact that most state regulators view interconnection charges as a way to transfer revenues from a premium niche market service to the LEC in order to subsidize residential telephone rates and support universal service goals.[13]

Interconnection to Long-distance Carriers

Wireless users want to be able to make and receive long-distance as well as local calls. Since the breakup of the Bell System in 1984, the LECs have been restricted to providing local service within geographic regions known as Local Access and Transport Areas (LATAs). Calls that cross a LATA boundary are considered long distance and must be handled by a long-distance carrier. In most cases, a wireless carrier first hands long-distance calls to the LEC, which in turn hands them to a long-distance carrier. Interconnection to the LEC is all that is needed for wireless users to be able to place calls to any telephone user across the nation.

However, in recent years, long-distance carriers have begun to connect directly to wireless networks, bypassing the LEC (see box 7-2).[14] Direct connections permit long-distance carriers to avoid paying *access charges* to the LEC. Access charges are essentially interconnection charges paid by

---

[10] The Communications Act of 1934 has been interpreted to require that regulators allocate the costs of providing telecommunications services among interstate and intrastate jurisdictions. The states, therefore, regulate the price of interconnection for intrastate calls, while the FCC regulates the price of interconnection for interstate calls. Because most calls from wireless phones are intrastate, the states are largely responsible for determining the interconnection costs of wireless carriers.

[11] Federal Communications Commission, Declaratory Ruling, op. cit., footnote 3, p. 15.

[12] Harry E. Young, *Wireless Basics* (Chicago, IL: Intertec, 1992), p. 90.

[13] Kennedy, op. cit., footnote 9, p. 46.

[14] For example, in the Washington, DC market, both MCI and AT&T have direct connections to the Southwestern Bell Mobile Systems (Cellular One) network.

BOX 7-2: Interconnection to Long-Distance Carriers

In most cases, wireless carriers hand off both local and long-distance calls to the local exchange carrier (LEC). The LEC then routes the long distance calls to a long-distance carrier (see figure 7-1). Increasingly, however, long-distance carriers are connecting directly to wireless carriers. The wireless carrier only routes local calls to the LEC, while long-distance calls are routed directly to a long-distance carrier (see figure 7-2). Although the link between the wireless network and the long-distance network is usually leased from the LEC, the LEC provides only simple transport and is not involved in setting up the call. In a few cities, long-distance carriers have bypassed the LEC entirely, using leased lines provided by new competitors to the LECs, called Competitive Access Providers.

FIGURE 7-1: Connection to Long-Distance Carrier Through Local Exchange Carrier

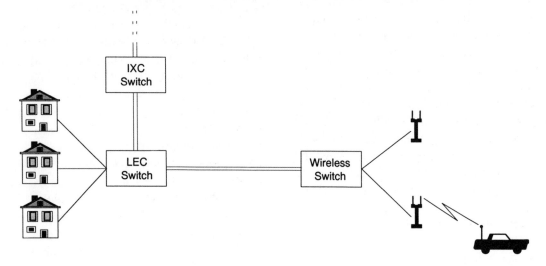

FIGURE 7-2: Direct Connection to Long-Distance Carrier

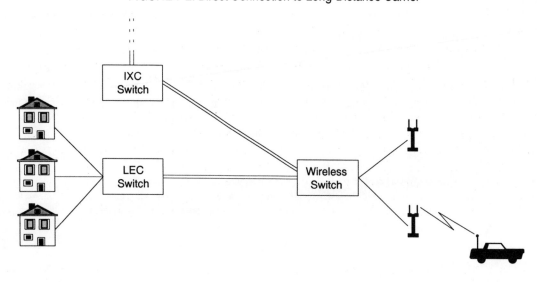

SOURCE: Office of Technology Assessment, 1995.

long-distance carriers whenever they receive traffic from a LEC, even when the calls originate on a wireless network. Long-distance carriers can avoid the access charges if they connect directly to the wireless carrier. In some cases, long-distance carriers pass on the access charge savings to wireless customers in the form of discounted long-distance calling.

Avoiding access charges, which can account for 40 to 50 percent of the cost of a long-distance call, is one reason for the recent interest shown by long-distance carriers in wireless communications. If the long-distance carriers can reach their customers without going through the LEC, they can cut access costs or put pressure on LECs to reduce the rates. However, these efforts raise questions about both the structure of local telephone rates and universal service. The access charge system was designed as a way to continue the Bell System's revenue transfer from long-distance to local service in the post-divestiture environment. If more long-distance carriers connect directly to wireless carriers—reducing the LECs' access charge revenue—they may undercut the system of subsidies that supports universal service.

■ Wireless/Wireline Interconnection Issues

Current rules for wireless interconnection focus on ensuring that wireless carriers are able to interconnect to the LEC. Now, however, existing and, especially, new wireless carriers are becoming concerned about the terms of interconnection agreements. First, new wireless entrants are worried that the present practice of negotiated interconnection agreements makes it possible for the LECs to discriminate among wireless carriers.[15] Second, as wireless technology becomes more efficient, interconnection charges will become a more significant fraction of wireless carriers'

overall cost structure. There may have to be reductions in interconnection charges if wireless carriers are to provide a mass market service or compete with the LEC in the market for local exchange service.

Nondiscriminatory Interconnection

In most states, interconnection charges are determined through negotiations between the LEC and the wireless carrier. In the early years of cellular service, several cellular carriers complained that the LECs were not negotiating in good faith or had not granted them the type of interconnection they requested.[16] However, in recent years the number of disputes has declined substantially. This may be due, in part, to the fact that the interconnection rules have been clarified by the FCC and are now well established. It may also be due to the fact that, in most markets, the second cellular carrier is no longer a small independent company, but is often part of a large company that is better equipped to negotiate with the LEC.

The cellular carriers have stated that they are generally satisfied with the current system of negotiated interconnection. However, many of the new PCS entrants are concerned that, despite the requirement that the LECs not discriminate, the established cellular carriers can obtain better terms because they are more familiar with the negotiation process.[17] The main problem for new entrants is that the agreements between the LECs and the cellular carriers are not made public in all states. It is difficult to enforce the nondiscrimination requirement without knowing the terms under which competing wireless carriers are obtaining interconnection.

One way to guarantee that all carriers obtain interconnection on the same terms is to require the filing of interconnection tariffs, as is done in New York and Florida. This protects new market en-

---

[15] See discussion in Federal Communications Commission, *Equal Access and Interconnection Obligations Pertaining to Commercial Mobile Radio Services (Equal Access NPRM)*, Notice of Proposed Rule Making and Notice of Inquiry, CC Docket No. 94-54 (1994), pp. 46-47.

[16] U.S. Department of Justice, Antitrust Division, *The Geodesic Network* (Washington, DC: 1987), p. 4.13.

[17] Federal Communications Commission, *Equal Access NPRM*, op. cit., footnote 15, p. 50.

trants unfamiliar with the interconnection negotiation process because all competitors have to obtain interconnection at the tariffed rate. The argument against tariffing is that it may not allow sufficient flexibility in the terms of interconnection. Moreover, the tariffing process can cause considerable delay before a new service can be offered by the LEC. Another option is to continue the present system of negotiated contracts, but require that the contracts be made available for public inspection.

Under current law, the FCC has limited ability to require states to use tariffs or require that contracts be made public. This is based on the division of jurisdiction in the 1934 Communications Act, which gives states primary jurisdiction over intrastate charges. If Congress decided that entry of new wireless providers would be facilitated by tariffing or public filing of the terms of interconnection agreements, it could provide guidelines on this issue. A public filing requirement that applies to LEC interconnection with all carriers, including wireless, is included in both S. 652 and H.R. 1555, the telecommunications bills currently being debated in Congress.

Local Exchange Competition

The cost of providing service, and the prices that wireless carriers charge, will significantly affect the role wireless technologies can play in the NII—whether they will remain providers of a relatively high-cost niche service (mobile communications) or whether they can broaden their appeal to compete in the market for local telecommunications services. The high cost of today's cellular service—and the correspondingly high prices charged to consumers—is primarily the result of inefficient analog technology. Increasing numbers of customers have been willing to pay these prices because of the value being placed on mobile communications.

New digital technology, however, will allow wireless networks to serve many more users at a lower cost per user (see chapter 3). As this happens, interconnection charges will become a larger fraction of the wireless carriers' overall cost structure and a more important determinant of the prices carriers can charge. The lower the interconnection charges, the lower the price at which wireless carriers will be able to provide service. The level of interconnection charges could even determine whether wireless carriers are limited to serving the mobile telephone market, for which consumers are willing to pay a higher price, or are also able to compete in the local exchange market. The cable companies, and others who view wireless local loop technologies as a way to compete in the local telephone services market, are arguing most strongly for reduced interconnection charges.

It is likely that some form of regulatory intervention would be required to reduce interconnection charges. Under the present system of negotiated interconnection agreements, wireless carriers could only obtain more favorable terms if they had equal bargaining strength. For the foreseeable future, however, wireless carriers will continue to be much more dependent on the LEC than the reverse.[18] Very few calls from LEC customers are to wireless users, while almost all wireless calls are to users of the landline network. Because of this imbalance, the LEC would have an incentive to maintain high interconnection charges even if wireless carriers were allowed to charge the LEC for completing calls.

As a result, regulators who want to bring interconnection charges down are faced with two difficult tasks. First, they may need to determine how much it costs the LEC to provide interconnection—a notoriously difficult task. Prices could then be set accordingly, allowing the LEC a reasonable margin of profit. Second, however, regu-

---

[18] For more discussion of this issue, see Rob Frieden, "Wireline vs. Wireless: Can Network Parity Be Reached?" *Satellite Communications*, July 1994, p. 20.

lators and policymakers must also determine the extent to which interconnection charges should continue to subsidize universal service. As the telecommunications industry has evolved from a single monopoly carrier into one with many participants, above-cost interconnection charges have been used to provide the LECs with revenues that subsidize local residential service. To reduce interconnection charges, regulators may need to find alternative funding sources to make up for the drop in revenues. The most common proposal for replacing interconnection charges as a source of subsidies is to create an expanded Universal Service Fund to which all carriers would contribute, and from which eligible carriers could withdraw funds to help them provide service.[19]

Regulators trying to encourage competition in the local telephone market will also have to determine whether to designate wireless carriers as *co-carriers* with the LEC. Today, although cellular carriers must pay the LEC to have wireless calls delivered to PSTN users, the reverse is not true—the LEC usually does not pay cellular carriers for completing calls that come from the PSTN. The FCC has stated several times that cellular carriers should be compensated for completing calls from the LEC, but most state regulators have chosen not to follow this recommendation.[20] In order to redress this imbalance, wireless carriers are petitioning states to be formally recognized as co-carriers. Co-carriage involves *mutual compensation,* in which each carrier compensates the other for calls completed. Today, most LECs only recognize other LECs, such as those with neighboring service areas, as co-carriers. Co-carrier status

would give wireless carriers greater bargaining power in negotiating with the LECs.

Another issue that potentially could affect the ability of wireless carriers to compete in local telecommunications markets involves the assignment of telephone numbers. While there has been long-standing concern on the part of cellular carriers that the LEC manages numbers in a way that disadvantages them, the issue is attracting more attention as existing area codes start to run out of numbers. When this happened in the past, area code regions were *split,* assigning part of the old area to a new number. But in recent years, LECs have proposed relieving the pressure for numbers by creating *overlay* area codes just for wireless carriers. Wireless carriers have argued that assigning different area codes to the LEC's potential (wireless) competitors could lead to discrimination in how different carriers (and their customers) are treated by the LEC.[21] The FCC has recently launched a proceeding to examine numbering issues in detail.[22]

In most respects, the interconnection issues that concern wireless operators are similar to those that concern new wireline competitors in the market for local telecommunications services. The primary difference is that wireless carriers have long had the right of interconnection, while state regulators have only recently begun to certify competitive local wireline carriers and grant them the interconnection rights they need to enter the market.[23] Regulators granted wireless carriers the right of interconnection more readily because they were seen as serving a separate, niche market (mo-

[19] The current version of S. 652, for example, specifies that only carriers designated as "essential telecommunications carriers" can withdraw from the fund. Section 104.

[20] Kennedy, op. cit., footnote 9, p. 44.

[21] See, for example, *Reply Comments of the Personal Communications Industry Association*, before the Federal Communications Commission, CC Docket No. 92-237, June 30, 1994, pp. 1-5.

[22] Federal Communications Commission, *Administration of the North American Numbering Plan*, Notice of Inquiry, 7 FCC Rcd 6837 (1992).

[23] See, for example, Richard L. Cimerman and Geoffrey J. Waldau, "Local Exchange Competition: Alternative Models in Maryland," in *Proceedings of the 22nd Annual Telecommunications Policy Research Conference*, Solomons, Maryland, Oct. 1-3, 1994, p. 221.

bile telephony) that did not threaten the local monopoly.

Under current law, the level of interconnection charges is primarily a state responsibility. Congress may choose to give more guidance to the states on the terms under which interconnection has to be provided. For example, both S. 652 and H.R. 1555, the telecommunications bills currently being debated in Congress, would require the LECs to treat all other carriers as co-carriers. In addition, if Congress determined that state regulation of interconnection charges was slowing the development of the wireless communications industry, it could either give the FCC a greater role or preempt the states entirely. However, a reduction in interconnection charges is likely to require coordinated action on the part of both state and federal regulators because these charges are entangled in the larger question of universal service subsidies.

INTERCONNECTION OBLIGATIONS OF WIRELESS CARRIERS

The question of whether a carrier should be obliged to interconnect with other carriers has been one of the constant themes of telecommunications policy debates over the past two decades.[24] Today, only the LECs have interconnection obligations. As a result of their control of the bottleneck local exchange, they are required to interconnect with long-distance carriers and with wireless carriers. A key issue is whether wireless carriers should have interconnection obligations of their own. In 1994, the FCC began examining whether some or all wireless carriers have sufficient market power to justify the imposition of interconnection obligations, or if, on a more fundamental level, interconnection obligations are required of all carriers in order to hold today's more fragmented and competitive "network of networks" together.[25] The interconnection obligations of wireless carriers are also being debated in Congress.

■ Interconnection with Long-Distance Carriers: *Equal Access*

As a result of the breakup of the Bell System in 1984, the relationship between wireline local and long-distance service providers changed. Current rules require LECs to provide "equal access" to all long-distance carriers—allowing wireline users to choose a preferred company to carry their long-distance calls. LATA boundaries define the limits of local service—whenever a call crosses a LATA boundary, it must be handed off by the LEC to the user's chosen long-distance carrier. The equal access rules were first applied by the Modified Final Judgment (MFJ) to the RBOCs after the breakup of the Bell System, and later extended by the FCC to apply to all other LECs.[26]

There are no FCC rules that require *wireless* carriers to provide equal access. However, the wireless affiliates of AT&T and the RBOCs are subject to consent decrees that require them to provide equal access, regardless of the fact that they are not required to do so under FCC rules. The restrictions on the RBOCs' cellular affiliates were imposed by the court that oversees the MFJ. The restrictions on AT&T were imposed as part of the settlement to an antitrust action brought by the Department of Justice (DOJ) when AT&T acquired McCaw.[27] Wireless carriers not subject to these consent decrees, such as GTE and Sprint, are not required to allow their customers a choice of long-distance carriers.

---

[24] See, for example, Gerald W. Brock, *Telecommunication Policy for the Information Age* (Cambridge, MA: Harvard University Press, 1994).

[25] Federal Communications Commission, *Equal Access NPRM*, op. cit., footnote 15.

[26] Ibid., pp. 6-7.

[27] U.S. Department of Justice, "Proposed Final Judgment and Competitive Impact Statement; United States of America v. AT&T Corp. and McCaw Cellular Communications, Inc.," notice, Federal Register 59 (165): 44158, Aug. 26, 1994.

Over time, more cellular systems have been converted to equal access. At first, all of the independent, or *A-side*, cellular carriers were free of the equal access restrictions, as were the *B-side* affiliates of GTE and other non-RBOC LECs. But in recent years, the RBOCs have begun buying A-side systems outside their home region. In Washington, DC, for example, the A-side system is controlled by an RBOC cellular affiliate. The court that oversees the MFJ has ruled that these out-of-region systems must also be converted to equal access. More recently, systems operated by the largest A-side carrier, McCaw, were required to convert to equal access after McCaw was acquired by AT&T. It has been estimated that over 60 percent of cellular customers are now served by equal access carriers.[28]

Implications of Equal Access Restrictions

The nature of the equal access restrictions imposed on a wireless carrier affects several aspects of its operations, including service packaging and system design and construction.

Bundled local and long-distance service

Unlike the wireless affiliates of the RBOCs and AT&T, carriers not subject to equal access rules do not have to give their customers a choice of long-distance carrier. They can even set up their own long-distance operation and funnel all of their customers' traffic to it, selling their customers a bundled package of local and long-distance service. Few wireless carriers have extensive long-distance networks of their own, but most resell long-distance service purchased at "wholesale" rates from one of the long-distance carriers.

Carriers that are allowed to sell bundled packages of local and long-distance service can market their services very differently from equal access carriers. They have the flexibility to create expanded "local" calling areas, much larger than a LATA, because they do not have to distinguish between intra- and interLATA calls.[29] They are able to incorporate the cost of the interLATA part of the call into the basic airtime charge, which applies to all calls within the larger calling area. Some carriers have even eliminated the concept of "long distance" entirely, offering calls to any location in the nation as part of the basic airtime charge.

On the other hand, equal access rules prevent the wireless affiliates of AT&T and the RBOCs from automatically funneling their wireless customers' traffic to their own long-distance operation. They must give their customers a choice of long-distance carrier. For many years, the RBOCs' cellular affiliates were, like their wireline telephone companies, prohibited from providing long-distance service at all. However, the court that oversees the MFJ recently approved a waiver request that allows the RBOCs' cellular affiliates to resell long-distance service, as long as they provide equal access and comply with several other restrictions. Both S. 652 and H.R. 1555 would codify and somewhat liberalize this exemption.

In general, the wireless affiliates of AT&T and the RBOCs may not offer wide-area "local" calling because the equal access rules require them to hand off interLATA calls to the customer's chosen long-distance carrier. However, there are several exceptions to this rule. The court that oversees the MFJ has often waived the equal access rules when it found that a "community of interest" crossed a LATA boundary and the RBOC's competitor was able to offer regional calling. The DOJ exempted AT&T from complying with equal access rules in those areas where the RBOCs are exempt, and also grandfathered 19 other systems operated by McCaw that crossed a LATA boundary.

The nature of the equal access restrictions under which a wireless carrier operates affects the configuration of the interconnection between it

---

[28] McCaw Cellular Communications, comments before the Federal Communications Commission, *Equal Access NPRM*, op. cit., footnote 15, p. 34.

[29] RBOCs can offer similar larger calling areas, but must get a waiver from the court to do so.

and a long-distance carrier. Long-distance carriers, if selected to provide wholesale long-distance service to a non-equal access carrier, nearly always arrange for a direct connection to the wireless carrier's switch. The volume of traffic is usually high enough to justify the cost of the leased line, especially when the savings on access charges are taken into account. When connecting to an equal-access carrier, on the other hand, long-distance carriers are more likely to connect through the LEC. Because the long-distance traffic is divided among several long-distance carriers, the volume of traffic is often insufficient to justify a direct link.

Impact of equal access on wireless system design

Equal access restrictions bring with them the requirement that wireless networks be designed to operate within LATA boundaries.[30] For example, they prevent a wireless carrier from connecting its switch to a cell site in a different LATA. Equal access rules would require that this link be open to competition from other providers of interLATA service. Because it is not technically feasible to design a wireless network in such a way that these internal operations are open to competition, wireless networks have to be contained within the LATA boundary. Non-equal access systems, on the other hand, can gain efficiencies by integrating functions across a wider area that includes several LATAs.

Because LATA boundaries were drawn with the landline network in mind, it has often been difficult to design wireless networks in a way that conforms to the LATA boundaries. One problem was that the FCC drew its cellular licensing map without regard to LATA boundaries. In many cases, cellular licensing areas include parts of more than one LATA, preventing an equal access carrier from serving the entire licensing area from a single switch, which may be the most efficient configuration. The court that oversees the MFJ has, on several occasions, granted waivers that permit the RBOCs to build networks that cross a LATA boundary.[31]

In addition, LATA boundaries and equal access have not been easily reconciled with the requirements of a mobile service. It is possible, for example, that a call will change from local to long-distance in mid-call if a user drives across a LATA boundary. Because it is technically impossible to transfer the call to the user's chosen long-distance carrier during this *intersystem hand-off,* the MFJ court has granted a waiver that permits RBOC wireless affiliates to continue these calls.

Finally, there may be significant advantages in network construction and operation, as well as other economies of scope, that may not be possible with continued segmentation of local and long-distance services. The cellular industry argues that users value large local calling areas. In addition, if a cellular carrier is reselling long-distance service, it can buy service at bulk rates that are cheaper than the retail rates that most individual users could obtain on their own. This has particular implications for satellite providers because it is likely that a call made by a mobile satellite system user will be headed outside the LATA in which the gateway is located. For this reason, satellite carriers intend to purchase long-distance service in bulk and then bundle it with the their usage charges at a flat per-minute rate, regardless of the destination of the call.

Proposed Changes to Wireless Equal Access Restrictions

In 1994, the FCC proposed requiring all cellular carriers to observe the equal access rules. In part, this proposal was intended to ensure that all companies in a competitive industry are subject to the same rules.[32] The FCC does not currently have the

---

[30] Kennedy, op. cit., footnote 9, pp. 102-108.

[31] Kennedy, op. cit., footnote 9, p. 106.

[32] Federal Communications Commission, *Equal Access NPRM,* op. cit., footnote 15, p. 20.

power to ensure competitive parity by removing the equal access restrictions from AT&T or the RBOCs because these conditions are a consequence of antitrust law and can only be modified by the courts or Congress. The FCC is only able to ensure competitive parity by imposing the equal access restrictions on the remaining wireless carriers. While preferring that competitive parity be achieved by removing their restrictions, the RBOCs supported this effort. The FCC has not yet acted on its equal access proposal.

Because of the problems associated with applying LATA boundaries to a mobile service, the FCC suggested that the larger Major Trading Areas (MTAs) be used instead of LATAs to distinguish between local and long-distance calls.[33] Long-distance carriers have opposed this proposal because it would reduce the amount of traffic considered to be long distance. The use of MTAs would also create competitive parity issues because the RBOCs' wireless affiliates would still be required to observe LATA boundaries, unless Congress or the courts altered the terms of the MFJ.

Wireless equal access has been an issue in recent congressional debates on revising the nation's telecommunications laws. Both S. 652 and H.R. 1555 would supersede the consent decree provisions that impose equal access restrictions on the wireless affiliates of AT&T and the RBOCs. Both bills would also require wireless carriers to allow their customers to reach all long-distance carriers. However, carriers could require their users to dial five-digit access codes to reach most long-distance carriers,[34] while reserving the more convenient "1+" access for calls routed through their own long-distance network. In the past, equal access has meant giving users the ability to *presubscribe* to their choice of 1+ carrier, as

they are able to do with their wireline telephone service.

In discussions concerning equal access rules, the key issue is whether wireless carriers have the ability to restrict competition in the market for long-distance service. Equal access rules were imposed on wireline LECs because their control over the local exchange *bottleneck* allowed them to also dominate the long-distance market. Wireless carriers, by contrast, do not control a bottleneck. The market for wireless communications has always been capable of supporting competition and has never been viewed as a natural monopoly. If there were several competing wireless carriers, there would be competition in wireless long distance even if each carrier did not offer a choice of other long-distance carriers.

To the extent that competition in the market for mobile telephone service is limited, it is because the FCC initially licensed only two cellular carriers. The DOJ imposed equal access restrictions on AT&T's cellular operations because it believed that AT&T would have sufficient market power, as one of only two cellular carriers in a market, to reduce competition in the market for cellular long-distance service.[35] The DOJ also required rigorous equal access restrictions as a condition of RBOC entry into the cellular long-distance market. Proponents of extending the equal access rules have pointed to the DOJ's actions to argue that these safeguards are required. However, the market for local mobile telephone services is about to become significantly more competitive with the entry of an ESMR carrier and three to six PCS carriers.

Conflicting Models

Although economic arguments may indicate that equal access requirements should not be imposed on wireless carriers, the sale of integrated local

---

[33] Ibid., p. 32.

[34] This is similar to the procedure by which users access long-distance carriers other than the one to which a payphone is presubscribed. The codes are of the form "10XXX," where the last three digits denote the carrier.

[35] U.S. Department of Justice, op. cit., footnote 27, at 44169.

and long-distance wireless service would be at odds with the telecommunications industry model that has been established over the past decade for the much larger wireline market. From a functional perspective, wireless can be used to provide access to a long-distance network in much the same manner as a wireline local exchange network. There is considerable pressure to structure the market so that long-distance carriers can sell service to wireless users in the same way that they sell to wireline users.

Without equal access, long-distance carriers cannot sell their service directly to end users, as in the wireline model. Instead, they have sell to the wireless carriers, who then resell the long-distance service to their customers as part of a bundled package. From the long-distance carriers' perspective, it is difficult to market services that can be used with both wireless and wireline access because there is no guarantee that their services would be accessible from all wireless carriers. In particular, *virtual private networks* that include volume discounts and custom features cannot necessarily be accessed from a corporate customer's chosen cellular carrier.

As the amount of wireless traffic grows, the conflict between the two models of the telecommunications industry could become more significant. Long-distance carriers have been the main supporters of the FCC's equal access proposal, preferring to sell directly to end users rather than ceding control over the packaging of services to the networks that originate the call. However, both AT&T and Sprint have acquired wireless interests of their own and may have an interest in permitting a greater degree of bundling. Long-distance carriers that have wireless access networks of their own would have a competitive advantage over long-distance carriers that do not.

Aside from economic considerations, another set of arguments in favor of equal access relies on the general NII concept of a network of networks. According to this argument, the future telecommunications infrastructure will be made up of many different networks, and users should be able to choose their telecommunications services from many different providers, mixing and matching as needed. They should not have to switch wireless carriers in order to change their long-distance service, for example. For this to be possible, all networks would have to interconnect, regardless of market power.

■ Interconnection of Wireless Carriers

Today, calls between customers of different wireless carriers are almost always routed through the local exchange network. Because the LEC is required to interconnect with all wireless carriers, it provides a common link between them. However, in the same way that a wireless carrier can circumvent the LEC and connect directly to a long-distance carrier, it can also choose to connect directly to another wireless carrier. This configuration avoids the interconnection charges that would have to be paid if the traffic were routed through the LEC. Direct connections are used only rarely, however because the volume of wireless to wireless traffic is usually too small to justify the cost of the leased line.

In 1994, the FCC proposed that wireless carriers be required to interconnect with other wireless carriers. Most wireless carriers opposed this proposal, arguing that interconnection through the LEC was sufficient to guarantee connectivity. They also pointed to the fact that there are relatively few direct connections between wireless carriers today. Others, however, argued that the amount of wireless to wireless traffic will soon increase, and that clear rules should be established now. In part, the FCC appeared to be concerned that purely voluntary interconnection arrangements would lead to a lack of connectivity or inefficiencies in network design.[36] After studying theissue, however, the Commission tentatively concluded that it would be premature to require wireless carriers to interconnect with other wireless carriers.[37]

A related question is whether roaming agreements (see chapter 3) should continue to be voluntary or if wireless carriers should be required to negotiate them. Today, it is in the interest of cellular carriers to negotiate roaming agreements with each other because all carriers benefit from being able to advertise wide area service and from the increased use of their systems. The cellular industry also voluntarily negotiated roaming agreements with a new provider of mobile satellite services, American Mobile Satellite Corp. (AMSC) allowing calls to be forwarded to users through AMSC's satellite network when they are outside cellular coverage areas (see chapter 3).

However, new wireless entrants have expressed concern that the incumbent cellular carriers will choose not to negotiate roaming agreements with them. Until there are PCS networks throughout the nation, new PCS providers might want to offer their customers a *dual-mode* phone that would use PCS-band service in their home market and cellular service when roaming. But it might be in the cellular industry's interest to refuse to negotiate roaming agreements, limiting their new competitors to isolated islands of service that could not compete with nationwide cellular roaming.

The location information that wireless carriers collect to facilitate roaming is also becoming increasingly valuable to other wireless and wireline carriers. There are many possible services that can be offered based on knowledge of a user's current location. For example, if LECs and long-distance carriers had access to cellular carriers' location information, they could deliver calls more efficiently and less expensively to roamers. Today, if a user is visiting another city and someone in that city wants to call them, the call is first sent to the user's home cellular system—incurring a long-distance charge to the caller. The cellular carrier determines that the user is roaming and then sends the call back to the LEC in the same city it came from—incurring a long-distance charge for the cellular subscriber. Thus, even if the two individuals are literally in the same building, the call must travel to the cellular user's home system and back again—turning an inexpensive call into a very expensive one. Ideally, local telephone companies and cellular companies could share information about roamers that would allow the visited LEC to deliver the call directly to the visited cellular carrier—eliminating all the unnecessary long-distance transfers and charges. In comments on the FCC's interconnection proceeding, a major interexchange carrier argued that it should be guaranteed access to information about its customers in the cellular industry's location databases.[38] The cellular industry believes that location information is proprietary and that it should not be required to share the information with other carriers.[39]

---

[36] "We ask commenters to focus on whether interconnection requirements would advance competition and encourage efficiencies and lower rates in the mobile services marketplace. We do not wish to encourage a situation where most traffic from one CMRS service subscriber must pass through a LEC switch for its traffic to reach a subscriber to another CMRS service, if such routing would be inefficient or unduly costly." Federal Communications Commission, *Equal Access NPRM*, op. cit., footnote 15, p. 54.

[37] Federal Communications Commission, *Interconnection and Resale Obligations Pertaining to Commercial Mobile Radio Services*, Second Notice of Proposed Rule Making, CC Docket No. 94-54, April 20, 1995.

[38] Federal Communications Commission, *Equal Access NPRM*, op. cit., footnote 15, p. 58.

[39] Ibid.

Zoning Regulations and Antenna Siting | 8

One of the most contentious issues facing the wireless industry today involves the location of transmitting antennas. The cellular and personal communications service (PCS) industries estimate that they will have to build 100,000 new antennas by the year 2000 in order to provide adequate mobile telephone service to the public.[1] Local communities, however, are increasingly opposed to the new antennas for aesthetic, health, and safety reasons, and are applying local zoning rules and municipal ordinances to force carriers to locate the antennas elsewhere or halt construction altogether.[2] In response to the increasing number and cost of these objections, two wireless industry trade associations petitioned the Federal Communications Commission (FCC) to nullify or preempt local regula-

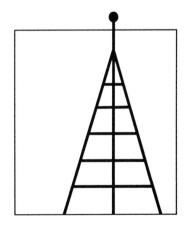

---

[1] Bob Roche, director of research, Cellular Telecommunications Industry Association, personal communication, May 31, 1995.

[2] Examples of reporting on this issue from local communities include: James Rush, "Towering Controversy: Expansion of Cellular Antenna Systems a Local, National Issue," *The Seattle Press,* vol. 10, No. 3, Apr. 12-26, 1995, pp. 1*ff*; Sandi Coburn, "Cellular One's Call Waiting," *Suburban News* (New Jersey), June 15, 1994, pp. 1, 14; Michelle De-Blase and Dina Masarani, "East Brunswick, Old Bridge Vote: Local Officials Urge Cellular Tower Limits," *Home News* (New Jersey), Sept. 30, 1994, pp. B1, B6; Norman O'Donnell, "Phone Trouble: Everyone Wants Cellular Phones, but Many Don't Want To Live Near the Antennas That Make Them Work," *Gannett Suburban Newspapers* (New Jersey), Aug. 24, 1994, pp. 1A, 2A; "Cellular Phones: West Hollywood, Cal., Denies Transmission Post," *EMF Litigation News*, November 1993, p. 535.

tions on antenna siting.[3] The FCC has not yet acted on these petitions. Local restrictions have also been a serious concern to the satellite broadcast industry, which has been fighting local rules on satellite receiving dishes for many years. At their foundation, these issues revolve around the question of which should take preeminence: federal policy or local law?

FINDING

The issue of federal preemption of local zoning and other regulations represents a battle between two valid, but conflicting, public policy goals. On the one side, federal policymakers, as set forth in the Communications Act of 1934, are trying to bring advanced communications services to the public. On the other side, communities and citizens are trying to preserve local control over their land and affairs—a long-standing tenet of American political culture. **In essence, the issues surrounding federal preemption of local regulations affecting antenna siting derive from ambiguous language contained in the Omnibus Budget Reconciliation Act of 1993—the legislation that established the Commercial Mobile Radio Service (CMRS).[4]** In that Act, Congress stated in part "...no State or local government shall have any authority to regulate the entry of or the rates charged by any commercial mobile service or any private mobile service, except that this paragraph shall not prohibit a State from regulating the other terms and conditions of commercial mobile services." Each side in the preemption debate has interpreted this passage as supporting its position.

Without additional information or clarification, congressional intent regarding preemption in the case of zoning and antenna siting remains unknown. This ambiguity is likely to cause continuing uncertainty until the FCC and appeals processes run their course. The Congress has not engaged in any debate or action on federal preemption of local regulations of wireless operations, and there is no information that could clarify what the Congress might think on this issue. As a result, attention is currently focused on the FCC, where the petitions for rulemaking have been submitted and the process of evaluating them is under way.

In responding to this issue, Congress has two primary options. First, it could let the FCC process run its course. The existing petitions for rulemaking, if accepted by the FCC, could result in a formal proceeding being established. This proceeding would doubtless receive considerable attention in the industry and in state and local communities, and there are indications that the FCC is looking at this issue carefully. The process would, however, take several years to wind its way through the FCC rulemaking process and the almost inevitable court challenges and appeals.

Secondly, Congress could make clear its intentions regarding the legislative language and offer a specific interpretation regarding local zoning and antenna siting—either by supporting it explicitly,[5] or by requiring states and local governments to resolve the antenna siting issues through negotiations with the wireless companies. A specific finding from Congress—either for or against preemption—would at least remove the uncertainty

---

[3] Cellular Telecommunications Industry Association, "Petition for Rulemaking," before the Federal Communications Commission, *In the Matter of Amendment of the Commission's Rule To Preempt State and Local Regulation of Tower Siting For Commercial Mobile Services Providers*, RM-8577, Dec. 22, 1994, and Electromagnetic Energy Association, "Petition for Further Notice of Proposed Rulemaking," before the Federal Communications Commission, *In the Matter of Guidelines for Evaluating the Environmental Effects of Radiofrequency Radiation*, ET Docket No. 93-62, Dec. 22, 1994.

[4] 47 U.S.C., sec. 332(c)(3)(A).

[5] Some leaders in the House of Representatives have already signaled that they now support preemption. See remarks by Rep. Newt Gingrich to Wireless '95 conference, New Orleans, LA, Feb. 1, 1995.

surrounding the issue, and allow the industry to move ahead with existing plans or pursue alternatives. Congressional action could also help clarify the issue of local restrictions on receive-only satellite dish placement, a matter that the FCC ruled on in 1986 when it partially preempted local regulations. [6]

BACKGROUND

The battle over antennas used to send and receive radio signals is not new, but its character is changing. In the 1980s, the fight was over local restrictions on the "big ugly dishes" used for receiving C-band satellite transmissions—pitting homeowner against homeowner or local zoning board. Today, although restrictions on satellite dishes remain contentious, the dispute has broadened as citizens and local governments have taken up positions against unwanted transmission towers used primarily to provide cellular telephone (and future PCS) services.

Wireless telephone service providers—cellular, PCS, and ESMR—are now in the process of establishing or expanding their networks. In order to deliver services, they have to place antennas in areas that will allow them to reach their customers. Sometimes these antennas can be located away from residential areas, but in other cases, engineering, topographical, or capacity considerations mean that antennas have to be located close to homes.

In the early days of cellular telephone system construction, it was relatively easy for companies to locate sites and build antennas. Property owners could be found who had little objection to antennas or base-station equipment, and many did not understand that their locations had value to carriers. Communities did not have ordinances limiting antenna siting or other characteristics of radio facilities. Furthermore, wireless carriers had more latitude in placing antennas; objections could usually be met by simply moving to another suitable site close by.

Today, cellular and PCS companies are having a much harder time siting antennas, both technically and politically. They are trying to erect new antennas to cover areas that currently have poor service, usually due to topography or cellular system congestion associated with high demand. Changes in cell structure and system architecture, however, are more difficult to make now because adjacent cells are already established. To function most effectively, antennas generally need to be located close to the center of their cells; as cells get smaller, the latitude for placement shrinks as well (see figure 8-1). [7] In a typical high-density area, where cells may be as small as one mile in diameter, this means that an antenna would ideally be located in a central four-city-block area. [8]

At the same time, despite the increasing reliance and value that many residents put on wireless communications, public opposition to these antennas is growing rapidly. Ironically, it arises most often, although not exclusively, in communities that have the highest per capita use of cellular telecommunications, notably wealthy suburban neighborhoods close to major metropolitan centers. Citizens often object to the antennas because they can be unsightly and bring down property values, and because they fear the possible health hazards associated with the radio waves the antennas emit (see chapter 11). Some question the need for or appropriateness of these new services. In a few cases, minor changes—

---

[6] Federal Communications Commission, *Preemption of Local Zoning or Other Regulation of Receive-Only Satellite Earth Stations*, 59 R.R.2d 1073 (1986).

[7] Ideally, the transmitter should be located at the center of the cell, but in any case should be located at a distance no more than one-fourth of the cell radius from the center. Cellular Telecommunications Industry Association, "Local Zoning vs. Wireless Communication: A Case for Federal Preemption?" briefing paper, (January 1995), p. 2.

[8] Jaymes D. Littlejohn, "The Impact of Land Use Regulation on Cellular Communications: Is Federal Preemption Warranted?" *Federal Communications Law Journal*, vol. 45, No. 2, April 1993, p. 250.

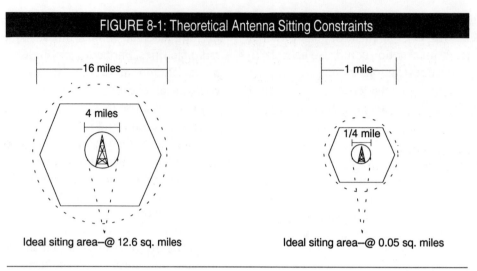

FIGURE 8-1: Theoretical Antenna Sitting Constraints

Ideal siting area—@ 12.6 sq. miles

Ideal siting area—@ 0.05 sq. miles

NOTE: Figure not drawn to scale.
SOURCE: Office of Technology Assessment, 1995.

planting bushes around equipment shacks or somehow disguising the antennas—are enough to satisfy citizen objections. In other cases, however, citizens want the antennas/towers moved so that they are less obvious or further away from populated areas—to lessen any possible health risks. And in some cases, citizens feel so strongly that no changes are acceptable; they seek to prohibit the tower/antenna altogether.

Citizen objections manifest themselves in restrictive zoning regulations or other municipal ordinances. This gives citizens' groups the ability to challenge the siting of each tower or antenna a wireless company wants to put up. They maintain that carriers can move their towers to other locations, but are usually unwilling to do so because it will cost them more money. There is also resentment among some citizens and public interest groups at the arrogant way they believe the carriers have treated their objections.

The process of challenging a particular antenna site, which can work itself out in both local zoning hearings and in court, is both time-consuming and expensive.[9] As a result, the wireless industry wants the federal government to preempt local and state regulations on antenna siting, so that they can move ahead with building their systems. They maintain that it is often not just a matter of cost, but of engineering requirements that dictates antenna placement. Early on, the industry received some support from FCC chairman Reed Hundt, who noted that local taxation, zoning, and other local restrictions could slow the widespread deployment of wireless technologies. In speeches to city and county organizations he encouraged them "to find a way to tolerate the presence of the new [PCS] equipment—relay stations and antennas—that this service requires."[10] To date, no general accommodations have been reached, and the issue has become highly politicized in many communities.

The satellite industry, meanwhile, is still fighting the battles first joined in the 1980s, when local restrictions on satellite dishes were put into place. Today, public zoning restrictions on satellite dishes are limited, but private homeowners' association rules or condominium covenants are permitted by the FCC. In addition, some commu-

---

[9] Estimates of the added costs to the wireless industry of local regulatory proceedings are not available.

[10] "Hundt Says Local Government Regs Could Slow Competition," *Telecommunications Reports*, Mar. 13, 1995, p. 24.

nities do not follow the guidelines specified in the FCC's preemption order, and since the FCC has limited enforcement resources, in these communities the law is ignored. The FCC has brokered discussions between the direct broadcast satellite industry and local government representatives on a blanket preemption of restrictions on direct broadcast service antennas.[11] As in the case of cellular and PCS antennas, the issue is not yet resolved.

■ Antenna Siting for Cellular and PCS Services

Antennas and base-station equipment for land-based wireless telecommunications systems vary in size and appearance depending on factors such as power output, frequency, topography, and expected usage. Engineering considerations determine both the number of radios needed per cell site (based on number of customers served) and the power levels of the radios—smaller cells use lower power. PCS base-stations, for example, may have a power output of up to 100 watts per channel—a typical site might have up to 30 channels, so total output might reach 3,000 watts if all channels were in use simultaneously.[12]

The equipment needed at each cellular or PCS base station generally consists of an antenna, radio transceivers, and the hardware needed to link to other cell sites or switches in the system. Because of differences in power levels and architectures, the equipment needed for individual cellular and PCS cell sites varies in size and configuration. For cellular base stations, antennas can be a small (3 to 4 feet) rod, a panel (4 to 8 feet tall and 1 to 2 feet wide), or a combination of rods and panels. In high-use areas, a complete antenna installation

may consist of 12 to 16 panels, located on a free-standing pole (up to 150 feet), a tall building, or another high structure (water towers, television antennas, etc.). In lower use areas, antennas can be mounted on smaller towers or even low-rise apartment buildings. The radio equipment for cellular telephone systems is usually housed in large trailer-sized (20' x 10' x 7') facilities equipped with air conditioners for peak-use cooling. PCS cell-site equipment consists of smaller whips and panels, and the radio hardware can be housed in a metal box about the size of a small refrigerator.

■ Siting Satellite Dishes

In the case of satellite dishes, local restrictions are aimed not at the large dishes used by companies to transmit programming to a satellite—these are usually located far from residential areas—but at the smaller (18 inches to 10 feet) dishes consumers use to receive programming at their homes. These antennas must be positioned so that they can easily receive signals from satellites. Depending on the consumer's exposure to the southern sky, and the landscaping and other physical structures present in the area, a customer may be able to put a dish in the backyard, on the roof, or in a place out of sight of neighbors.[13] Some customers, however, must put their dishes in their front yards or elsewhere in view of others in the area.

Some communities have zoning ordinances, or restrictive covenants, or other conditions that limit the type, placement, or appearance of these dishes, and some forbid their use altogether. Restrictions exist because residents object to the size or appearance of these dishes. In a few cases, developers make arrangements with cable companies to pre-wire communities, at the cable compa-

---

[11] Cellular Telecommunications Industry Association, "Reinventing Competition: The Wireless Paradigm and the Information Age," (February 1995), p. 13.

[12] Due to collocation of antennas, local effective radiated power levels may vary substantially.

[13] There are three generally available types of satellite dishes that correspond to different satellite frequencies and services: 1) large 8- to 12-foot diameter dishes, known as C-band antennas; 2) smaller dishes, about 3 feet in diameter, known as Ku-band antennas, used, for example, to receive broadcasts from Primestar; and 3) small dishes, about 18 inches across, known as direct satellite service (DSS) dishes, used to receive broadcasts from DirecTV and United States Satellite Broadcasting (USSB).

nies' expense, in exchange for restrictive covenants that are written into deeds or other community rules. Some communities restrict only satellite dishes of a certain size or those not camouflaged (a typical disguise is a patio umbrella), while others make no distinction at all, prohibiting even traditional television antennas.

In response to growing concerns that restrictive covenants would affect the health of the satellite industry, in 1986, at industry's urging, the FCC ruled that the only permissible local restrictions were those that were narrowly written; based on health, safety, or aesthetic concerns; and that did not discriminate against receive-only satellite antennas. All others restrictions would be preempted.[14] With this ruling, the FCC attempted to balance the interests of the industry and consumers in receiving satellite broadcasts with the interests of communities in local control of land-use and enforcement of health, safety, and aesthetic regulations.

In 1993, the satellite industry pressed the FCC to modify the 1986 order to clarify the types of local restrictions that would be prohibited.[15] The industry claimed that many communities were imposing "noncompliant" regulations that the FCC was powerless to oppose—in particular size and height restrictions—which, by their nature, single out satellite dishes, including lot size limitations, limits in commercial or industrial areas, and other placement or screening requirements, or any flat bans.[16] The FCC is currently considering modifications to the 1986 order.

GENERAL COURT GUIDELINES ON FEDERAL PREEMPTION

Politically, there are few issues that raise the ire of a small but vocal segment of the population more than federal preemption of states' rights and local regulations. The recent Supreme Court decision striking down federal restrictions on gun possession near public schools,[17] that reversed decades of Court rulings on use of the interstate commerce clause of the Constitution to accomplish federal goals, underlines the necessity of considering carefully the appropriate and justifiable division of regulatory responsibilities between the states and the federal government. When coupled with continuing concern about the health effects of electromagnetic radiation, the local control of antennas could become a very divisive issue for policymakers.

The issues surrounding federal preemption of local zoning laws regarding antenna siting are part of a larger conflict between federal policy and state laws. In general, the supremacy clause of the Federal Constitution says that federal law overrides, or can prohibit, exercise of state laws. General rules on preemption are impossible to formulate because of the diversity and complexity of circumstances.[18] As Supreme Court Justice Black wrote for the majority in *Hines* v. *Davidowitz*, the test to be applied in such cases is whether a state law "stands as an obstacle to the accomplishment and execution of the full purposes and objectives of Congress."[19]

---

[14] Federal Communications Commission, "Preemption of Local Zoning or Other Regulation of Receive-Only Satellite Earth Stations," report and order, 47 CFR Part 25, *Federal Register* 51(31):5519-5527, Mar. 14, 1986.

[15] Satellite Broadcasting and Communications Association of America, "Reply Comments," before the Federal Communications Commission, *Preemption of Local Zoning Regulation of Satellite Antennas*, Report No. DS-1311, July 12, 1993.

[16] Ibid., pp. 9-12.

[17] *United States* v. *Lopez*, No. 93-1260, decided Apr. 26, 1995.

[18] Ronald D. Rotunda and John E. Nowak, *Treatise on Constitutional Law: Substance and Procedure*, 2nd ed., vol. 2, sec. 12.1, pp. 62-63. It should be noted that there is no mention of preemption in the Constitution itself.

[19] 312 U.S. 52, at 67 (1941).

Thus, congressional intent to preempt state law is the principal element of a preemption claim, and finding congressional intent when it is otherwise not explicitly expressed has been the task of the courts. Where no explicit congressional intent can be found, the courts have labored to balance state and federal interests to avoid conflicting regulation at the different levels of government. In general, the Court has given greater deference to state and municipal regulations that concern traditionally local issues—such as zoning, health, and safety measures—even while attending to the facts of each case considered on its own.[20] In other cases, although federal preemption has been granted by the courts with some ease, there seems to be increasing reluctance to allow it. One indication of this reluctance was shown when, in 1987, President Ronald Reagan issued an executive order directing that federal preemption should be sought:

> ... only when a statute contains an express preemption provision or there is some other firm and palpable evidence compelling the conclusion that the Congress intended preemption of the state law, or when the exercise of State authority directly conflicts with the exercise of Federal authority under the Federal statute.... Any regulatory preemption of State law shall be restricted to the minimum level necessary to achieve the objectives of the statute pursuant to which the regulations are promulgated.[21]

This order confirmed the trend evident in the Supreme Court, that had, by that time, begun to show increasing reluctance to usurp state and local law.[22]

THE CASE FOR FEDERAL PREEMPTION

The legal issue of land-use regulation and wireless telecommunications has been framed in terms of: 1) whether Congress's intent that new wireless services be quickly and comprehensively rolled out means that it intended that state and local land-use regulations be preempted, and 2) whether the FCC has the authority to preempt state and local regulations that impede the development of commercial mobile radio services (CMRS).

In building its case for preemption, the industry argues that Congress and the FCC have determined that development of nationwide wireless telecommunications services is a policy objective of the United States, citing language from the FCC's own rulings:

> We [the FCC] expect cellular to become an important communications tool, the extensive use of which can be of significant benefit to the American economy and to the more general public interest, and we are accordingly anxious to have it implemented as quickly as possible.... We believe that cellular is important enough to the public interest to warrant special attention to avoid delays.[23]

In order to meet this goal, wireless carriers maintain that they must be free to build towers where they are needed and not be subject to long local procedures that delay implementation. They argue that preemption is needed if services are to be deployed as quickly and widely as possible.

In the Omnibus Budget and Reconciliation Act of 1993, which amended section 332 of the Communications Act,[24] Congress said that "[n]o State

---

[20] Rotunda and Nowak, op. cit., footnote 18, sec. 12.3, p. 73.

[21] Reagan, R. R., President, United States, "Executive Order No. 12612—Federalism," (Oct. 26, 1987), secs. 4(a), (c), reprinted in 52 FR 41685 (1987).

[22] Rotunda and Nowak, op. cit., footnote 18, sec. 12.4, p. 76.

[23] Federal Communications Commission, "Public Mobile Radio Services," final rule, 47 FR 10,018, 10,033 (1982), cited in Littlejohn, op. cit., footnote 8, p. 259.

[24] This amendment streamlined all commercial mobile radio services into one regulatory framework. Public Law 103-66, Aug. 10, 1993.

or local government shall have any authority to regulate the entry of or the rates charged by any commercial mobile service or any private mobile service." States may only regulate "other terms and conditions."[25] The industry argues that only a narrow reservation of authority was reserved for state and local governments over telecommunications activities in order that "[s]tate and local governments may not lawfully bar entry, create regulatory disparities or introduce significant inefficiencies in the production of CMRS through zoning and other similar regulation."[26] By this, the wireless industry asserts that: 1) Congress tacitly allowed federal preemption, because zoning regulations introduce inefficiencies in the establishment of CMRS services, and 2) given the FCC's long-standing commitment to efficiency as a major criterion in regulating radio services, the FCC should preempt local zoning regulations.[27]

In carrying out congressional mandates, questions have arisen regarding the authority of the FCC to preempt local regulation. Under the interstate commerce clause, as developed through various court cases dealing with telecommunications regulation,[28] the FCC has regulatory authority over telecommunications that have interstate connections. This discretionary power generally covers any system connected to the public switched telephone network, including cellular telephony and new PCS. Preemption proponents argue further that the FCC has jurisdiction over equipment that is used in providing wireless services, such as antenna siting where heights and locations can affect service delivery. They note that the Court of Appeals for the District of Columbia Circuit has held that:

> If the [1934 Communication] Act's goal of providing uniform, efficient service is ever to be realized, the Commission must be free to strike down the costly and inefficient burdens on interstate communications which are sometimes imposed by state regulation.[29]

To date, however, the FCC has not decided whether it should act on this issue. Although it can strike down regulations that restrain interstate telecommunications activities, it is not required to do so, nor does it mean that sweeping national preemption is necessary. Until such a determination is made by the FCC or Congress, each challenge to local laws and regulations (each individual siting) must be argued by the cellular carriers on an individual basis.[30] Because each local proceeding could take many months, this could slow service deployment or upgrades, add significantly to the network's start-up costs, and slow earnings of wireless operators.[31]

---

[25] 47 U.S.C., sec. 332(c)(3)(A). OTA found no legislative history in this regard.

[26] Cellular Telecommunications Industry Association, op. cit., footnote 3, p. 7.

[27] Littlejohn, op. cit., footnote 8, pp. 259-261.

[28] Ibid., pp. 253-256, citing *Puerto Rico Telephone Company* v. *FCC*, 553 F.2d 694, 698 (1st Cir. 1977), which determined that the FCC could prohibit the private branch exchange (PBX) rule as it, in effect, encroached on the FCC's authority over interstate commerce, and relied on *Ambassador, Inc.* v. *United States*, 325 U.S. 317 (1945), which affirmed that the FCC's jurisdiction "extends to 'interstate wire communication from its inception to its completion.'"

[29] *National Association of Regulatory Utility Commissions* v. *FCC*, 746 F.2d 1492, 1501 (D.C. Cir. 1984), cited in Littlejohn, op. cit., footnote 8, p. 256.

[30] Littlejohn, op. cit., footnote 8, p. 256.

[31] For examples of local opposition to cellular antennas that wireless companies say show significant added costs or other burdens, see McCaw Cellular Communications, Inc., "Comments," before the Federal Communications Commission, *In the Matter of Amendment of the Commission's Rule To Preempt State and Local Regulation of Tower Siting For Commercial Mobile Services Providers*, RM-8577, Feb. 17, 1995, pp. 10-19, and Southwestern Bell Mobile Systems, Inc., "Comments," before the Federal Communications Commission, *In the Matter of Amendment of the Commission's Rule To Preempt State and Local Regulation of Tower Siting For Commercial Mobile Services Providers*, RM-8577, Feb. 16, 1995, pp. 8-15.

THE CASE AGAINST FEDERAL PREEMPTION

Opponents of preemption argue that state and local rights, including regulating the power output of facilities in their jurisdictions, must be preserved because they are the appropriate loci for protecting public health, safety, and welfare.[32] They object to antennas on several grounds: antennas can be obtrusive and may have unacceptable visual impacts on neighborhoods, which lowers property values; there may be health hazards from electromagnetic radiation emitted from antennas close to residences and schools; and without local regulations tailored to local conditions, antennas may be poorly constructed or unsafe.

■ Local Control

Preemption opponents argue that there is a limitation to the FCC's power when matters pertain exclusively to local or intrastate matters.[33] Under sec. 332 (c) (3) of the Communications Act:

> ... no State or local government shall have any authority to regulate the entry of or the rates charged by any commercial mobile service or any private mobile service, except that this paragraph shall not prohibit a State from regulating the other terms and conditions of commercial mobile services.

Opponents argue that this exception permits them to continue to regulate antenna placements under local zoning laws because zoning falls under "other terms and conditions," and is not related to "entry of or the rates charged by" CMRS providers. In their view, while it may be more costly or difficult to establish service quickly, CMRS providers can, nevertheless, establish service. The Cellular Telecommunications Industry Association's (CTIA) position that any regulation is an obstacle to entry is overly narrow, opponents argue.[34] Opponents of preemption point to tests of federal preemption involving amateur radio antenna regulations, as decided in *Guschke* v. *City of Oklahoma City*.[35] This case determined that despite general federal encouragement of amateur radio as socially important, that finding alone was not sufficient to warrant federal preemption of local regulations.

Furthermore, where the relevant market for service is local, as it is with many wireless services, communities argue that they have the right to decide what costs and benefits they are willing to sustain, as long as there are no substantial impacts on other areas. If local costs are raised by local restrictions, and these costs are not borne by other communities, then it could be argued that preemption is an unnecessary intrusion.[36]

Communities feel that opening the door to federal preemption of local zoning and land-use restrictions may result in other intrusions:

> This attempt at preemption by the cellular phone industry with the cooperation of the FCC is a blatant attack on our communities that is more of a threat and at a lower level of morality than any neighborhood drug dealer... If this preemption is allowed it will open the door for the federal government to attack any and all zoning

---

[32] "Local Groups Oppose Radio Tower Preemption Request," *Telecommunications Reports*, Feb. 20, 1995, p. 45.

[33] *Louisiana Public Service Commission* v. *FCC*, 476 U.S. 355 (1986), sec. 2(B), cited in Natural Resources Defense Council, "Comments," before the Federal Communications Commission, *In the Matter of Amendment of the Commission's Rule To Preempt State and Local Regulation of Tower Siting For Commercial Mobile Services Providers*, RM-8577, Feb. 16, 1995, p. 5.

[34] Natural Resources Defense Council, op. cit., footnote 33, p. 3.

[35] 763 F.2d 379 (10th Cir. 1985), cited in Littlejohn, op. cit., footnote 8, p. 260.

[36] According to the economic analysis of rights, as articulated by Ronald Coase, for an efficient economic outcome to be achieved, it matters little which party bears the economic burden of ameliorating a noxious or objectionable condition. In the case of antenna siting, either the wireless company or the local residents pay for making antenna siting less objectionable, but in end the cost of service will be the same. The fact that costs can be arbitrarily allocated means that some basis for deciding must be determined. For a discussion of Coase's Theorem, see Charles Fried, *Right and Wrong* (Cambridge, MA: Harvard University Press, 1978), pp. 81-107.

regulations in all of our communities whenever a wealthy and powerful industry group with an influential lobby sees those regulations as an obstacle to increased profit... At a time when there is so much talk in Washington, D.C. about taking back our neighborhoods there is a clear example here of us losing those very neighborhoods to big business.[37]

■ Health, Safety and Aesthetic Concerns

In addition to arguments concerning the legality of preemption, opponents further argue that the safety of radio emissions has not been fully established, and that local zoning and other regulations are appropriate measures to take in order to protect public safety (see chapter 11 for more discussion of health issues). Aesthetic concerns undoubtedly lie at the core of many objections to antennas, but these are harder to argue for without running afoul of charges of inconsistency, beauty being in the eye of the beholder.[38] As a practical matter, aesthetics is generally formally given as a reason for restricting antenna siting in cases where obvious historical or other design considerations are at stake in a community.

The Natural Resources Defense Council notes that section 332 (a) of the Communications Act[39] directs the FCC to take action after considering whether such action will "promote the safety of life and property." It argues that local zoning regulations are designed to protect public health, and that preempting them could harm the public. Communities claim that this language provides them with legitimate grounds for regulating or prohibiting the placement of antennas within their boundaries. Until a consensus on the safety of

broadcast antennas is established, they will continue to have the right to limit placements.

The industry counters that health concerns are used arbitrarily and capriciously by communities to delay or prevent antenna installations:

Despite overwhelming and uncontroverted evidence that the extremely low power emissions or radio frequencies from properly designed and constructed antenna sites fall well below every state and federal exposure limitation, (usually by factors of 500 to 3000 percent), the unfounded health and safety concerns of local citizens are most easily appeased by simply rejecting applications and letting the courts overturn the decision—at great expense and costly delay for the commercial mobile service provider.

Health and safety claims are also often a subterfuge for underlying and unreasonable "aesthetic" concerns. In most typical communities telephone poles, water towers, broadcast towers and microwave relay sites proliferate, yet zoning boards often find that mobile antennae poles and towers violate vague "aesthetic" standards included in local zoning codes. Were the same standards to be applied to other forms of communications these communities would have no telephone service, no radio service, no television service and no utilities.[40]

Regarding the aesthetics of satellite dishes, the FCC has held that local regulations do hold sway in some cases:

State and local zoning regulation or other regulations that differentiate between satellite receive-only antennas and other types of antenna facilities are preempted unless such regulations (a) have a reasonable and clearly defined

---

[37] See George Curtis of Seattle, WA, "Comments," and R. James Pidduck, of Edmonds, WA, "Comments," before the Federal Communications Commission, *In the Matter of Amendment of the Commission's Rule To Preempt State and Local Regulation of Tower Siting For Commercial Mobile Services Providers*, RM-8577, Feb. 14, 1995 and Feb. 17, 1995.

[38] See, for example, Town of Greenburgh, NY, "Local Law on Temporary Moratorium on the Establishment of New Commercial Antenna," 1995, and Abby Gilbert of Washington, DC, "Comments," before the Federal Communications Commission, *In the Matter of Amendment of the Commission's Rule To Preempt State and Local Regulation of Tower Siting For Commercial Mobile Services Providers*, RM-8577, Feb. 12, 1995.

[39] 47 U.S.C., sec. 332 (a).

[40] Cellular Telecommunications Industry Association, op. cit., footnote 7, p. 5.

health, safety, or aesthetic objective; and (b) do not operate to impose unreasonable limitations on, or prevent, reception of satellite delivered signals by receive- only antennas or to impose costs on the users of such antennas that are excessive in light of the purchase and installation cost of the equipment.

Regulation of satellite transmitting antennas is preempted in the same manner except that state and local health and safety regulation is not preempted.[41]

These issues will likely continue to be contentious for the foreseeable future, given their pervasive scope, and because they pit national objectives for quick and inexpensive service provision against deeply held beliefs, traditions and laws concerning local land use regulation. Some basis must be given for deciding who will bear the costs of antenna siting; this would seem to be the primary responsibility of the Congress.

---

[41] 47 CFR 25.104.

Wireless Technologies and Universal Service | 9

O ne of the most important contributions that wireless technologies can make to the emerging National Information
Infrastructure (NII) is to support and extend the provision
of communications services to all Americans. The main
purpose of the Communications Act of 1934 was:

> to make available, so far as possible, to all the people of the
> United States a rapid, efficient, Nationwide and worldwide wire
> and radio communications service with adequate facilities at rea-
> sonable rates.[1]

The term "universal service" has come to mean widespread availability of basic telephone service at affordable rates. Today, 93.8
percent of U.S. households have telephone service, down somewhat from the all-time high of 94.2 percent, recorded in 1993.[2]

Policymakers are concerned with both providing telecommunications service to households that do not have it and with maintaining universal service during the transition to a more
competitive market. Wireless technologies can contribute to universal service goals by providing unserved users with access to
service and/or by allowing customers to be served at lower costs
than with wireline technology. However, policymakers also recognize that the definition of universal service will evolve to include more advanced communication and information services. If
wireless technologies are to play a continuing role in supporting

---

[1] 47 U.S.C. 151.

[2] Federal Communications Commission, "Telephone Subscribership in the United
States," April 1995, table 2.

universal service, they will have to keep pace with the capabilities of wireline systems.

FINDINGS

- **Wireless technologies can provide access to telecommunications services in areas where wireline service is not available.** The first component of universal service is physical access—the availability of service regardless of location. Although most households in the United States have access to wireline telephone service, in some parts of the nation it is difficult or impossible to deliver service with wireline technologies because of high cost, difficult terrain, or geographic barriers. But radio waves can cross water, canyons, and other obstacles, providing telephone service to homes that would otherwise remain unserved. In addition, broadcast and satellite technologies are the only means available to deliver video programming and other advanced services to some parts of the nation.

- **Wireless technologies may be able to serve some homes at lower cost than wireline technologies.** With wireline technology, the cost to build a copper loop depends on the distance from the telephone company's *central office* to the home. In sparsely populated rural areas, where many homes are far from the central office, it can be very expensive to provide wireline telephone service. Wireless local loop systems, which connect homes to the telephone network through a radio link, may be less expensive than long rural copper loops. With wireless technology, the cost to serve a home is less dependent on distance from the central office.

If wireless proves to be a lower cost alternative in rural areas, it would allow for a reduction in the industry cross subsidies currently needed to keep rural telephone service affordable. Federal policies have long supported the use of these subsidies to extend universal service to rural areas, and as a result, telephone penetration in rural areas no longer lags behind that of the cities. However, the system of subsidies is being threatened by the transition to a competitive telecommunications industry, in which consumer prices are expected to be driven closer to the actual cost of providing service. Deploying a less expensive technology would allow for a reduction in subsidies for rural telephone service while keeping prices affordable.

- **Despite the potential cost advantage of wireless technology, it is premature to conclude that it can eliminate the need for rural telephone subsidies.** Few households currently have wireless telephone service. The new digital technologies that will allow for low-cost wireless local loops are only now being introduced. Production economies have not been achieved, and final prices are not yet set. For this reason, determining the cost—both system capital cost and subscriber equipment cost—of different levels of wireless service (basic voice through interactive broadband) is difficult. Moreover, it is not clear whether wireless technology can maintain a cost advantage while providing the high-speed two-way video and data services that may be required as the definition of universal service evolves.

Even if wireless systems can provide lower cost alternative telephone service in rural areas, a broader portfolio of policies will still be required to support affordable telephone service for low-income users in both urban and rural areas. Wireless technology may provide a way to keep rural telephone service affordable, while reducing the subsidies, but there are still millions of users in both urban and rural areas who cannot afford telephone service even at current rates. If anything, cities with a large low-income population have a more acute universal service problem. The deployment of wireless technology is unlikely to make telephone service significantly more affordable for these low-income households. Special programs such as Lifeline and LinkUp America, which subsidize users directly, will likely have to be maintained.

In order to more fully explore the potential of wireless technologies in helping meet evolving

NII and universal service goals, **Congress could support experimentation with wireless technologies by rural telephone companies.** The use of wireless to provide basic telephone service in rural areas is unproven, and there are many uncertainties. Pilot projects or demonstration projects could help to establish whether wireless is, in fact, a viable option and also help determine the applications in which wireless can be used most effectively.

Congress could also direct the Federal Communications Commission (FCC) to determine whether additional spectrum should be allocated to wireless loop service in rural areas. In the seven years since the Commission last examined this issue, wireless technology has advanced considerably and interest in rural wireless has grown. Some local exchange carriers believe that the current allocation is insufficient and have urged the Commission to allocate additional spectrum.

THE ROLE OF WIRELESS TECHNOLOGIES IN UNIVERSAL SERVICE

The current concept of universal service entails the provision of basic telephone service at affordable rates. Wireless systems, both terrestrial and satellite-based, have certain advantages, including coverage and a different cost structure, that may allow them to support universal service by improving access in areas that have no telephone service and/or by lowering the cost of service. Terrestrial "wireless local loop" systems broadcast from a tower to the homes in the surrounding area; the range can be up to 20 miles or more. The signals are received by an antenna mounted either on a pole near the house or on the outside wall of the house, and then connected by wire to a telephone inside the house. Telephone service can also be delivered via satellite, although satellite service is usually more expensive than terrestrial wireless service.

▌ Extending Service to Unserved Populations

The first component of universal service is physical access—the requirement that service be available. In the United States, there are very few areas that have no telephone service. The long effort to bring telephone service to rural America has been largely successful. However, a small number of households remain unserved because the wires needed to provide service do not reach them.

Households without physical access are generally in areas where wireline technology is not viable, due to prohibitive cost, difficult terrain, or a geographic barrier such as a river or mountain. The data on unserved households is unreliable, but one group estimated that there were approximately 150,000 households in areas where there was no certified telephone company and about 330,000 households in areas where there was a telephone company but no service was available.[3] Another survey found about 500 to 2,000 unserved customers in Colorado, mainly in mountainous regions.[4]

There have been several estimates of the number of rural households that could be served with wireless technology, either because they have substandard telephone service or because they are without telephone service. The last time the FCC examined the issue of rural radio, in 1987, one survey found that 7,731 subscribers, scattered among 138 telephone companies, could be served or upgraded through radio loop technology.[5] However, the petitioners who initiated the FCC proceeding estimated the nationwide total of eligible subscribers at approximately 900,000 by counting

---

[3] Rural Radio Task Force, comments before the Federal Communications Commission, "Petition for Rulemaking to Establish Basic Exchange Telecommunications Radio Service," CC Docket No. 86-495, May 9, 1986, pp. 14-16.

[4] George Calhoun, *Wireless Access and the Local Telephone Network* (Boston, MA: Artech, 1992), p. 185.

[5] Federal Communications Commission, *Basic Exchange Telecommunications Radio Service*, Report and Order, CC Docket No. 86-495, 3 FCC Rcd 215 (1988).

households that were without telephone service or had four- or eight-party-line service.[6] Finally, a study by Bellcore estimated that 213,000 to 246,000 households could be served by radio.[7]

There is a clear role for wireless technologies in serving these remote and difficult locations. Bell Atlantic, for example, serves a household on an island in the James River with terrestrial wireless technology.[8] In Nevada, in the Antelope and Reese Valleys, 50 residential customers who did not have service will soon receive it from a cellular company.[9] Wireless technologies can also be used for temporary installations that do not justify the construction of a wireline network, for emergency restoration of service, and to provide interim service until wireline facilities have been constructed.

Although most installations of wireless local loops have relied on terrestrial technology, satellites may offer another option in especially remote areas. Universal access is inherent in the use of satellite technology—once the satellite has been launched, any location within its footprint can get service. In Alaska, satellites have played a key role in delivering service to remote villages for many years. US West has launched a trial in which Very Small Aperture Terminal (VSAT) equipment is used to provide telephone service to 43 Wyoming customers.[10] New mobile satellite services may offer telephone and more advanced services

to fixed users in isolated areas. (See chapters 3 and 5.)

Finally, wireless could provide service to those who have no permanent home. For example, four to five million migrant farmworkers, who usually have limited acces to a telephone, could use wireless—if service was less expensive.[11] Currently, the Census Bureau's statistics used to measure telephone penetration do not count the use of mobile telephone service if it is used instead of wired service to a home.[12] But a small number of people may already be using a mobile phone as their primary phone.

∎ Increasing Affordability

Physical availability is only one component of universal service. Service must also be affordable. In some applications, wireless technologies could support universal service goals by delivering telephone service at a lower cost than wireline technologies. Until recently, this would have seemed unlikely—there are no more than a few thousand households in the United States that get their telephone service over a wireless link, mainly in remote and hard-to-reach areas. According to some published figures, however, the cost of a wireless local loop has dropped to between $800 and $1,200, which is comparable to the average cost of a copper loop in the United States.[13] And in areas

---

[6] Federal Communications Commission, *Basic Exchange Telecommunications Radio Service*, Notice of Inquiry, CC Docket No. 86-495, 2 FCC Rcd 326 (1987).

[7] Federal Communications Commission, *Basic Exchange Telecommunications Radio Service* Report and Order, op. cit., footnote 5.

[8] Personal Communication, Donald Brittingham, Bell Atlantic, Mar. 20, 1995.

[9] "Nevada PSC OKs Programs for Service to Remote Areas," *Telecommunications Reports*, vol. 61, No. 1, Jan. 9, 1995, p. 11.

[10] "US West Deploys USATs for Rural U.S. Telephony," *Telecommunications,* Americas Edition, vol. 28, No. 4, April 1994, p. 8.

[11] Some of these workers already spend $40 or more per week on long distance calls to their families, but the added cost of wireless subscriptions put cellular out of their reach. Based on OTA interviews with migrant workers and migrant health professionals.

[12] Jorge Schement, Alex Belinfante, and Larry Povich, "Telephone Penetration 1984-1994," in Proceedings of the 22nd Annual Telecommunications Policy Research Conference, p 4.

[13] Terry Sweeney, "Lenders Backing Wireless Loops," *CommunicationsWeek International,* Dec. 12, 1994, p. 3. See also, Bruce Egan, "Economics of Wireless Communications Systems in the National Information Infrastructure," unpublished contractor report prepared for the Office of Technology Assessment, U.S. Congress, Washington, DC, November 1994.

that are sparsely populated or have difficult terrain, the cost of a copper loop can easily reach as high as $2,000 to $5,000, making wireless solutions much more attractive.[14]

New digital technologies are the primary driver behind low-cost wireless loops. (See chapter 3.) Reductions in the cost of wireless local loop systems are also being driven by the explosive growth in demand for mobile telephone service. Because the equipment used to provide fixed wireless service is similar to that used for mobile service, fixed users can piggyback on the technology advances and declining cost of mobile technology. As mobile service becomes more widely used and the price of equipment drops due to economies of scale, fixed wireless services will also become less expensive.[15]

Impact of Wireless Technology on Rural Subsidies

Background

Wireless loops may play an important role in reducing the cost of providing telephone service in rural areas. One of the characteristics of wireless technology is that the cost to serve a home does not depend on whether the home is close to the transmitter or far away, as long as it is within range. With wireline technology, on the other hand, the cost to serve a home depends directly on its distance from the central office. In sparsely populated rural areas, homes are located further apart, requiring long, expensive loops dedicated to each customer. For the most remote customers, even terrestrial wireless technologies may be too expensive—if a cell site serves a very small number of households, for example. In such cases, satellite technology may be the only cost-effective option.

Because of these high costs, telephone penetration rates in rural areas of the United States were much lower than in the cities for the first half of this century. To remedy this situation, federal and state regulators developed policies designed to make rural telephone service more affordable. The Rural Electrification Administration (REA—now the Rural Utilities Service) offered low interest loans, provided technical support, and also helped with the formation of cooperatives in areas where commercial companies chose not to provide telephone service. But the more important policy tool was the subsidization of rural telephone service with revenues transferred from customers in lower cost urban areas. It has been estimated that about $5.5 billion flows from urban to rural users to maintain rural telephone rates comparable to those in urban areas.[16]

One subsidy mechanism that is used to keep rural telephone rates low is rate averaging, by which regulators require that carriers charge both urban and rural customers the same rate. As a result, rural users are charged less than it costs to serve them, while urban users pay more in order to provide the necessary subsidies. Rate averaging is the primary tool used by the larger local exchange carriers, the Bell Operating Companies, to provide affordable service in their rural territories. These companies serve a diverse customer base of rural and urban customers and can successfully transfer costs from one group of customers to another.

Most of rural America, however, is served by small independent telephone companies—some serving only a few hundred households—that op-

---

[14] A. Javed, P. O'Kelly, K. Dick, and M. Lucey, "Wireless Technology Evolution and Impact on the Access Network," in Proceedings of the 1994 Conference on Personal Wireless Communications, p. 12.

[15] In general, systems developed specifically for wireless loop applications provide a higher level of voice quality than those based on modified versions of mobile technologies. Many of today's mobile technologies are designed to deliver voice quality lower than that of wireline systems, trading off quality for the advantages of mobility and increased capacity. Achieving better voice quality adds to the cost of the system.

[16] Telecommunications Industries Analysis Project, "Apples and Oranges: Differences Between Various Subsidy Studies," Oct. 10, 1994, p. 2.

erate only in high-cost areas and have few offsetting low-cost loops. The FCC tries to ensure that these small companies can deliver affordable telephone service by subsidizing them with revenues from a Universal Service Fund. The money paid into this fund comes from the long distance carriers, who contribute about one cent of every dollar of their revenues. All local telephone companies with loop costs more than 15 percent above the national average are eligible to withdraw from the Fund. The higher their loop costs, the more funds they can withdraw. In 1993, about $750 million was transferred from the long distance carriers to high-cost local telephone companies.[17]

Proposed changes to the Universal Service Fund could encourage small telephone companies to look for lower cost loop technologies. Under current rules, telephone companies withdraw from the Universal Service Fund in proportion to their loop costs. As a result, they make an adequate return on investment, regardless of whether they have used the most efficient technology. The FCC is currently examining whether it is possible to base subsidies on a projected reasonable cost to serve an area, based on *proxy factors* such as population density or terrain type.[18] In the past, high-cost assistance based on proxy factors was rejected in part because the data was more difficult to assemble or verify than simple loop cost.[19]

The system of subsidies has largely been successful; telephone penetration rates in rural areas no longer lag behind those in urban areas. However, there is a concern that the subsidy flows will be more difficult to maintain in a deregulated and competitive environment. For example, a Bell Operating Company that priced urban service above cost in order to subsidize rural users could find its rates undercut by a new competitor that served only the urban market. As competition drives prices closer to cost, those who have benefited from the existing system of cross-subsidies—primarily rural users—may see their rates rise. One organization of rural telephone companies estimated that their subscribers' monthly bills would increase by about $12 per month.[20]

Although there is ongoing debate about the extent to which higher rural prices would cause users to drop off the network, Congress has indicated a desire to maintain a balance between urban and rural rates. Both S.652 and H.R. 1555, the telecommunications bills currently being debated in Congress, state that consumers in rural and high-cost areas should have access to telecommunications services at the same rates as urban consumers. One way to achieve this objective would be to find a subsidy scheme that is compatible with a competitive market. Mechanisms to accomplish this have been the subject of much discussion, but there is, as yet, no consensus on the best solution.

The impact of wireless systems

The promise of wireless technology is that it would provide a way to keep rural rates affordable while at the same time reducing the reliance on subsidies from urban users. Because of its cost structure and the advent of digital technology, wireless technology may be able to serve some sparsely populated rural areas at about the same

[17] Federal Communications Commission, *Amendment of Part 36 of the Commission's Rules and Establishment of a Joint Board*, Report and Order, CC Docket No. 80-286, note 4, Dec. 23, 1993.

[18] Federal Communications Commission, *Amendment of Part 36 of the Commission's Rules and Establishment of a Joint Board*, Notice of Inquiry, CC Docket No. 80-286 (1994), p. 22.

[19] Ibid., p. 23.

[20] OPASTCO, "Keeping Rural America Connected," p. ES-4, 1994.

cost per household as a densely populated urban area. The objective of equal urban and rural rates could then be achieved without cross-subsidies.[21] For those who see no simple way to continue the rural subsidies in a competitive environment, or view the continuing debate over universal service as an impediment to the transition to a more competitive telecommunications industry, the prospect of a technological fix is attractive.

Even if wireless were found to be a lower cost option, however, it would probably be deployed on a piecemeal basis. Nearly all households, even in rural areas, already have wireline telephone service. Wireless technology would be used initially to bring service to the small number of households that currently have none or to provide for new growth in rural areas. It may also be used to upgrade substandard loops, but only about 3 percent of the existing copper loops are rebuilt each year. As a result, it will take some time before the cost structure of the rural telephone network would change enough to allow for a reduction in subsidy requirements.

Most studies that show wireless making a dramatic impact on the cost of rural telephone service assume that the network is being built from scratch.[22] In fact, in countries that are building their telecommunications infrastructure for the first time, wireless is often the technology of choice. Fixed cellular access systems have been deployed in over 40 countries,[23] primarily in developing countries such as Indonesia, India, and the Philippines, but also in Spain and in central Europe. The market for wireless local loop equipment has been estimated at about $4 billion over the next three years, and provides an important export opportunity for U.S. manufacturers.[24]

Low-Income Populations

Among the 6.2 million Americans who do not have telephone service, low income is the primary predictor. For example, of households on welfare, 27.9 percent lack telephones.[25] Now that policymakers have succeeded in bringing telephone service to rural America and in equalizing urban and rural rates, they are beginning to concentrate on bringing telephone service to these low-income populations. If anything, universal service concerns are at least as great in urban areas with significant low income populations as in rural areas—the focus of universal service policy initiatives in the past.

Despite its potential cost advantages, however, wireless technology is unlikely to lower the cost of telephone service sufficiently to make it more affordable for low-income populations. It may help keep rural telephone rates close to urban rates at lower subsidy levels, as noted above, but it will not dramatically lower the average cost of telephone service in the United States. Although wireless probably has a cost advantage over copper when used for rural or longer suburban loops, it is, at best, comparable in cost to copper when used for the much larger number of short urban loops. In addition, any savings from a reduction in rural subsidies paid by urban users would be spread across a very large number of households and reduce the average urban bill only slightly.

---

[21] ". . . .the public interest is unquestionably served when basic telephone service can be provided in a more cost effective manner — particularly in rural areas which generally require universal service subsidies to keep rates for local service affordable." US West comments before the Federal Communications Commisison, *Allocation of Spectrum Below 5 Ghz Transferred from Federal Government Use* , ET Docket 94-32, Dec. 19, 1994, p. 6.

[22] See, for example, Hatfield Associates Inc., "The Cost of Basic Universal Service," July 1994.

[23] Jean-Philippe Haag, "Fixed Cellular Solutions for Wireless Access," *Telecommunications,* vol. 28, No. 12, December 1994, p. 57.

[24] Sweeney, op. cit., footnote 13.

[25] Schement, op. cit., footnote 11.

Because wireless technology will make telephone service more affordable only in a limited number of applications, it cannot, by itself, dramatically increase current levels of penetration. The lack of telephone service among low income groups is a complex problem whose solution will almost certainly require the continuation of federal and state programs that address the affordability question more directly. One such program reduces monthly subscriber charges (the Lifeline Service program), while another provides for reduced installation charges (the LinkUp America program). Over the past decade, states that have pursued aggressive federally supported assistance policies have shown the greatest increase in penetration among households below the poverty level.[26]

POLICY ISSUES

Wireless has considerable promise as a tool for maintaining and expanding universal service, especially in rural areas. But the use of wireless technologies in fixed applications is still rare; OTA was unable to determine the number of households whose telephone service is provided with wireless technologies, but it is probably no more than a few thousand. It is premature to assume that the deployment of wireless technology can eliminate the need for a rural subsidy program. Moreover, it is uncertain whether wireless technology can maintain a cost advantage while providing the high-speed two-way video and data services that may be required as the definition of universal service evolves (see below). However, federal policy should make available sufficient spectrum for the potential of wireless in rural areas to be explored.

■ Wireless Technology and the Evolving Definition of Universal Service

Wireless technology can provide today's definition of universal service—"basic" voice telephone service. As technology advances and users' needs change, however, the requirements for universal service are expected to broaden; perhaps to include high-bandwidth services such as image transfer and video. The telecommunications bills currently being debated in Congress, for example, define universal service as an evolving level of services. Both S.652 and H.R. 1555 envision that the FCC would periodically determine which services should be provided at affordable rates to all Americans, including those in rural areas.

Wireless technology already plays an important role in providing one-way video services, although they are not part of the current definition of universal service. For example, while 96 percent of U.S. households currently have access to cable television, 4 million households remain unserved. Most of these are in areas where constructing cable systems would be prohibitively expensive.[27] By contrast, at least one or two channels of broadcast television is available in 99.5 percent of households, and over 1 million households in areas without cable service get service from large C-band satellite dishes. Most recently, high-powered direct broadcast satellites (DBS) have brought multichannel video to unserved areas at a price that is competitive with cable rates in urban areas.[28]

In the future, the definition of universal service is likely to include two-way data communications capability that would allow subscribers to access the Internet or online services. Most terrestrial wireless access systems currently allow data to be

---

[26] Schement, op. cit., footnote 11, p. 11.

[27] Federal Communications Commission, "Broadcast Television in a Multichannel Marketplace," June 1991, p. 71.

[28] Beth Murphy, "Rural Americans Want Their DirectTV," *Satellite Communications*, March 1995, p. 30.

transmitted at 9,600 bits per second, the speed of a moderately good wireline modem, to access on-line services and for other applications. Some of the newer systems designed specifically for wireless local loop systems offer even higher fax and data transmission rates. In some respects, wireless may be better able to provide advanced services than the existing wireline network. In rural areas, deteriorating copper loops may not be able to support high-speed fax and data transmission, and it may be less expensive to install a new wireless loop than to rebuild an aging copper loop.

It is unclear, however, whether wireless will be able to match all of the new services that will be provided over advanced wireline networks and still maintain its cost advantage in more than the most difficult to reach locations. Both S.652 and H.R.1555 would require that the services available to urban and rural users be reasonably comparable. In the cities, there is growing interest in a wireline technology known as Integrated Services Digital Network (ISDN) that offers a 128,000 bit per second data stream to and from the home. Both telephone companies and cable companies are also beginning to upgrade urban networks with fiber and coaxial cable to provide high-bandwidth services. No existing wireless access technology can match these capabilities, although the proposed Spaceway and Teledesic satellite systems would provide high-speed data communications services. (See chapter 5.)

∎ Spectrum Availability

Spectrum allocations determine the viability of wireless services—whether they can be offered at all, their capabilities, and the cost of the service. For example, the amount of spectrum allocated

determines whether fixed wireless service is limited to basic telephony, or can also carry high-bandwidth information-age services such as interactive multimedia or video. The band in which the spectrum is allocated also affects the economics of the service. Lower frequencies are especially useful because the signal propagates further, allowing more households to be covered from the same tower and decreasing the cost per household.

For wireless to provide the services that constitute the universal service package, sufficient spectrum must be made available. Today, only a limited amount of spectrum is available for fixed voice services—almost all of the spectrum that is allocated for wireless telephony is restricted to mobile applications. The only spectrum available to serve fixed users is allocated to a service called BETRS (Basic Exchange Telecommunications Radio Service), which was established in 1987 by the FCC. But because demand for the service was uncertain, the FCC did not create an exclusive frequency allocation and allocated only a small number of channels.[29] In addition, the FCC only allows carriers with Personal Communications Service (PCS), cellular, or Specialized Mobile Radio (SMR) licenses to serve fixed users on an "incidental" or "ancillary" basis.[30] Their customers may choose to employ their mobile phones in a fixed application, but the network has to be designed primarily to serve mobile users.

In large part, the restrictions on the provision of fixed services by mobile service providers are due to concerns about competition. Competition in the provision of local telephone service has historically been limited by the belief that such service was actually a natural monopoly most effectively pro-

[29] The FCC allocated 26 frequencies in the 450 megahertz band to BETRS on a co-primary basis. In the cities, these frequencies are used for a mobile telephone service, but the FCC reasoned that in rural areas, where BETRS would be more useful, they are often vacant. BETRS *Report and Order*, op. cit., footnote 5. In 1988, the FCC also permitted the use of cellular frequencies for BETRS, but in practice only the 450 megahertz band has been used.

[30] "There is only a limited amount of spectrum for these new PCS services, and fixed service uses generally can be accommodated by other means or in other frequency bands. Therefore, the primary focus of PCS will be to meet communications requirements of people on the move." Federal Communications Commission, *Amendment of the Commission's Rules to Establish New Personal Communications Services,* Notice of Proposed Rule Making and Tentative Decision, ,7 FCC Rcd 5689 1992.

vided by only one carrier. Most states still limit competition in the local telephone service market (although this is changing), and a broad grant of permission to cellular or PCS carriers to provide fixed as well as mobile service might have been seen as sanctioning competition in the local exchange market. In creating the BETRS service, the FCC was careful to note that it would only grant authority to provide BETRS to companies that were either certified local exchange carriers or had some other form of permission from the state to provide local exchange service.[31]

As state barriers to local exchange competition begin to come down, the FCC has the option to allow mobile services providers to provide fixed service. In one survey of small telephone companies, 32 percent believed that wireless would be a competitor.[32] Noting that the PCS frequencies are unlikely to be fully utilized for mobile services in rural areas, the FCC recently indicated that it is willing to consider waiver requests to use PCS fre-

quencies to provide fixed services.[33] However, this position was stated in passing in an unrelated proceeding, and there is still considerable uncertainty about which uses of the PCS spectrum are permitted. The FCC will need to clarify its position regarding wireless fixed telephone services before full competition can emerge in the local telephone market.

The FCC also has the option to allocate additional spectrum specifically for wireless local loop applications.[34] Several local exchange carriers recently requested that the FCC allocate spectrum transferred from the federal government to wireless local loops. However, under most of these proposals, the wireless local loop spectrum would only be available to the incumbent local exchange carrier. As the telecommunications industry becomes more competitive, it is unlikely that the FCC could exclude other carriers from competing for this spectrum.

---

[31] Federal Communications Commission, *Basic Exchange Telecommunications Radio Service*, Report and Order, op. cit., footnote 5, p. 217.

[32] Western Alliance, *Universal Service in the Nineties*, p. 14.

[33] Federal Communications Commission, *Allocation of Spectrum Below 5 Ghz Transferred from Federal Government Use*, First Report and Order, op. cit., footnote 20.

[34] United States Telephone Association comments before the Federal Communications Commission, ET Docket No. 94-32, Dec. 19, 1994, p. 3.

Privacy, Security, and Fraud | 10

A s wireless technologies become more widely used and more closely integrated into the National Information Infrastructure (NII), concerns about privacy, confidentiality, the security of communications, and protection from fraud will become increasingly important (see box 10-1).[1] Although laws that address such issues do exist, users of wireless technologies generally have less assurance of confidentiality and protection from fraud than do users of traditional wireline systems. This is due to the fact that most radio transmissions are much easier to intercept than those transmitted over a wireline system. The extent to which the public is aware of these problems is unclear, but among radio enthusiasts the open nature of radio signals has long been recognized, and is the basis of the popular pursuit of scanning or recreational eavesdropping.[2]

Until recently, privacy violations and fraud affected a relatively small number of users and technologies. Today, as wireless communications systems proliferate and the number of radio communication devices expands, the problems are becoming more severe—the worst of which is theft of service through fraud. Concerns about the confidentiality and security of wireless data transmission, for example, are rising as more companies turn to

---

[1] OTA has done several studies of aspects of telecommunications privacy and security. See U.S. Congress, Office of Technology Assessment, *Information Security and Privacy in Network Environments,* OTA-TCT-606 (Washington, DC: U. S. Government Printing Office, September 1994) and U.S. Congress, Office of Technology Assessment, *Protecting Privacy in Computerized Medical Information,* OTA-TCT-576 (Washington, DC: U. S. Government Printing Office, September 1993).

[2] Scanners have their own magazine, *Monitoring Times,* which has a circulation of 30,000.

BOX 10-1: Definitions

Many of the terms used in this chapter to discuss privacy and security have ambiguous meanings, and are used in various ways by different people.[1] In this report, OTA uses the following definitions:

- *Confidentiality* refers to the nondisclosure of information beyond an authorized group of people.
- *Privacy* is distinguished from confidentiality in that privacy refers to the balance struck between an individual's right to keep information confidential, and society's right to have access to that information for the general welfare. Privacy laws codify this balance, and also provide for some level of individual control over information about themselves.
- *Security* refers generally to the protection individuals desire against unauthorized disclosure, modification, or destruction of information they consider private or valuable. Security is maintained through the use of safeguards, which can be implemented in hardware, software, physical controls, user or administrative procedures, and the like. In practice, security and safeguards are often used interchangeably.
- *Fraud* refers to the use of deception to gain something of value, such as someone using another's telephone account number or other identifier to steal telephone service.

---

[1] For more detailed discussion of these definitional issues, see U.S. Congress, Office of Technology Assessment, *Information Security and Privacy in Network Environments,* OTA-TCT-606 (Washington, DC: U. S. Government Printing Office, September 1994), pp. 26-29, 82-83.

SOURCE: Office of Technology Assessment, 1995.

---

wireless technologies to meet their data communication needs. The use of radio technologies in the context of the NII is especially problematic because the vulnerability of the radio link to eavesdropping also exposes the wireline portion of public voice and data networks to privacy and security violations and fraud, and in ways that are difficult to guard against. This chapter examines the problems of privacy, security, and fraud in today's wireless networks, and discusses possible technical, regulatory, and administrative solutions.

FINDINGS

Wireless technologies invite privacy and fraud violations more easily than wireline technologies due to their broadcast nature. The privacy implications of widespread use of mobile wireless technologies are potentially serious for both individuals and businesses. There will be a continuing need to guard against eavesdropping and breaches of confidentiality, as hackers and scanners devel-

op ways to listen in and track wireless communications devices.

- It is unclear how successful efforts to address privacy and security concerns regarding wireless telecommunications have been. Laws designed to protect wireless telephone users, while potentially helpful, may not go far enough, and enforcing them is difficult. Likewise, the success of the efforts of wireless service providers to combat fraud and provide secure communications is hard to measure. Technical changes may make systems more secure than they are today, but each time new security measures are implemented, criminals find new ways to "beat the system." For the most part, industry implements technical changes that frustrate fraud and prevent violations of personal privacy. However, it is unlikely that wireless fraud will ever be completely eliminated.
- The true extent of service theft through fraud in the wireless (primarily cellular) industry is un-

known, but is estimated to directly cost the industry $482 million per year. Indirect costs may range as high as $8 billion per year. Unfortunately, this cost is distributed across all paying wireless customers in the form of higher bills. Customers can help protect themselves from fraud through vigilant scrutiny of their wireless telephone bills, but it is unclear how well the general public understands its vulnerability or the extent and cost of wireless fraud. Greater public awareness—through education and warnings provided by wireless service providers and equipment manufacturers—could help combat the problems.

■ Wireless systems, coupled with improved location identification technologies, may make it easier to track people's movements. In the course of listening in on a conversation or intercepting a data communication, an eavesdropper may be able to determine the location of the user. Location information is a particular concern to individuals, especially when it can be gathered in the normal course of wireless telecommunications operations.[3] Businesses using wireless systems for voice and/or data communications may be monitored for purposes of industrial espionage. Treatment of location information in law is not yet consistent.

■ Options

If Congress feels that wireless privacy, security, and fraud are problems, it could consider three principle options:

1. Congress could amend the U.S. Code to make possession of scanning equipment and number-altering software illegal.[4] Currently, possession of specialized scanners and software is not illegal—only its purchase and use with intent to defraud.

2. Congress could require cellular carriers and equipment manufacturers to give explicit warnings about the possibility of fraud and breaches of privacy in service agreements, instruction manuals, bills, or other service agreements; on handsets in the form of labels; and elsewhere to help educate consumers.

3. Congress could consider authorizing increased funding of the Electronic Crimes branch of the Secret Service, and of the enforcement division of the Federal Communications Commission, to combat wireless crimes. The Secret Service estimates that its electronic crimes enforcement effort would be at optimum staffing levels with 50 more agents, which would cost an estimated $4.5 million.

CONFIDENTIALITY AND PRIVACY

People using wireless communication systems—for either voice or data applications—may incorrectly assume that because their cellular telephone or portable computer operates roughly like their wireline counterparts that they are subject to the same privacy laws and possess the same safeguards. But there have been numerous widely publicized cases of eavesdropping on and recording of cellular telephone calls, including those of prominent political or society figures, such as Virginia Governor Douglas Wilder, and Princess Diana of Wales. Both the mayor and police chief of New York City reportedly have had their telephone calls monitored. Businesses routinely warn their employees not to conduct sensitive business on cellular telephones.[5]

Telecommunications privacy and security have been the subject of gradually evolving law and

---

[3] See Internet posting Subject: Does GSM track the physical location of a phone?, Date: 20 April 1995 08:32:19 +0200, From: mobile-rg@dxm.ernet.in, To: cellular@dfv.rwth-aachen.de, Message-ID: <9504200632.AA02651@lorien.dfv>.

[4] 18 U.S.C., sec. 1029 (a).

[5] Milo Geyelin, "Cellular Phone May Betray Client Confidences," *The Wall Street Journal*, Sept. 1, 1994, p. B1.

regulation since the early days of telephony (see box 10-2).[6] Telephone communications are generally protected against unauthorized listening or recording under the Communications Act of 1934 and other privacy statutes, principally the Electronic Communications Privacy Act of 1986 and the Communications Assistance to Law Enforcement Act of 1994. Fraudulent use of someone's telephone accounts is prohibited under the criminal code concerning access device fraud.[7]

There are two main types of information that merit protection in the wireless context: 1) the contents of a call or transmission and 2) the location of the sender or recipient. The privacy of call contents is easily understood, and has generated the most concern and regulation. Privacy of location information, however, is a relatively new concept, and may pose unusual management and social challenges.[8]

■ Privacy of Transmission Contents

As a practical matter, listening or scanning are generally not prosecuted, particularly when the contents of intercepted transmissions are kept confidential and when not used for a commercial purpose by the unauthorized recipients. This degree of privacy is sufficient for many people, such as those who use cordless telephones, but is nevertheless troublesome for those who desire confidentiality comparable to that of traditional wireline telephones. This relative insecurity of wireless telecommunications is responsible in part for interest in technological safeguards to protect confidentiality.

There are some security-protecting features of mobile communications, however, that make widespread and intrusive wireless monitoring less likely. While scanners can pick up conversations fairly easily, finding any particular one is difficult. It is even harder in networks with many simultaneous conversations and where one or both of the participants is mobile. Calls are handed off from cell site to cell site, making it hard to track a specific conversation for very long. Despite large investments in technologies that could pick out individual conversations from all those passing through the public switched networks, even the government, much less private individuals or organizations, still cannot do this well.[9]

Wireless data network providers, such as RAM and Ardis, claim that their systems are inherently more secure than analog cellular telephony, because of their digital formats, and error-checking and correction protocols. Data are typically transmitted in digital packets, each containing an address instructing that packet where to go and in what order. Eavesdropping would require intercepting the right packets, identifying the header

---

[6] James E. Katz, "U.S. Telecommunications Privacy Policy," *Telecommunications Policy*, vol. 12, December 1988, p. 354.

[7] 18 U.S.C., sec. 1029.

[8] Because wireless telecommunications systems are typically interconnected to other telecommunications networks, privacy of wireless signals can be compromised in either the wireless or the wireline portion of a transmission. Privacy also may be compromised by someone scanning the frequencies used for the wireless portion of a cellular call; in this case, the wireline portion of the call will also be compromised. The base station or the wireline system itself may be physically tapped as well. This section will focus only on attacks on the wireless portion of a call.

[9] Unclassified information on government surveillance capabilities is difficult to obtain. Public statements by current and former intelligence officials can give some indication of these capabilities, as in this report of a presentation given by former National Security Agency head, Adm. Bobby Inman: "Inman [pointed out to an MIT seminar] that current cellular phones are difficult to monitor because "there's no technology that can sweep up and sort out phone conversations" despite very large investments in this. He drew an analogy to a case where he had to inform President Carter that an insecure dedicated private land line to the British Prime Minister had been compromised. Inman told Carter that the nature of the public phone system, with its huge volume and unpredictable switching, would have made using a pay phone more secure." Internet posting to Red Rock Eater listserver, Date: Wed, 23 Nov. 94 09:54:12 EST, From: lethin@ai.mit.edu (Rich Lethin), Subject: Admiral Inman visits MIT.

BOX 10-2: Wireless Communications Privacy and the Law

The legal status of the privacy of wireless communications has evolved over time. Since most wireless signals can be received by anyone with a radio or scanner tuned to the correct frequency, they are inherently less secure than their wireline counterparts—undermining any reasonable expectation of privacy. Congress has, however, established limitations on the right of people to receive or intercept wireless transmissions. These limitations have grown more extensive and explicit as wireless telecommunications systems have become more widely used.

Historically, the struggle over the privacy of communications has been a battle between an individual's right to privacy and the legitimate needs of law enforcement to conduct surveillance (wiretapping, interception) in the investigation of crimes. Striking a balance in this area has proven difficult for the courts and Congress as wired and wireless communication technologies have advanced—new technologies made old assumptions, decisions, and regulations about privacy and surveillance obsolete. In fact, for the first 70 years of this century, the specific implications of privacy and wiretapping laws for wireless services (and vice versa) generally were not even considered because the public generally did not use radio systems to communicate with one another.

The first general set of communications privacy limitations are found in the Communications Act of 1934.[1] The act made the intercepting or divulging of private communications, by whatever medium, illegal, except by authorized communications company employees or on lawful demand by law enforcement officers.[2] In 1967 the Supreme Court ruled in *Katz* v. *United* States and *Berger* v. *New York*[3] that certain wiretapping operations violated the Fourth Amendment protection against unreasonable search and seizure. Largely in response to these cases and to law enforcement concerns about its ability to conduct wiretapping operations, Congress passed the Omnibus Crime Control and Safe Streets Act of 1968.[4] Title III of this act tried to strike a balance between individual privacy rights and law enforcements' needs, and set forth the conditions under which law enforcement could intercept private communications. Subsequently, some courts found that the protections of the Act against unauthorized interception generally did not apply to radio-based communications, while others protected some radio communications.[5]

As wireless technology developed and came into more widespread use, the special problems of privacy in a wireless environment became clearer—especially in the case of cordless and cellular phones. Early court cases limited an individual's reasonable expectation of privacy when using a wireless phone, holding that such calls were exposed to many people who could easily listen in—intentionally or by accident.[6] The Electronic Communications Privacy Act (ECPA) of 1986 extended the privacy provisions of Title III to cellular telephones, most pagers, and other electronic communications, including electronic mail, but specifically exempted cordless phones from privacy protections.[7] The Act also made the disclosure of protected communications illegal. In response to concerns about increased monitoring of cellular telephone calls, legis-

---

[1] Ch. 652, Title VII, sec. 705, 48 Stat. 1064, 1103 (June 19, 1934), codified at 47 U.S.C. sec. 605 (a).

[2] In *Nardone* v. *United States*, 302 U.S. 379, 380-81 (1937), the Supreme Court ruled that section 605 of the Communications Act generally prohibited interception and subsequent disclosure of wire communications. In the middle third of this century, however, law enforcement authorities continued to use wiretaps, and the number of court cases over wiretaps arising in the 1930s and 1940s makes it clear that section 605 prohibitions did not end the practice of wiretapping.

[3] 389 U.S. 347 (1967); 389 U.S. 41 (1967).

[4] See especially Title III, Pub. L. 90-351, June 19, 1968; 82 Stat. 197.

[5] *State* v. *Delaurier*, 488 A.2d 688 (R.I. 1985). In *United States* v. *Hall*, however, the court held that a transmission between a mobile telephone and a landline telephone was protected, but a call between two mobile telephones was not. 488 F.2d 193 (9th Cir. 1973).

[6] See *United States* v. *Hoffa*, 436 F2nd 1243 (7th Cir. 1970).

[7] Pub. L. 99-508, Oct. 21, 1986; 100 Stat. 1848.

(continued)

BOX 10-2 (cont'd.): Wireless Communications Privacy and the Law

lation banning the manufacture or import of scanning devices capable of receiving cellular frequencies was passed in 1992.[8]

The Communications Assistance to Law Enforcement Act of 1994 (CALEA) finally extended to cordless telephones and wireless data communications systems—including wireless local area computer networks—the same protections cellular telephones enjoyed.[9] In several cases since 1986, the courts had found that users of cordless phones had no objectively reasonable expectation of privacy—as cordless telephones operate in readily accessible public spectrum used by a variety of unlicensed devices—and could be intercepted without a wiretap authorization.[10] By 1994, however, the use of cordless phones had become ubiquitous, and lawmakers found that the public believed their cordless phone calls were as private as a wired telephone—when, in fact, they were not. Responding to this sentiment, Congress made a legislative determination that such communications should be protected.

Conceptually, the limitations on intercepting wireless communications fall into two groups: those involving possession of scanning or listening devices, and those involving the actual receiving, using or divulging the contents of transmissions.

As noted above, the manufacture or import of cellular frequency scanning equipment is illegal. However, legitimate scanners (used to monitor police, fire, emergency and other public radio services, and manufactured without the ability to monitor cellular frequencies) can easily be adapted to receive cellular frequencies; information on how to make such adaptations is easy to acquire, and kits to make such adaptations are not banned and may be purchased legally. Even prohibiting all scanners outright is not sufficient to prevent scanning: nearly any cellular telephone call can be picked up using another cellular telephone.[11] It is estimated that there are over 5 million scanning units in the United States today; a unit typically costs $300 or less. Thus, possession of scanners or equivalent equipment capable of listening to cellular telephone calls is difficult to prevent; such devices are essentially available on the open market, and are widely used recreationally by some radio enthusiasts.

Apart from possessing a scanner or receiver, unauthorized and intentional listening to cellular and cordless telephone calls is also illegal, regardless of the frequencies monitored, as is divulging or making use of their contents.[12] *Inadvertently* received transmissions, such as when someone is scanning the spectrum for some legitimate purpose, may not be divulged or published either, and the person receiving such transmissions is enjoined from benefiting in any way from the communication. Broadcasts intended for use by the general public, such as communications to ships, airplanes, amateur or citizens band radio are not prohibited.

---

[8] Pub.L. 102-556, Title IV sec. 403(a), Oct. 28, 1992; 47 U.S.C., sec. 302a (d). The law denies authorization of equipment that can receive transmissions in the cellular telephone frequencies, of equipment that is capable of being altered to receive such transmissions, or that can convert digital signals in those frequencies to analog voice audio. The U.S. manufacture or importation of such devices is also illegal. In addition, under a different statute, 18 U.S.C., sec. 2512, the export, import, manufacture, assembly or possession of equipment whose primary function is the surreptitious interception of private electronic communications, including wireless transmissions, is illegal, and violators are subject to fines and/or five year prison terms.

[9] Pub.L. 103-414, Oct. 25, 1994; 108 Stat. 4279.

[10] See, e.g., *United States* v. *Smith*, 978 F.2d 171 (5th Cir. 1992); *United States* v. *Carr*, 805 F.Supp. 1266 (E.D.N.C. 1992).

[11] Some old television sets with UHF tuners can be tuned to cellular frequencies because these frequencies were allocated from the upper portion of the UHF band, channels 70 to 83.

[12] Two statutes apply in this general area. Under 47 U.S.C., sec. 605 (a) violators are subject to fines and/or months imprisonment, for the first conviction, and may be subject to civil damages as well, unless the court finds that the person was unaware of the violation, when damages may be reduced to a fine only. For violations involving commercial advantage, the penalties are fines and/or two years imprisonment for a first offense, and fines and/or five years for subsequent offenses. Under 18 U.S.C., sec. 2511(1), violators are subject to fines and/or a five-year prison term; first offenders are only fined.

SOURCE: Office of Technology Assessment, 1995.

codes, and then reassembling them, probably requiring weeks of work per message, and consequently the results in most cases would not be available in real time.[10]

Several different methods are being used or developed to make wireless networks more secure. Special modulation formats may be used. If signals are encoded in some way, an eavesdropper must have decoding equipment as well. Numerous techniques for encoding are undergoing testing or already deployed. In the future, digital transmission schemes, which were developed to make more efficient use of limited radio spectrum, may also make transmissions more secure. In addition, signals can be encrypted. Both types of technologies are discussed below.

Transmission Schemes

Analog cellular and other traditional radio systems typically transmit information over a single channel in what is known as "circuit switched" transmission. That channel is dedicated to the user for the duration of the call. The technologies are relatively simple and inexpensive, but they use radio spectrum inefficiently. They are also easy to listen in on—once a call has been found, a scanner can lock onto it until the conversation ends, or one of the parties leaves the cell and drops the channel.

New digital communications systems, such as time division multiple access (TDMA) or code division multiple access (CDMA) use spectrum much more efficiently because they break conversations into digital bit streams in order to carry more conversations simultaneously over the same amount of spectrum (these systems are described in more detail in chapter 3). These separate fragments are reassembled by the receiver and presented to the listener as a complete and intelligible conversation. These techniques also make transmissions more difficult to intercept. Without knowing what the disassembly scheme is, an

eavesdropper will hear only unintelligible noise. Thus, digital transmission schemes are desirable for reasons of both economy and security.

TDMA and CDMA differ considerably, however, in the degree of security and efficiency they provide. With TDMA, conversations are broken into segments based on a timing scheme. Each user of a channel is "assigned" one of three time slots by the cellular base station equipment. The time sequences must be known in order to separate out all the conversations occurring on that channel, and to reassemble any particular transmission. This is a straightforward technical task, but it is more difficult and costly to do than monitoring a comparable analog cellular conversation.

CDMA transmission schemes are based on a different principle, known as "spread spectrum." Instead of assigning a time slot on a single channel, CDMA uses many different channels simultaneously, and the network assigns a code to each fragment of a conversation like an identifying label. The receiver recognizes the specified code, sent at the beginning of the transmission, selects all transmissions with this code, and reassembles them into a coherent whole. CDMA is also inherently more difficult to crack because the coding scheme changes with each conversation, and is given only once at the beginning of the transmission. Receivers lacking the proper code to intercept will only hear digital noise.[11] Keeping track of codes is a demanding signal processing task, and it is not likely that eavesdroppers will have the technical or financial wherewithal to monitor CDMA traffic in the near future. Thus, monitoring transmissions on CDMA systems is considerably more difficult than with TDMA and far harder than with analog systems, providing a greater degree of security. However, since the technical standards for both TDMA and CDMA are open and published, they are theoretically susceptible to attack.

---

[10] Ellis Booker, "Is Wide-Area Wireless Secure?" *Computerworld*, vol. 26, No. 39, Sept. 28, 1992, p. 59.

[11] The inherent properties of this scheme explain its attractiveness to and use in the military.

Encryption

Additional security can be provided by a variety of separate encryption schemes. Voice encryption has been used since the 1920s for military use.[12] Commercial products have been available since the 1970s, and a few companies make such products today. Total sales of encryption products now number only a few thousand a year. Some cellular companies offer encryption services, but they are not widely used.

Encryption systems can use either analog or digital techniques. Analog systems manipulate analog wave forms by splitting and inverting the voice signals using ordinary filters. A harmonic signal is injected into the output, resulting in harmonic distortions. These encrypted signals are transmitted, and the reverse process is used to reconstruct the communication. Further encryption can be achieved by varying some of the parameters of the signal-splitting and harmonic distortion, but voice quality may suffer as more distortion is introduced. Companies manufacturing such systems claim that they cannot be decoded in real time, but they admit that they could be recorded and broken later. Nevertheless, these systems can provide a high level of security, but cost from $300 to $1,000 per unit (two units are needed—one for each end of a communication).

Digital encryption systems work by manipulating digitized voice signals. The data representing voice speech are compressed and processed to pass through only phonemes or speech elements (which are reconstructed by the receiver using special software). The digital bitstream is further manipulated using bit substitution, permutation, and other techniques. The encrypted data can be further scrambled, as noted above, with the use of digital transmission systems, which break the bitstream into packets and are coded and displaced in time. Such manipulations incur little or no cost in signal quality, because digital data can be accurately reproduced, and error-checking and correction techniques applied. Voice encryption schemes based on RSA, an encryption algorithm thought to be extremely secure, are on the horizon, and promise a level of privacy protection that is thought to be unassailable.[13] The main constraint with all encryption is the slow speed of processing and the lag that occurs if signals take too long to pass through the system. As signal-processing hardware and software improve, greater levels of security may become available, but the ability of decrypters is also likely to improve as well. To date, most voice encryption devices are bulky and inconvenient, and do not enjoy much consumer or carrier acceptance.

■ Privacy of Location

A new aspect of wireless networks is uncertainty about and concern for *privacy of location*, where a caller's location can be hidden to a certain extent from the network and from the recipient of the message. By the same token, location information is necessary, at least to the level of a sector within a cell, for the switching equipment to be able to successfully connect users.

This feature contrasts markedly with wireline networks where location of the parties is unambiguous, especially to the system operator, but also most likely to the correspondents. The ambiguity of wireless is likely to lead to a series of new issues for wireless users. Much of our common understanding of business, law, and social behavior is based on assumptions about the unchanging nature of place and people. With widespread deployment of wireless technologies, this is less likely to be the case. Assumptions about boundaries, jurisdictions, and proximity are challenged by mobility and ambiguous location information. People will likely develop strategies to uncover the location of users and to hide themselves from others.

---

[12] Material on voice encryption drawn from Dan Sweeney, "The Wages of Fear: Marketing Cellular Encryption," *Cellular Business*, vol. 9, No. 13, December 1992, pp. 58-66.

[13] Red Rock Eater listserver, op. cit., footnote 9.

Unlike wireline networks, wireless networks typically do not know the precise location of the parties to a transmission. This uncertainty varies depending on the type of system: satellite systems have the largest "granularity" of coverage because they are typically broadcasting either to whole continents or large regions. Cellular and other terrestrial networks have much smaller areas in which signals can be received and transmitted, with a maximum of about 20 miles for cellular systems. Future personal communication services (PCS) will use cells covering even smaller areas, perhaps only a few hundred yards in diameter. Location identifying techniques must confront the fact that while it is simple to identify a particular transmitter used by someone with a wireless device, the area that transmitter serves may be quite large or difficult to search, thereby making precise location difficult to determine.

A number of services already exist to address location concerns, and there will be implications associated with this inherent ability. Tracking people and things may be easier in the future with both Global Positioning Satellite (GPS) (see box 4-3) and non-GPS systems using lightweight and inexpensive receivers and radios. In trucking logistics, for example, wireless technologies have helped produce significant improvements in services for firms such as UPS and Federal Express, which now depend on such technologies to conduct their business.[14] Vehicle location services such as Lo-jack and Teletrak are already well established or are under development.

Cellular telephones are actually in operation more than most users think (if the phone is turned on, but not actually being used). To monitor the state of the network and be able to respond quickly when calls are made, the main cellular controlling switch periodically "pings" all cellular telephones. This pinging lets the switch know which users are in the area and where in the network the

telephone is located. This information can be used to give a rough idea of location, down to the level of a cell, or cell sector, or even smaller areas, depending on the system used.

With the prospective launch of PCS systems, with cell areas typically smaller than those of cellular telephone systems, it may be possible to specify particular areas in which a PCS phone may operate. Parents might use this to control the movements of their children, or administrators the movement of their employees. If a user strays from the approved area, a message might be sent, "Get back home now!" Such services would be inexpensive to provide, because they are a byproduct of the normal operation of this type of technology.[15] As yet, however, there has been no demand for such services.

A wireless user's location can also be calculated by using a combination of signal strength, angle of return, time delay and synchronization, in somewhat the same way that a person can infer distance by seeing or hearing an object with two eyes or ears. Technology developments in location identification for emergency 911 services with wireless systems will undoubtedly improve the ability of wireless service providers to locate individual users. These methods can be fairly accurate, particularly when used together, and they are likely to improve in the near future (see discussion of emergency 911 services in chapter 3). Law enforcement services already can locate an emitter to within six feet, if given sufficient time and resources, possibly in as little as a half hour.[16] (This level of detail would be the result of significant effort, for example, in serious fraud or drug investigations.)

Techniques are likely to be found that enable people to hide themselves from wireless networks and other people. Mobility allows users to contact others from any location; if they move quickly

[14] Frank Erbrick, UPS Vice President for Operations, OTA Advisory Panel meeting, May 12, 1994.

[15] Scott Schelle, vice president for operations, American Personal Communications, Inc., OTA Advisory Panel meeting, May 12, 1994.

[16] Interview with U. S. Secret Service officials, Dec. 12, 1994.

enough, it will be difficult to trace them. Simply turning off the handset will serve in many cases (but will also make the phone unusable for receiving calls).

One area of growing concern is how information about personal location and behavior could be gathered and used by a range of large information systems, such as electronic payment systems, credit card and other credit reporting, telecommunications transaction records, health record systems and the like.[17] The Communications Assistance to Law Enforcement Act forbids wireless carriers from divulging location information to anyone, except to law enforcement authorities with a proper warrant.[18]

The issues of personal information-gathering and disclosure are beyond the scope of this report. They generally do not involve matters of wireless telecommunications technologies, with one exception: the Intelligent Transportation System (ITS), formerly known as the Intelligent Vehicle Highway System (IVHS). The inherently mobile nature of transportation, and the reliance of ITS designers on wireless telecommunications for some aspects of the system, raises the issue of privacy protections.[19] Some analysts have argued that:

Many of these technologies involve surveillance of the location and behavior of identified vehicles and/or people, and the collation of such data for further use. These and other aspects of IVHS technologies raise concerns amongst the community, and have delayed adoption of some systems.[20]

[S]ome proposed designs require the system to collect vast amounts of data on individuals' travel patterns, thus raising the potential for severe invasions of privacy. To make social choices about IVHS, it is necessary to reason about potentials for authoritarian uses of an IVHS infrastructure in the hypothetical future.[21]

The design of such systems or subsystems needs to carefully considered with privacy concerns in mind.

■ Location and legal jurisdiction

Many aspects of the law are predicated on geographic location. To a certain extent, wireless telecommunications confound such geography-based distinctions, because with cellular telephones, boundaries (local or state, and to a limited extent, international) can be broached. With satellite-based communications, boundaries are essential-

---

[17] GSM systems reportedly know the location of all phones within 10 meters, and that the three closest cell sites track the phone at all times, to enable smooth hand-offs from one cell to another. Continuous location data could easily be recorded, even for many users, without posing an undue data burden—one observer estimates that 1 million users, tracked every 10 minutes to one square meter, for one year, would generate about 510 gigbits of uncompressed data, well within the data processing capability of most business and many personal computers. See Internet post, Date: Thu, 20 Apr 1995 08:32:19 +0200, From: mobile-rg@dxm.ernet.in, To: cellular@dfv.rwth-aachen.de, Subject: Does GSM track the physical location of a phone?, Message-ID: <9504200632.AA02651@lorien.dfv>.

[18] Public Law 103-414, sec. 103 (a)(2), Oct. 25, 1994, 108 Stat. 4281.

[19] For example, see Don Phillips, "Big Brother in the Back Seat? The Advent of the 'Intelligent Highway' Spurs a Debate Over Privacy," *The Washington Post*, Feb. 23, 1995, pp. D10-D11.

[20] Marcus Wigan, "The Influence of Public Acceptance on the Reliability of the Potential Benefits of Intelligent Vehicle-Highway Systems," *Information Technology & People*, special issue on "Identification Technologies and Their Implications for People," vol. 7, No. 4, 1994, pp. 48-62.

[21] Philip E. Agre and Christine A. Harbs, "Social Choice About Privacy: Intelligent Vehicle-Highway Systems in the United States," *Information Technology & People*, special issue on "Identification Technologies and Their Implications for People," vol. 7, No. 4, 1994, pp. 63-90.

ly meaningless. Work on transborder data flows has attempted to address this problem, but its resolution is unclear. The Internet also poses similar problems of geographic location, jurisdiction, and the law.[22]

CELLULAR AND OTHER WIRELESS FRAUD

With widespread use of wireless telephony has come widespread theft of service by fraudulent means. The true extent of cellular telephone fraud is unknown, but the number of attempted fraudulent calls may run as high as 3 million per month.[23] The Cellular Telephone Industry Association (CTIA) estimates that fraud amounts to about $482 million a year, based on estimates of out-of-pocket costs to companies for customer-identified calls for which the company reimburses customers.[24] Other analysts believe the cost is substantially higher. The government has no independent estimate of the extent of wireless telephone fraud.

For wireless technologies to enjoy the same public acceptance as wireline telecommunications, they will probably need to provide similar levels of security from fraud and misrepresentation. Fraud increases service costs for both businesses and consumers, and may make wireless less competitive than wireline services. Cellular customers ultimately pay for cellular phone fraud in the form of higher costs because companies pass these costs along to consumers.[25] It is also costly for law enforcement agencies to enforce fraud statutes, and it fosters the expansion of criminal activities, both directly and indirectly. Fraudulent phones are frequently used in the commission of other crimes, and hinder law enforcement efforts against those criminals.

This section will discuss cellular telephone fraud and how it is committed. It will also describe some of the technical and organizational cost-benefit tradeoffs the industry has made that shape the incidence of fraud. Finally, technical measures that might be taken to limit fraud in the future will be addressed briefly. The focus is on cellular telephones because currently experience widespread fraud. Although the pirating of satellite television signal is still a problem, it is not addressed here. The heyday of pirating is long since passed, and with the introduction of new digital transmission and encryption systems, fraud is expected to drop further.

■ Tumbling and Cloning

Cellular telephone fraud is conducted through what is known as "tumbling" and "cloning." Understanding how these work requires a brief description of how a cellular telephone identifies itself to the cellular network, and how billing is managed.

Every cellular telephone has a unique electronic serial number (ESN), "burned in" on a chip by the manufacturer. FCC regulations require that every phone have a unique ESN. In addition, every cellular telephone subscriber is issued a mobile identification number (MIN) when the phone is assigned a telephone number and activated by the service provider. For example, when a subscriber buys a cellular telephone at a retail store, the service provider assigns a telephone number from a batch of numbers provided by the local telephone

---

[22] Dan L. Burk, "Transborder Intellectual Property Issues on the Electronic Frontier," Arlington, forthcoming in vol. 5, *Stanford Law & Policy Review*, available at URL gopher://gopher.gmu.edu:70/00/academic/colleges-depts-insts-schools/ law/working/dburk2.

[23] Susan Kumpf and Nora Russell, "Getting the Jump on Fraud," *Cellular Business*, vol. 9, No. 10, October, 1992, p. 24.

[24] "Secret Service, CTIA Crack Down on Cellular Fraud," *Telecommunications Reports*, vol. 61, No. 15, Apr. 17, 1995, p. 32. Cellular telephone firms are unwilling to give an accurate accounting of cellular telephone fraud to CTIA. Telephone toll fraud generally may be as much as $8 billion per year, with international toll fraud comprising 65 to 80 percent of the total. Dan O'Shea, "Security Products Abound, But Is Toll Fraud Too Tough?" *Telephony*, vol. 225, No. 9, Aug. 30, 1993, pp. 7, 13.

[25] Because cellular companies are unregulated, there are no public ratepayer issues with cellular fraud.

monopoly, and records both the MIN and the ESN as an associated pair.

When a call is initiated, the phone transmits its ESN and MIN to the cellular switch. This is done over a signaling channel, reserved for setting up a call between the handset and the switch. If the two match, then the call is permitted to proceed and a voice channel is opened. If a call is made outside the regular service area, the remote cellular company relays the ESN/MIN pair to the home company or to a regional database to check whether the number is valid (the negative number list), in accordance with an industry standard, IS-41. If it is authenticated, the call is permitted to go through. The air time and roaming charges are forwarded to the home company at the end of the call, and the two companies settle up periodically to clear outstanding balances.

With traditional analog cellular systems, "tumbling" is quite simple. A fraud perpetrator (or "bandit," the preferred term) randomly or sequentially changes the ESN and/or the MIN after each call. Because the cellular switch takes some time to verify each number, some proportion of calls may get through the system before the system denies access. Tumbling is currently not very prevalent because cellular operators have installed systems that can defeat it fairly easily. When GTE installed its pre-call validation system in December 1991, 25 percent of attempted fraudulent calls were denied connection. Other cellular carriers have even higher levels—for example, up to 61 percent by Ameritech Mobile Systems in Detroit, MI.[26] Once the technology is deployed, bandits typically move on to other forms of fraud.

"Cloning" works a bit differently. Cloners pick up ESN/MINs on busy streets or highways with scanning equipment that is legally available, although their use for this purpose is illegal.[27] The devices typically monitor cellular signaling channels, and display broadcasted ESN/MIN pairs. Cloners record these number pairs, and send them to other cities, whose carriers may be unable or unlikely to verify that the number is in use elsewhere or was so recently used in another place as to be fraudulent. In the remote city, a participant in the fraud scam uses a standard personal computer or laptop with legally available software to reprogram the ESN/MIN in a cellular telephone, which can be done with existing external connectors to the phone.[28]

This phone is then either sold or used by someone wanting to make free calls or who does not want to be traced, either by law enforcement agencies who might have a wiretap order on a known number or by the telephone company for billing purposes.[29] Because a fresh number has not yet been identified as fraudulent in the negative number list, checking that database will not prevent fraud the first time it is tried. Depending on whether the original owner of the stolen number notices the charges on the bill, and how often the databases are updated, a cloner may be able to use the cloned phone for some time and run up a substantial bill. Real-time access to subscriber lists and activity records between companies handling calls is available in some markets for the purpose of defeating such scams. Industry officials esti-

---

[26] Kumpf and Russell, op. cit., footnote 23, pp. 24-25.

[27] These scanners are legitimately used by technicians in servicing cellular telephone equipment. They are designed to work within a very short range, about 10 to 15 feet. However, it is a simple matter to make them receive over a larger area by boosting the power. These scanners are readily available, including by mail-order.

[28] Phones could be made unreprogrammable, but there are legitimate reasons to keep them reprogrammable. One is the ability to change the number if the service provider changes, without having to change phones. Another is to allow changes in case the phone is compromised by a cloner.

[29] Some reports put the street price of a cloned phone at $300, with a guarantee to replace it if the number is turned off. Michael Meresman, "The Phone Clone Threat," *Mobile Office*, vol. 5, No. 11, November 1994, p. 62.

mate that, by the end of 1995, up to 70 percent of the U.S. carriers will have this capability.[30]

Today, if the customer notices fraudulent charges and notifies his or her company, the company will remove the charge, pay the long distance charges, reimburse costs to the remote company if roaming has occurred, and absorb the loss. Companies have done this since beginning operations in the early 1980s, but are under no legal obligation to do so.

∎ Call Selling

Call selling is an illegal activity conducted with cloned cellular telephones. In the view of CTIA, this may be a greater revenue drain on firms than simple cloning. In essence, in a call selling operation, perpetrators set up their operation in a hotel room or an apartment with a number of cloned cellular telephones. They advertise informally to immigrant communities, among others, that they will sell calling time to their home countries significantly below international rates. The defrauders not only do not pay for the use of the telephones, but they also receive cash payments for their use. Immigrant communities are willing to spend a significant portion of their monthly income to call overseas, and are typically looking for ways to reduce their calling costs.

Such fraud operations are highly profitable, less risky and much less physically dangerous than other types of organized crime, such as drug trafficking. As a result, some law enforcement officials believe that cellular fraud will continue to grow significantly in the future.[31] Cloners move quickly to break new protection schemes, often succeeding within six months of their introduction.[32] The switch to digital technologies will offer users some protection, but analog systems will

continue to operate and be susceptible to fraud for many years.

∎ Law Enforcement

Altering the ESN/MIN pair of cellular telephones by counterfeiting these numbers is covered by the same statutes as credit card or currency counterfeiting, in that fraudulent means are used to gain access to the telecommunications system.[33] Thus, identifying and arresting perpetrators of cellular fraud is primarily the responsibility of the U.S. Secret Service, which has primary federal jurisdiction over fraud. State and local law enforcement officials are also involved to some extent, as well as the U.S. Drug Enforcement Administration, U.S. Customs Service, and the Federal Bureau of Investigation, depending on what other crimes are perpetrated using a cellular telephone. The Secret Service has recently put 20 of its 1,200 agents through the Electronic Crimes Special Agent Program, which prepares them for all types of electronic crimes, including wireless fraud.

Fraud investigation usually begins when a subscriber or carrier identifies some suspicious activity—for example, a rapid increase in traffic at a particular cell site. The carrier will then locate the source of activity using radio triangulation techniques, and will turn this information over to the Secret Service, who will attempt to get a warrant and make an arrest. The cities with the most cellular fraud are New York, Los Angeles, and Miami, but some of the recent large cellular phone fraud operations have been outside these three centers: in late 1991 and early 1992, over 57,000 calls were made in 19 days by Palestinians in the West Bank and Gaza to other countries in the Middle East via Phoenix, AZ, in a three-way calling scam.[34] Because the most costly element of cellular tele-

---

[30] Ibid, p. 64.

[31] Ibid, pp. 60-69.

[32] Tom McClure, CTIA Fraud Taskforce head, interview, July 5, 1994.

[33] 18 U.S.C., sec. 1029.

[34] Anthony Ramirez, "Theft Through Cellular Clone Calls," *The New York Times*, Apr. 7, 1992, p. D-1.

phone fraud is international calling, companies are beginning to offer international service only to those customers who specifically request it, about 5 percent of all cellular subscribers.

A number of technical efforts are under way to combat cellular (and by extension, other wireless telephone) fraud. Handsets can be made more secure and difficult to clone,[35] and cellular switches can be equipped with database and signal processing equipment and software to detect fraud and stop it there. Carriers are adopting personal identification numbers (PINS) that must be entered manually by the subscriber before a call can be completed, as is done with electronic bank cards.[36] The disadvantage of this method is that customers have to key in additional numbers, making calling less convenient.

Call screening systems with fast database and call pattern-recognition software are also being deployed. These systems work by monitoring the past activity of a particular subscriber. If new activity does not fit the established pattern, the calls are flagged and the owner of the phone is contacted to confirm unusual use. AirTouch, NYNEX, and Bell Atlantic Mobile have all begun to use these services within the past two years, and report reductions of up to 75 percent in stolen minutes.[37]

Experiments are also under way with systems that would identify the particular electronic signature of individual phones (each phone has slightly different electronic characteristics due to variation in the electronic value of components, which gives each phone a distinctive and identifiable profile).[38] Digital technologies will also make cloning more difficult. However, digital encoding schemes are known and can be broken, given enough time and computing power, even though the equipment to pick out numbers is more costly.

In fact, digital telephone standards IS-41 and the Global System for Mobile Communications (GSM) provide one such digital scheme. Cellular telephones would be programmed with a secret number that would never be transmitted. During call setup, the handset would prompt the cellular switch to transmit back to the handset a one-time number. The handset would then generate a one-time response based on its own secret number and the transmitted number to validate the call to the cellular switch. Since one of the two numbers lies in the carrier's database and changes with each call, and the other number is never transmitted, each number is unique and impossible to reverse-calculate.[39] Next-generation digital cellular telephones could perform this validation function easily, but existing analog telephones could not without expensive retrofitting.

It appears that cellular telephone fraud could be minimized by technical means, if the costs of

---

[35] Originally, the ESN was to be unprogrammable, a permanent part of the phone. However, cellular handset resellers resisted marketing such handsets, because the cellular carriers (in general unrelated to the resellers) charged the resellers for establishing service, making accounting changes, and the like. Resellers insisted on programmable cellular telephones, which the carriers ultimately did not oppose, primarily because the carriers depend heavily on the resellers to market their system and provide customer service. Some observers believe that this business dynamic between resellers and carriers is responsible for the technical configuration of cellular phones, which is inherently less secure than an ESN that is not reprogrammable. Internet posting to Telecom Digest, coyne@thing1.cc.utexas.edu, Subject: Re: Bell Atlantic Mobile Joins the PIN Crowd, Date: 10 Jan 1995, 20:12:46 GMT, Organization: the University of Texas at Austin, Austin, Texas.

[36] This service configuration was introduced in late 1994 by NYNEX Mobile Communications and Bell Atlantic Mobile. The ESN/MIN pair is transmitted over the reverse signaling channel, while the PIN is sent over the voice channel. Cloners are unlikely to be listening to both channels simultaneously or be able to associate the two numbers. If the PIN is compromised, the subscriber can simply get a new PIN by phone, rather than a whole new ESN/MIN, which is much more costly. Other companies have used variations on the PIN concept.

[37] Meresman, op. cit., footnote 29, p. 62.

[38] Ellis Booker and James Daley, "Cellular Carriers Gain New Fraud-Detection Weapon," *Computerworld*, vol. 27, No. 44, Nov. 1, 1993, p. 71.

[39] Meresman, op. cit., footnote 29, p. 32.

stopping it were lower than the level of fraud, and if users would be willing to forego the convenience of simple number dialing. Law enforcement officials and other industry observers agree that the problem is tractable. With more competitors in the marketplace for PCS, the ability of carriers to pass along these fraud costs will be limited. Carriers will likely have a greater incentive to limit costs by more vigorously limiting fraud. They could press equipment manufacturers for handsets that contain unclonable technologies, to overcome the weakest link in the wireless security chain. As new technology is deployed the problem will diminish. However, industry officials believe that analog phones will be used in North America for a number of years, and will undoubtedly be targeted by bandits because they are inherently less secure. It is likely that the fraud problem will decrease, but it is unlikely that it will disappear altogether. Bandits are notorious at learning new techniques to defraud operators and subscribers, and will likely continue their efforts with new technologies.

The level of effort the Secret Service devotes to wireless fraud is difficult to indicate in dollar amounts. Agency officials told OTA that the Secret Service would only handle major fraud cases. Because there are technical fixes to much of the fraud activity, it appears industry will have to deal with lower level criminal activities on its own. The Secret Service sees its primary role as identifying new fraud techniques, and then working with industry (which is itself conducting an extensive antifraud program) to develop countermeasures to combat those techniques. The agency is satisfied that carriers have been cooperative in responding to suggestions by law enforcement; changes suggested by the Secret Service usually are made within three or four months.

The Secret Service and the industry agree that easy availability of scanners capable of picking up ESN/MIN pairs, and software used in altering ESNs, contributes to law enforcement's problem in policing fraud. Although *sales* of scanners are illegal[40]—other than to an employee, agent, or contractor of a cellular carrier or government employee with specific need—their *possession* is not. The FCC is formally responsible for enforcement of this provision in the law, but has few resources to do so. In fact, scanners are readily available through retail electronics stores and mail-order companies. These scanners are intended to be used for bench-testing only. They are supposed to comply with FCC rules limiting their range to 15 feet, but this limitation is easily defeated by extending the devices' antennas. Under current law, a scanner is only illegal if it is used with intent to defraud,[41] which is difficult to prove. Possession of or sale of ESN-altering software is currently not illegal. Penalties for cellular fraud include prison terms of up to 15 years, and fines up to $250,000.

Law enforcement and the industry would like to make the *unauthorized possession* of a scanner illegal, thereby closing what they consider to be a significant loophole in the current law. They would also like to make illegal the production, use, or trafficking in software used to alter ESNs. They argue that such legislation would also spread the burden of law enforcement to more agencies, enabling better enforcement.

■ Consumer Protection

Consumers are not well informed about cellular fraud, its frequency, its methods of perpetration or means of identifying it. Many consumers do not receive itemized bills, and have no way of verifying billing accuracy.[42] Service agreements, owners' manuals, and bills themselves usually do not warn users about the possibility of fraud. As noted above, wireless companies will generally absorb

---

[40] 47 U.S.C., sec. 302(a).

[41] 18 U.S.C., sec. 1029(a).

[42] Many companies charge a supplementary fee to provide itemized bills.

the cost of fraud that consumers identify. But un-identified fraud costs are borne by the user, and all fraud is reflected in higher costs to all customers. While service providers are moving steadily to combat fraud once it is found, they may not be alerting their customers to its possibility. Despite efforts to inform them, many users believe that cellular telephones are as secure as, and operate in the same manner as, traditional wireline telephones. Clearer warnings that this is not the case may be in the public's best interest.

Health
Issues 11

Over the past several years, concerns have been raised about the potential health risks of portable cellular telephones and emissions from radio antennas. These concerns are rapidly becoming one of the most controversial issues surrounding the widespread use of wireless technologies. Although some research on possible adverse health effects has been conducted, it has not been conclusive—government, industry, and academic researchers agree that it is not yet possible to say with certainty whether the devices or the antennas do or do not pose a risk to human health or how serious any risk may be. As a result, the long-term issues surrounding the health and safety effects of cellular telephones and other wireless devices remain unresolved. In the face of this uncertainty, the debate over the safety of wireless devices and systems is likely to become an important public policy problem as concerned citizens take their concerns to state and federal policymakers and regulators.

OTA did not conduct an indepth assessment of the possible health effects associated with radio communication devices and systems. Nor did it exhaustively review and critique the health effects research conducted to date. Such an endeavor is properly the focus of an additional, more narrowly focused study. Rather, this chapter presents only a general overview of the research performed to date, and discusses the controversy that surrounds these issues.

FINDINGS

The debate over the safety of wireless systems is characterized by high emotion and heated rhetoric—on all sides. Picking through the rhetoric and separating fact from fiction will be extremely difficult for lawmakers and regulators as the controversy continues.

The findings presented below are based on the general state of research as it exists in early 1995. As more studies are completed, issues may become clearer; although evidence gathered to date and the experiences of other public health-related issues—including the controversies over electric power lines and tobacco—indicate that resolution of these issues could be years or even decades away.

- **Scientific research to date has found no conclusive evidence that low power microwave radio communication signals adversely affect human health. However, currently available scientific information is insufficient to conclude that there are no long-term adverse health effects—either from hand-held wireless communication devices or from towers.**[1] Because of the paucity of data on biological and health effects, and the ambiguity in the results of research conducted so far, neither public interest advocates nor industry have made a clear and convincing argument sufficient to prove their case. All parties agree that more research is needed to determine whether there could be any health effects from long-term exposure to radio frequency (RF) radiation at the power levels used by wireless communications devices, what they might be, and how serious a risk they could pose. Specifically, additional research will be required as new technologies are developed that use differ-

ent frequencies, power levels, and transmission formats.
- **Public concern persists over many forms of radiation, including nonionizing electromagnetic radiation.**[2] The willingness of the public to give credence to anecdotal reports of radiation-induced human health risks is an enduring phenomenon. Maintaining the public's trust and confidence in technologies associated with radio waves demands extraordinarily high levels of responsible scientific work and policy development. Given the character of public concern over many types of hazards in the environment, the technical complexity of new wireless systems, the difficulty the public has in understanding the complex results of scientific research, and the likelihood of many more radio devices working at new frequencies and with new technologies, it may be prudent for the federal government, including Congress, to continue to monitor technology and industry developments and the ongoing research into wireless health issues.

Industry has taken some steps to address public concerns, and is making substantial funds available for research. However, especially in health-related areas, it may be difficult for the public or policymakers to trust that industry-funded research will always be conducted in an objective manner. **Some continuing federal role—as an overseer of**

---

[1] U.S. Food and Drug Administration, "Talk Paper" on cellular telephone safety, 1993; U.S. Federal Communications Commission, Office of Engineering and Technology, "Information on Human Exposure to RF Fields from Cellular Radio Transmitters," 1994; Institute of Electrical and Electronics Engineers, "Position Statement on RF from Portable and Mobile Phones and Other Devices," 1992; U.S. Congress, General Accounting Office, *Status of Research on the Safety of Cellular Telephones*, GAO/RCED-95-32 (Washington, DC: November 1994), pp. 3-4, 15; Mark Fischetti, "The Cellular Phone Scare," *IEEE Spectrum*, vol. 30, No. 6, June 1993, pp. 43-47; "Cellular Phone Industry Research Group Sees Need for 'Basic Information in All Areas'; Proposals Under Review," *Microwave News*, September/October 1994, pp. 9-10; Scientific Advisory Group on Cellular Telephone Research, *Interim Status Report: Potential Public Health Risks from Wireless Technology: The Development of Data for Science-based Risk Management Decisionmaking*, Nov. 29, 1994, p. 4; "SAG Chairman Comments on Significance of Research Agenda; Proud of Group's Track Record," *Cellular Telephone Update*, vol. 2, No. 1, Fall 1994, p. 2.

[2] Although "radiation" is the preferred technical term when discussing radio wave emissions from wireless transmitters, radio communication radiation should be clearly differentiated from the harmful ionizing and particulate ("hard") radiation associated with nuclear energy. These two types of radiation are not the same. Public concern about all forms of electromagnetic radiation may be fueled by a misunderstanding of the technical terms involved.

industry-funded work, as a participant in the research and testing process, or in mounting its own research program—may be desirable to assure research integrity and to maintain high levels of public trust and confidence in these technologies.

A vigorous federal government role is particularly important given the difficulties in evaluating technologies that have not yet reached large-scale deployments. As wireless technologies become more ubiquitous, unanticipated interactions or consequences may appear. What appears to be a negligible or unknown problem in the lab or at reduced scale may turn out to have significant effects when widely deployed, as was the case with lead paint and asbestos.[3] **Long-term monitoring of the effects of radio frequency exposure on humans may be necessary to avoid surprises and persistent public uncertainty.**

THE CONTROVERSY SURROUNDING HEALTH EFFECTS

The debate over the possible health effects from the radio waves used by cellular telephone and other mobile communications systems is intensely polarized. On one side, some citizens and a few researchers are firmly convinced that such radio waves pose a substantial health risk to public health. They believe that cellular phones should be redesigned or banned and that construction of new radio transmitters and antennas, especially those needed for cellular and future personal communications services (PCS) systems, should be restricted and perhaps even stopped. (Radio interference with medical devices such as pacemakers is addressed in chapter 12.) On the other side, equipment manufacturers and service providers maintain that there is no credible evidence that their products and services threaten human health. Without clear and definitive proof of harm, they argue that the development of new systems (and

expansion of existing systems) should continue. Both sides have evidence—scientific studies, statistical records, and anecdotal reports—they believe supports their case. The result is a confusing and often conflicting body of scientific and medical literature.

In disputes like this, identifying and evaluating risk to the public is often difficult. Many elements contribute to understanding risk, and often these are confused, misinterpreted, or misrepresented. In many cases, the elements become divisive public policy issues as different groups with different perspectives battle over what is legitimate, acceptable, and "true," and what is not. In situations where individuals cannot avoid exposure—as in the case of radio waves—it is the role of government through the regulatory and policy process to decide what level of risk is acceptable and to enact the necessary provisions to protect public health. To focus government resources and policy efforts most effectively, it is important for policymakers and regulators to understand the different stages involved in evaluating this risk.

The first step in assessing this type of risk is establishing *causality*—what effects are due to what causes, and how certain is the relationship between them. Disputes can arise between different parties claiming that effects are or are not associated with particular causes, and disagreements frequently center on the adequacy of the science that supports a particular position. This is true with radio wave radiation and its effects on animal tissues. High-power microwave radiation, for example, is known to produce thermal effects (heating), but the possible nonthermal effects of radio waves, which include changes in cell membrane permeability, cell metabolism, or on genetic material, are more contentious. A few researchers have found some such effects, but results are still considered tentative, and the mechanisms causing them are not well understood.

---

[3] George Brandon, "Pulling Together an Electromagnetic Field Defense: Defendants Need a Coordinated Strategy for the Mass Tort Some Call the 'Asbestos of the '90s,'" *The National Law Journal,* Aug. 1, 1994, p. B19.

The second element in assessing risk is demonstrating *harm* from the effects. Even if a cause and an effect can be positively linked, this does not necessarily mean that harm results. Making this connection is at the heart of current debates over the safety of radio communication systems. In the case of radio waves' effects on animal tissues, this means that any observed *biological* effects need to be clearly linked to observed *health* problems. Heating effects have been shown to cause adverse health reactions, but *not* at the low power levels used by today's cellular telephones. Determining harm is more difficult with nonthermal effects—which might affect basic cell functions that are only now beginning to be understood—and will be the subject of long debate.

In any case, some people will view any biological effects as harmful, whether or not there are any actual impacts on health. Fundamentally, an assessment of risk and one's reaction to it is quite subjective and personal. For example, many people are afraid to fly, although airline fatalities are rare. On the other hand, automobile safety receives far less public scrutiny, even though tens of thousands die annually from highway accidents.

In trying to evaluate the possible harm from radio communication systems, different groups disagree over what *standards of proof* should be used to determine safety or harm—that is, what proof is adequate to prove or disprove potential adverse health effects. One view requires *proof of no harm* before a technology is deployed. This approach is generally taken, for example, by the pharmaceutical industry and the U.S. Food and Drug Administration: firms must show, through extensive self-funded testing, that a new drug has few significant known adverse effects when used as prescribed.

An alternative approach is to permit a technology to be deployed, under certain guidelines, until it can be shown convincingly that negative effects result, or *no proof of harm* (note word order difference from above). In this case, experimentation is not limited to test groups in experimental settings, but also takes place among the public where a technology can be fully and vigorously evaluated in real-world conditions. For example, software producers expect bugs in early releases of their products because they know they cannot completely test programs and applications on their own beforehand.[4]

Most technologies fall somewhere between these two positions: initial experimentation is extremely limited in scale and scope, often confined solely to the laboratory. Next, the technology or product is subjected to more rigorous evaluation to see if hazards exist. After a period of controlled testing and evaluation, standards may be issued by the relevant technical body, such as the Institute of Electric and Electronics Engineers (IEEE). These standards may be accepted by government regulators, and become enshrined as substantial benchmarks guiding general and large-scale use and deployment of the technology or product.

If new information about hazards or other negative effects later comes to light, the standard may be changed with the agreement of the standards bodies and regulators. Changes at this stage may be difficult due to the institutional interests surrounding the status quo and the changing standard of proof required to attend to problems. With technologies or products such as asbestos, lead paint, or tobacco that come to be seen as hazardous, the firms that manufacture them have, in many cases, successfully resisted efforts to label them as bad for health, despite steadily mounting evidence to the contrary.

Another issue in determining harm is the *integrity* of the process by which research is conducted, including that of the people performing the work. If research is conducted in a way that raises questions of bias or poor quality, then such work will fail to settle questions about cause and effect, as well as potential hazards. Charges of bias, ignoring contrary evidence, or slipshod research meth-

---

[4] This difficulty in testing before full-scale release poses particularly acute problems for systems that operate highly reliably the first time, but cannot easily be subjected to real-world tests, such as antiballistic missile system software.

ods may be unfounded, but nevertheless must be taken seriously. Failure to demonstrate good faith or adherence to good scientific practice in the process by which information is gathered and evaluated may lead to continuing controversy. The makeup of research teams, lack of financial or other ties to firms with a stake in the outcome, fair and open evaluation of research proposals and research results, open publication of results or other public reporting requirements, participation by all interested parties, regardless of their affiliation—all these contribute to the integrity of the research process. These factors are also essential to reducing public concerns about research bias, and to increasing public trust and confidence in the technologies or products in question.

In the face of inconclusive and ambiguous evidence, different groups have different reactions. Opponents of widespread deployment of cellular and PCS facilities, and those claiming that cellular telephones promote cancer, argue that the industry should be held to the "proof of no harm" test. Without convincing proof of their safety, some people believe that antennas and towers should be restricted or moved and phones should be redesigned or prohibited altogether, even those that conform to current safety guidelines. The wireless industry, on the other hand, argues that there has been no proof of harm to date, and that changes in standards and use of the technologies should occur only when substantial and persuasive proof of harm is demonstrated. The industry also argues that it is funding research into biological and health effects, and that this research will help settle disputes about the safety of microwave radio frequency technologies. Compromise between these two groups will be very difficult, because their reactions to uncertainty are based on diametrically opposed philosophies—stop until safety is guaranteed or keep going until harm is proven—and both hold up different standards of proof.

Faced with a technical and policy controversy such as this, policymakers have difficult choices to make. If a technology is already being widely used, as is the case with many wireless technologies, using a "proof of no harm" standard is unrealistic. Television broadcasting towers, public safety radios, cellular towers and antennas, and hand-held cellular telephones have been deployed for years, and are used by tens of millions of people. Stopping these systems until definitive testing can be done is not realistic in today's political climate. However, finding out about possible harm through monitoring and active research is a viable option. Identifying early indications of effects or harm is in the public interest, even if short-term costs are high. Research to determine cause-and-effect relationships, and to ascertain the extent to which and under what circumstances harm may ensue, is essential. Some researchers also suggest that those concerned about possible hazards from electromagnetic radiation practice "prudent avoidance," which is avoidance of emissions where it is economically, operationally or physically easy to do so.[5]

BIOLOGICAL AND HEALTH EFFECTS OF WIRELESS TELECOMMUNICATIONS

Cellular and other radio communications devices should be distinguished from low frequency electromagnetic fields found around electric power lines. Electric power systems in the United States operate at a low frequency of 60 cycles or hertz (Hz) and at high power, while cellular telephones operate at much higher frequencies, 800 to 900 megahertz (MHz), and at extremely low power levels. New PCS systems will operate at even higher frequencies, 2 gigahertz (GHz) and still lower power levels. Researchers have established that the effects of electromagnetic radiation vary greatly with frequency and power levels, and empirical work over the last several decades has been

5 See U.S. Congress, Office of Technology Assessment, *Biological Effects of Power Frequency Electric and Magnetic Fields,* OTA-BP-E-53 (Washington, DC: U.S. Government Printing Office, May 1989), pp. 77-80 for a discussion of prudent avoidance in the context of electric power line electromagnetic radiation and potential human health effects.

conducted to determine safe levels at various combinations.[6] Because of this variability, however, effects found at one level are not generalizable to other frequency/power combinations—independent research must be conducted.

■ Research Is Inconclusive

While considerable research has been conducted on the effects of electromagnetic fields generally, very little work has yet been done on the possible health effects of exposures in the specific frequency and intensity ranges generated by wireless communications devices and systems. A particular weakness in the existing literature is the lack of research on the impact of long-term exposures.

The data that does exist paints an ambiguous picture. Some—but not all—research conducted on cells and animals suggests that exposures to fields with characteristics similar to those generated by cellular phones may cause behavioral and biological effects, including abnormal cell growth and increased incidence of malignancies.[7] The results of other studies involving claimed links between radio waves and cancer are inconsistent and difficult to interpret.

> [GAO] has concluded that [no] research has been completed on long-term human exposure to low levels of radiation specifically from portable cellular telephones. Research findings on exposure to other sources of low-level radio-frequency radiation are inconclusive. Some laboratory studies show that biological effects can occur when animals and cells have undergone extended exposure to low-level radio-frequency radiation; others do not. Scientists at FDA and EPA said that existing research does not provide enough evidence to determine whether portable cellular telephones pose a risk to human health.[8]

There are two fundamental issues concerning radio-frequency electromagnetic radiation and human exposure. The most obvious is the thermal or heating effect of such radiation on tissue. It is well known that high-power radio waves will generate heat in exposed tissues. Microwave ovens, high-powered radars, and other high-power microwave devices, for example, radiate energy—a small portion of which is absorbed by body tissues. The rate at which this energy is absorbed is called the specific absorption rate (SAR). Absorbed energy raises the temperature of the tissues through the excitation of water molecules (the typical microwave oven operates at about 600 watts at 2450 MHz). The higher the power level the more heat is generated at a given distance for a given sample, and the higher the frequency, the more of the incident energy is superficially absorbed.

The thermal effects of radio communication devices are generally not considered harmful. Wireless devices are required to comply with well-established standards governing human exposure to electromagnetic radiation. These standards incorporates a substantial safety factor as a cushion against unanticipated effects or exposure in unusual situations. As a result, researchers have been unable to measure heating of tissue at the low power levels used by hand-held cellular telephones. Microwaves do not penetrate metal, so shielding against them is fairly straightforward. In addition, power densities decline rapidly with distance from the source, so exposure can be reduced by lowering the power level and maintaining proper distances from operating antennas.

The second, and more controversial, issue is the possibility that RF radiation may cause non-thermal effects, including changes in genetic

---

[6] For recent reporting on low-frequency power effects, see Tekla S. Perry, "Today's View of Magnetic Fields," *IEEE Spectrum,* vol. 31, No. 12, December 1994, pp. 14-23. High frequency standards are dealt with in Institute of Electrical and Electronics Engineers, *IEEE Standard for Safety Levels with Respect to Human Exposure to Radio Frequency Electromagnetic Fields, 3 kHz to 300 GHz,* IEEE C95.1 1991, approved by IEEE Sept. 26, 1991, approved by the American National Stardards Institute Nov. 18, 1992, (New York: Institute of Electrical and Electronics Engineers, November 1994).

[7] See U.S. Congress, General Accounting Office, op. cit., footnote 1, pp. 29-31, for a brief review of this literature.

[8] Ibid, p. 3.

BOX 11-1: Origins of Recent Concern About Brain Cancer and Cellular Telephones

Public concern about low-power, high-frequency radio devices such as cellular telephones has its origins in a wrongful death lawsuit filed in April 1992, by David Reynard against his cellular telephone company, alleging that his wife's frequent and prolonged use of her cellular telephone contributed to her death by brain cancer. The story was first reported in the Ft. Lauderdale *Sun-Sentinel*, and received widespread attention following an interview with Reynard by Larry King on the CNN television network in January 1993.

News of the suit led to a significant drop in the stock prices of cellular companies and led to efforts by the companies to assure the public that cellular telephones are safe. While there was broad public concern at the time about the safety of the devices, committed users apparently were unwilling to forego use of the phones: cellular telephone subscription rates and usage did not significantly drop during this time. The case was dismissed on May 17, 1995, for lack of evidence meeting Florida's standards for admissibility.[1] There are currently seven other cases pending on the safety of cellular telephone use.

---

[1] *H. David Reynard, et al.*, v. *NEC Corp., et al.*, "Order," in United States District Court, Middle District of Florida, Tampa Div., case no. 94-825-CIV-T-21E. See also John Schwartz, "Court Call Favors Cellular: Judge Throws Out Claim of Link to Brain Cancer," *The Washington Post*, May 20, 1995, p. A2.

SOURCE: Office of Technology Assessment, 1995.

---

structure, the changes in the permeability of cell membranes, and disturbances in cell metabolism. These nonthermal effects theoretically could occur at lower power levels and under different modulation schemes than would be necessary to generate thermal effects. Much research in this area remains to be done, as government, industry and the academic communities agree. While there is no evidence that low-power, high-frequency radio signals *cause* cancer in cells, the possibility has been raised that such low-power radio waves could *stimulate* the growth of cancerous or precancerous cells, although early evidence is very weak (see box 11-1). Some preliminary evidence

of microwave effects on DNA has also been reported, but not yet confirmed.[9]

■ Exposure Standards Are Still Being Debated

To protect people from harmful exposure to high levels of electromagnetic energy, the Institute of Electric and Electronics Engineers (IEEE) developed standard IEEE C95.1, which was revised and adopted by IEEE in September 1991 and approved by the American National Standards Institute (ANSI) in November 1992.[10] Essentially, the standard says that devices operating between 100

---

[9] Henry Lai and Narenda Singh, "Acute Low-Intensity Microwave Exposure Increases DNA Strand Breaks in Rat Brain Cells," *Bioelectromagnetics,* vol. 16, spring 1995, forthcoming. See report in "Microwaves Break DNA in Brain; Cellular Industry Skeptical," *Microwave News,* vol. 14, No. 6, November/December 1994, pp. 1, 11-13.

[10] Institute of Electrical and Electronics Engineers, *IEEE Standard for Safety Levels with Respect to Human Exposure to Radio Frequency Electromagnetic Fields, 3 kHz to 300 GHz,* IEEE C95.1 1991, approved by IEEE Sept. 26, 1991, approved by the American National Stardards Institute Nov. 18, 1992 (New York: Institute of Electrical and Electronics Engineers, November 1994). These standards are based on several decades of biological and radiological work, particularly on the question of electromagnetic radiation and cancer. For the most recent verion of the standard, promulgated in 1991 and 1992, the standards committee had 14 biological evaluation working groups, with 125 scientists, physicians, and engineers drawn from academia, the private sector, and government. Similar standards have been adopted by other organizations as well.

MHz and 450 MHz are within permissible limits if they radiate less than 1.4 watts, and the radiating structure is at least one inch from the body.[11] At higher frequencies, the permitted power levels drop: for example, at 1500 MHz, the limit is 0.4 watts. Most hand-held telephones used in the United States operate at no more than 0.6 watts. Mobile telephones (installed in cars) are permitted to emit up to 3 watts because car phone antennas are installed outside vehicles away from close human contact. These levels are considerably below the 4 watt per kilogram energy absorption threshold identified in the scientific literature as the lowest level at which adverse effects due to heating had been noted and replicated. In a December 1992 report, IEEE concluded that "prolonged exposure at or below the levels recommended in these guidelines is considered safe for human health."

The exposure limits in the standard were derived from work done by the U.S. Navy and the IEEE before 1960, and reviewed and revised every five years, according to ANSI policy. Because of this historical foundation, the standard principally addresses concerns about the thermal effects of microwave radiation. Nonthermal effects, while reportedly discussed in the standards committee deliberations, are not directly addressed by the ANSI/IEEE standard, in part because little research on them had been done when the standard

was last revised.[12] Too little is known about the mechanism(s) by which nonthermal effects operate to set standards for exposure, presuming harmful nonthermal effects exist. As the IEEE standard document notes:

> Biological effects data that are applicable to humans for all possible combinations of frequency and modulation do not exist. Therefore, this standard has been based on the best available interpretations of the extant literature and is intended to prevent adverse effects on the functioning of the human body[13]. . . .

> Research on the effects of chronic exposure and speculations on the biological significance of nonthermal interactions have not yet resulted in any meaningful basis for alteration of the standard. It remains to be seen what future research may produce for consideration at the time of the next revision of this standard.[14]

Disputes over biological and health effects revolve around the continued acceptability of this standard as new research is performed.[15] As of spring 1995, the FCC was still considering whether to adopt the C95.1-1992 standard for *all* devices operating at microwave radio frequencies. Analog cellular telephones are presently exempt from testing under FCC rules because of their low power levels. However, the FCC indicated in 1994 that PCS phones would be subject to testing and SAR level limitations unless their maximum power

---

[11] This is a conventional way of stating the levels permitted under the standard, expressed in terms of what levels the emitting devices may have. The standard actually says nothing about emitting devices, but specifies exposure levels for humans, and is considerably more complex and detailed: it covers a wide range of frequencies (from 3 kHz to 300 Ghz), and power levels, measured as electric field or magnetic field strength or power density, depending on the frequency range. Compliance with the IEEE/ANSI standard also requires that, at cellular phone frequencies, actual exposure for the general public (measured by the specific absorption rate) not exceed 0.08 watts per kilogram whole-body average or 1.6 milliwatts per kilogram peak exposure in any one gram of tissue over 30 minutes. The maximum power density level is 0.57 milliwatt per square centimeter of tissue for over the whole body. These levels are somewhat different for other radio devices, such as ESMR, PCS or police radios. See Mark Fischetti, "The Cellular Phone Scare," *IEEE Spectrum*, vol. 30, No. 6, June 1993, pp. 44, 46.

[12] IEEE notes that most reports of biological effects have dealt with acute exposures at relatively few frequencies rather than with chronic exposures, and its work reflects this data base. The cutoff date for the literature review on which the standard depends was December 1985, with some carefully selected exceptions. See Institute of Electrical and Electronics Engineers, op. cit., footnote 6, p. 26-27.

[13] Institute of Electrical and Electronics Engineers, op. cit., footnote 6, p. 21.

[14] Ibid., p. 24.

[15] Louis Slesin, publisher of *Microwave News*, is a careful exponent of those advocating increased attention to biological effects of high-frequency, low-power electromagnetic radiation on humans. See for example, "Cellular Phones: Why the Health Risk Can't Be Dismissed," *Microwave News*, vol. 13, No. 1, January/February 1993, pp. 1, 11-12.

BOX 11-2: Statistics and large numbers

In 1991, there were approximately 17,600 deaths caused by brain cancer in the United States and about 514,300 cancer deaths overall. The cancer rate, between five and six deaths per 100,000, has not changed significantly over the past decade.[1] In a population of 180 million adults 20 years old and above, there are about 20 million cellular telephone users, or about 11 percent of the adult population. Mathematically, one would expect about 1,956 cellular telephone users to get brain cancer, independent of any specific cause. The National Cancer Institute, a part of the National Institutes of Health, estimated that there would be 350 new cases of brain cancer among cellular telephone users in 1993.[2] It is unknown how many actual cases occurred, since data on cancer and cellular telephone use is not yet available.

The lesson in these numbers is that, just because someone uses a cellular telephone and gets cancer, there is no reason to assume it is the phone that *caused* it. Because the numbers are so small, it would be difficult to distinguish cancer due to cellular telephones from other possible causes. If it were scientifically proven that cellular telephone users contract cancer at rates above the average, all other things being equal, it might be concluded that cellular telephones had a role to play. But even this is difficult to say with certainty because so many factors contribute to the incidence and growth of cancer.

---

[1] Letter from Dr. F. Kristian Storm, Professor, Departments of Surgery and Human Oncology, University of Wisconsin, Comprehensive Cancer Center, to Rep. Edward Markey, Feb. 2, 1993.

[2] Mark Fischetti, "The Cellular Phone Scare," *IEEE Spectrum,* vol. 30, No. 6, June, 1993, pp. 43-47.

SOURCE: Office of Technology Assessment, 1995.

---

output was less than 0.1 watt and a 2.5 centimeter separation was maintained between the user and any radiating structures.[16] The standard has been endorsed by the cellular industry and the FDA's Center for Devices and Radiological Health, but EPA, the National Institute for Occupational Safety and Health and others have objections.[17]

■ Research Activities

Research into the possible health effects of radio communication devices and systems is under way in a variety of institutions, including work sponsored by the cellular telephone industry. Questions have been raised about the potential bias of such work,[18] but these concerns appear to have been addressed.[19] Planned research may provide some answers to recently raised questions about the health effects of wireless telecommunications.

Research is concentrated in epidemiology, dosimetry, toxicology, and clinical studies. Through statistical studies of large populations, epidemiological studies seek to determine whether the occurrence of a disease can be associated with characteristics of people or their environments (see box 11-2). Dosimetry studies attempt to develop appropriate models of exposure relevant to human use of cellular and other wireless telephone use. Laboratory studies use controlled experi-

---

[16] *Microwave News*, vol. 14, No. 5, September/October 1994, p. 8.

[17] *Microwave News*, vol. 14, No. 3, May/June, 1994, p. 13.

[18] U.S. Congress, General Accounting Office, *Status of Research on the Safety of Cellular Telephones*, GAO/RCED-95-32 (Washington, DC: November 1994).

[19] Letter from Dr. George Carlo, Chairman, Wireless Technology Research, to Mr. Keith O. Fultz, Assistant Comptroller General, Resources, Community and Economic Development Division, U.S. General Accounting Office, Apr. 10, 1995.

ments with cell tissues or animals to ascertain the biological effects of particular radio-frequency emissions. These types of studies, epidemiological and laboratory, are necessary to assess whether there is a health risk to the population.

Two major research programs are being conducted in the United States. In the first, Motorola, a major manufacturer of cellular telephones and switching equipment, is funding a number of studies, some of which are published in the peer-reviewed literature. The other major research program is a three-to-five year effort, estimated to cost upward of $25 million, funded by the cellular telephone industry using an unrestricted deposit-only escrow fund that may be increased as research questions are refined.[20] This effort is overseen by Wireless Technology Research (WTR) (formerly the Scientific Advisory Group (SAG)),[21] and will support a number of multidisciplinary studies in epidemiology, cell cultures, test models, and genetics.[22] Both analog and digital transmission formats will be examined at power levels and frequencies used by current cellular systems, as well as those of proposed PCS. The resulting scientific work is subjected to review through an independent peer-review board coordinated by the Harvard University School of Public Health's Center for Risk Analysis.[23] Results will be submitted for publication in the scientific literature.

Research on cellular telephone health effects is also being conducted in Europe, although differences in transmission frequencies, power levels, and waveforms make it difficult to know the applicability of research findings in the United States. In the United Kingdom, the National Radiological Protection Board is developing computer models to characterize the fields induced in the human head by hand-held devices. Both German Telkom and the Research Association for Radio Applications—a consortium of manufacturers and cellular providers—are sponsoring behavioral and health effects research in Germany. The European Commission commissioned a study of thermal and nonthermal health effects from wireless device emissions in late 1994. The study is being conducted at the Center for Personkommunikation at Aalborg University, Denmark.

The credibility of industry-funded research depends on an open process, extensive peer and government review, adherence to accounting and auditing standards, no-strings-attached funding, appropriate research questions and methods, and timely disclosure of research results. For the CTIA-sponsored effort, the peer-review panels and the research itself are funded through an escrow account to provide for strict independence. GAO (see below) questioned whether the research efforts conducted under the cellular industry program could be considered truly objective and credible; the WTR established a new nonprofit administrative structure to manage the research funds and altered its funding and supervisory structures to respond to GAO's concerns.[24] Government funds might be contributed to the effort,

---

[20] Interview with Scientific Advisory Group (now Wireless Technology Research) staff members, March 29, 1995.

[21] Membership of the Scientific Advisory Group consists of Dr. George L. Carlo, of the Health & Environmental Sciences Group, Ltd., and George Washington University; Dr. Ian Munro, of CanTox, Inc.; and Dr. Arthur W. Guy, University of Washington, Seattle. On Mar. 31, 1995, the SAG became Wireless Technology Research, LLC.

[22] Scientific Advisory Group on Cellular Telephone Research, "Potential Public Health Risks From Wireless Technology: Research Agenda for the Development of Data for Science-Based Decisionmaking," (Washington, DC: Scientific Advisory Group on Cellular Telephone Research, Aug. 25, 1994).

[23] Details of Wireless Technology Research and associated activities can be found in *Wireless Technology Update*, its organization newsletter published in Washington, DC.

[24] Letter from Dr. George Carlo, Chairman, Wireless Technology Research, to Mr. Keith O. Fultz, Assistant Comptroller General, Resources, Community and Economic Development Division, U.S. General Accounting Office, Apr. 10, 1995.

but the WTR believes that bureaucratic and budget constraints make this unlikely.

■ Government Initiatives

The General Accounting Office (GAO) completed a short study of research performed on the safety of analog cellular telephones in November, 1994. The report notes that no one federal regulatory agency in the United States has responsibility for wireless communications device emissions; EPA has overall responsibility for advising the government on EMF exposures, the FDA establishes standards for devices that emit radiation, and the FCC approves wireless communications

devices for use and assures that their emission levels meet safety standards.

The study also concluded that little research on the health effects of wireless telecommunications devices on humans is planned by the federal government, with the exception of an epidemiological study by the National Cancer Institute to be completed in 1997 or 1998. In 1984, the Environmental Protection Agency convened an interagency working group on electromagnetic frequency radiation, composed of scientific specialists. The Food and Drug Administration is establishing an oversight group that includes policy specialists as well.[25]

---

[25] Members include the U.S. Food and Drug Administration, the U.S. Environmental Protection Agency, the National Telecommunications and Information Administration, the National Institutes of Occupational Safety and Health, and the Occupational Safety and Health Administration.

Electromagnetic Interference and Wireless Devices | 12

As new radio devices and wireless systems proliferate, particularly at low power levels and in nontraditional applications, and with the increasing numbers of other passive electronic devices in society, radio frequency interference among them may become an increasing problem. As devices become smaller, people are increasingly likely to carry and use them in situations unanticipated by designers. Nonradio electronic devices such as personal computers have not necessarily been designed to be immune from wireless telecommunications emissions, and can also cause interference to radio receivers.[1] This chapter discusses how wireless devices and systems may interfere with each other as well as with other electronic equipment and identifies some possible solutions.

FINDING

Interference between different wireless systems and between wireless systems and other electronic devices is potentially serious, but also is amenable to technical and regulatory solutions. Wireless devices can cause interference to electrical components and vice versa, and as new generations of digital radio equipment become widely used, these problems may increase in the short term. However, installation of lower power microcells, improved shielding, and electrical design techniques can usually mitigate most interference problems. In cases where other solutions are not feasible, carefully targeted use restrictions may be required.

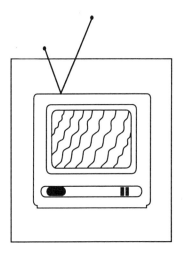

---

[1] Causes of interference include high clock rate timing pulses used in computers, video games, etc., and their harmonics.

BACKGROUND

Any short length of wire in an electronic circuit or in an integrated chip can act as an antenna when exposed to radio waves and give rise to electric currents that may interfere with the normal operation of the circuit.[2] This potential electromagnetic interference (EMI) is an inherent property of radio or television transmissions, electric motors, and household switches, as well as natural phenomena such as lightning, aurora borealis, and sunspot activity.[3] To protect against it, shielding—either in a metal case or special shielded wire—or better circuit design is necessary. Most of the time designers anticipate problems, and build devices not subject to interference when used as directed. However, there are cases in which devices are not shielded adequately against EMI, many involving medical devices.

While many of the reported EMI problems involve older analog radio transmitting devices, the wireless industries are increasingly turning to digital transmission formats to improve quality and increase capacity. This transition may pose new EMI problems because digital equipment may interact with other devices in unpredictable ways. For example, digital Global System for Mobile Communications (GSM) handsets and Time Division Multiple Access (TDMA) telephones emit higher strength peak electric fields than do analog telephones, while maintaining the same average power levels.[4] This scheme results in better transmission and reception at a lower average power output—extending battery life—but it may also cause greater interference than analog phones. The increasing use of spread spectrum, including Code Division Multiple Access (CDMA) technologies, has also led some engineers to predict that, with a large number of users, interference between competing devices may make the systems unusable.

INTERFERENCE WITH MEDICAL DEVICES

Medical devices can be affected by interference from radio devices, including cellular telephones, and this has recently become a public issue.[5] Pacemakers, apnea monitors, blood-gas pumps, hearing aids, wheelchairs, and electronic imaging devices have reportedly been interrupted or interfered with in the presence of cellular telephones or other radio devices.[6] In some cases, deaths have occurred, though none have been attributed to cellular telephones. In spring 1995, pacemaker wear-

---

[2] K. J. Clifford, et al., "Mobile Telephones Interfere with Medical Electrical Equipment," *Australian Physical & Engineering Sciences in Medicine,* vol. 17, No. 1, 1994, p. 23.

[3] EMI effects increase with power and decrease with distance.

[4] Stewart Fist, "GSM and TDMA digital phones," April, 1994, unpublished manuscript.

[5] Jeffrey L. Silberberg, "Performance Degradation of Electronic Medical Devices Due to Electromagnetic Interference," *Compliance Engineering,* fall 1993, pp. 25-39; "Cellular Telephones and Radio Transmitters: Interference with Clinical Equipment," *Technology for Respiratory Therapy,* vol. 14, No. 5, November 1993; Tom Knudson and William M. Bulkeley, "Stray Signals: Clutter on Airwaves Can Block Workings of Medical Electronics," *The Wall Street Journal,* vol. 223, No. 116, June 15, 1994, pp. A1, A12.

[6] Some documented illustrative examples:
   • A fetal heart beat detector picked up radio and CB broadcasts and static instead of heart beats.
   • A ventilator malfunctioned due to interference from a guard's walkie-talkie.
   • A user of a powered wheelchair had moved to a new home and was showing his friends, also in powered wheelchairs, around the neighborhood. While moving up a hill, the user heard clicking noises and took his hand off the joystick. The wheelchair made a sudden about turn and headed down hill at high speed. The wheelchair would not respond to further movement of the joystick. The wheelchair continued down the hill for about 25 yards, veered left, and went over a cliff. The user suffered a broken hip and several other injuries. His friends' wheelchairs were from a different manufacturer and were not affected. The wheelchair user's new home is several miles away from a radio station and three blocks from a major interstate highway.
   • An external defibrillator/pacemaker stopped pacing when an ambulance attendant used a hand-held transmitter too close to the patient. The patient was not resuscitated.
   These examples are taken from Jeffrey L. Silberberg, op. cit., footnote 5, pp. 25-39.

ers were warned not to use new digital cellular phones because of interference problems.[7]

Specific problems have surfaced with new digital mobile telephones and hearing aids. Time division digital transmissions can produce loud audio tones in some hearing aid models and other analog audio devices from up to 100 feet away. The tones can reach 130 dB—the sound of an airplane taking off as heard by a person standing on the runway.[8] The interference lasts as long as the hearing aid is close to the digital phone, but returns to normal when the phone is turned off or moves out of range.

Shielding can reduce the amount of interference hearing aids encounter, but there are limits to what shielding can be done. There are three types of hearing aids, those worn in the ear, outside the ear, and in a pocket and attached by wire. Hearing aids worn in the ear, by far the most popular, are least amenable to shielding, because they are already very small; hearing aids worn in the pocket are most susceptible to EMI, but can be easily shielded.

There are about six million hearing-aid users in the United States today, and the number is projected to increase as the baby-boomer generation ages. It is not known what types of hearing aids (in-ear, on-ear, or pocket), or how many (one or two ears) are used, nor is it known how many hearing-impaired people use cellular telephones. The projected cost of retrofitting hearing aids to eliminate interference is unknown; this may not be feasible given their small size and life span.

The potential for EMI has long been studied and understood by radio engineers and medical technologists, and a substantial body of technical work and engineering expertise exists. Like other forms of electromagnetic interference, shielding devices against electromagnetic radiation and controlling the output levels of emitting devices are the two main ways compatibility is attained. Another is the proper installation and spacing of medical equipment to minimize the potential for interaction.

Standards have been set for both transmitting devices and for shielding of computing and medical devices, based on both lab testing and field experience. Voluntary standards were promulgated in 1979 by the Food and Drug Administration (FDA) specifying that medical equipment should be protected against interference up to seven volts per meter between 450 and 1000 MHz.[9] A more recent standard issued by the International Electrotechnical Commission, one of the main standards' bodies in this area, relaxes suggested permitted exposure to three volts per meter in the frequency range from 26 to 1000 MHz.[10] The Association for the Advancement of Medical Instrumentation, a voluntary standards body in the United States, has convened a committee to address EMI problems.[11] Table 12-1 gives the FDA's 1994 draft suggestions on the minimum distance that should be maintained between transmitters of various power outputs and medical devices with various amounts of shielding.

[7] Mark Landler, "Cellular Phones May Affect Pacemakers," *The New York Times*, Apr. 29, 1995, p. B1; John J. Keller, "Cellular Phones May Affect Use of Pacemakers," *The Wall Street Journal*, Apr. 29, 1995, p. B1. Medtronic, Inc., a major pacemaker supplier in the United States, advises pacemaker users to turn off portable phones placed in breast pockets, hold phones ten inches away from the chest, and use the phone on the opposite side of the body from where the pacemaker is implanted.

[8] Michael Ruger, attorney, Baker & Hostetler, Washington, DC, personal communication, Feb. 17, 1995; "TDMA Mobile Phones Accused of Interference," *Microwave Engineering Europe*, March/April 1993, pp. 16-17.

[9] U.S. Food and Drug Administration, Bureau of Medical Devices, "Electromagnetic Compatibility Standard for Medical Devices," BMD Publication No. MDS-201-0004, Oct. 1, 1979.

[10] International Electrotechnical Commission, *Medical Electrical Equipment, Part 1: General Requirements for Safety, 2. Collateral Standard: Electromagnetic Compatibility—Requirements and Tests*, 1993.

[11] Knudson and Bulkeley, op. cit., footnote 5, p. A12.

Power rating of radio source and example sources	Immunity level of medical device					
	Unknown (assume 0.1 volt/meter)	1 volt/ meter	3 volts/ meter	10 volts/ meter	20 volts/ meter	40 volts/ meter
0.07 watt Microcell cellular phone	4.6 meters	0.5 meter	0.3 meter	0.3 meter	0.3 meter	0.3 meter
0.01 watt	5.5	0.6	0.3	0.3	0.3	0.3
0.1 watt Wireless computer equipment	17.4	1.7	0.6	0.3	0.3	0.3
0.6 watt Portable cellular phone	42.6	4.3	1.4	0.4	0.3	0.3
1 watt	55.0	5.5	1.8	0.6	0.3	0.3
3 watts Transportable cellular phone	77.8	7.8	2.6	0.8	0.4	0.3
5 watts	123.0	12.3	4.1	1.2	0.6	0.3
10 watts	173.9	17.4	5.8	1.7	0.9	0.4
20 watts	246.0	24.6	8.2	2.5	1.2	0.6
50 watts	388.9	38.9	13.0	3.9	1.9	1.0
100 watts State police radio Amateur radio	550.0	55.0	18.3	5.5	2.8	1.4
1,500 watts Amateur radio	2.1 km	213.0	71.0	21.3	10.7	5.3
100 kilowatts FM broadcast TV stations ch. 2-6	17.4 km	1.7 km	579.8	173.9	87.0	43.5
316 kilowatts TV stations ch. 7-13	31 km	3.1 km	1.0 km	309.2	154.6	77.3
5 megawatts TV stations ch. 14-69	123 km	12.3 km	4.1 km	1.2 km	614.9	307.5

TABLE 12-1: Proposed Electromagnetic Interference Protection Distance, in Meters

To find the minimum recommended protection distance between a medical device and a transmitter from this table, first locate the value in the top row that is closest to the RF immunity of the medical device. Then follow that column down to the row corresponding to the rated power of the transmitter. The entry in that cell of the table is the minimum recommended protection distance [in meters] between that medical device and that transmitter.
SOURCE: U.S. Food and Drug Administration, "EMI Protection Distance," draft, Aug. 15, 1994.

However, with the growing number of both radio and medical devices and their shrinking size, more interference is likely to occur. Because transmission equipment can rarely be altered to reduce interference, regulators think the best solution is for device manufacturers to pay close attention to shielding, working in consultation with the designers and manufacturers of emitting devices.[12]

[12] U.S. Congress, General Accounting Office, *Electromagnetic Interference with Medical Devices*, GAO/RCED-95-96R (Washington, DC: Mar. 17, 1995).

Other measures by themselves may not be sufficient. For example, proposals have been made to restrict the use of wireless devices in hospitals and clinics, but the ubiquity and small size of such devices makes policing difficult. Moreover, health care is becoming more decentralized with sensitive medical equipment increasingly housed in homes and outpatient clinics. Mobile care-givers, in turn, are becoming more reliant on wireless communications to interact with doctors and technicians at hospitals in other locations. This evolution in care-giving requires that medical equipment and wireless communications exist side-by-side. Users of medical or radio devices are generally unaware of field strengths, frequencies, the position, or in some cases even the *presence* of electromagnetic radiation. Warnings, when they do exist, rarely tell users what to do beyond "avoid electromagnetic interference."

Clearly, incorporating shielding into medical devices early in the development process is essential. Other measures may provide some help in minimizing interference problems: promulgating strong standards, limiting radio devices in well-identified areas, and providing good consumer education of the dimensions of EMI.

■ Regulatory and Legislative Initiatives

In October 1994, the Subcommittee on Information, Justice, Transportation, and Agriculture of the House Government Operations Committee held hearings on medical device interference from wireless and cellular devices.[13] The Federal Communications Commission (FCC) and FDA have primary oversight responsibilities for this area, and have consulted frequently on design and standards issues. However, legislative interest in this issue appears to have precipitated action in the industry to address EMI problems. For example, the Cellular Telecommunications Industry Association (CTIA) and the Health Industry Manufacturers Association have jointly funded a Center for the Study of Wireless Electromagnetic Compatibility at the University of Oklahoma to study medical device interference. This center convened a Forum on Electromagnetic Compatibility in September 1994, which discussed these issues.

The Hearing Aid Compatibility Act of 1988 required that all telephones be made compatible with hearing aids by 1991.[14] However, in a concession to the cellular telephone industry, the act excluded mobile phones. The act did permit the FCC to revisit the issue at a later date, with the presumption that new technologies would be made compatible with hearing aids. The FCC has determined that PCS equipment will be exempt from compliance with the act, noting that U.S. operators who choose GSM will use a different frequency than their European cellular counterparts, that few hearing-aid users will be affected, and that cost-effective solutions to mitigate interference are available.[15] There is some concern in the hearing-aid users' community that PCS operators will choose GSM as their standard. The FCC has convened an advisory committee to examine this issue.

INTERFERENCE WITH AIRCRAFT CONTROL SYSTEMS

Although there are no documented cases of civilian airline crashes caused by cellular telephone or other interference, electronic devices may pose

---

[13] U. S. Congress, House of Representatives, Committee on Government Operations, Subcommittee on Information, Justice, Transportation and Agriculture, 103d Congress, "Do Cellular and Other Wireless Devices Interfere with Sensitive Medical Equipment? Are Pacemakers, Hearing Aids, Apnea Monitors, Blood Pumps and Other Sophisticated Medical Devices Affected by Outside Electromagnetic Interference (EMI) from Cellular And Other Wireless Devices?" photocopied hearing statements, various witnesses, Oct. 5, 1994.

[14] 47 U.S.C., sec. 610. FCC regulations on hearing aid compatibility can be found in 47 CFR, sec. 68.4.

[15] Letter from Hon. Reed Hundt, Chairman of the FCC, to Sen. Bob Packwood, Chairman, Committee on Commerce, Science, and Transportation, Subcommittee on Communications, Apr. 12, 1995.

problems to airplane control systems.[16] Of the approximately 100 reported cases of alleged interference, about one-third appear to have some validity, according to technical experts.[17] FAA regulations hold the airline companies responsible for setting policies on the use of portable electronic devices; given the difficulty in assuring safe operation under all operating conditions, all airlines have decided to prohibit the use of any electronic devices during take-offs and landings.[18]

Inside an aircraft, radio transmitters, such as cellular telephones, can induce transient currents in wires and even be amplified in the aluminum airframe, because any unshielded metal can act as an antenna. CTIA is currently testing cellular telephones in planes to certify their safe use on the ground. (In addition, cellular telephone use on commercial aircraft in flight is not allowed because a single cellular telephone at even moderate altitudes would tie up many terrestrial cellular base stations simultaneously, since many base stations could be "seen" simultaneously by an airborne cellular telephone.)

A potential problem with American Mobile Satellite Corp.'s (AMSC's) transportable telephone is that it will operate at a frequency adjacent to that used by the Global Positioning Satellite (GPS) system, which will serve as the basis of the new generation of air traffic control systems in the United States. Operating such a telephone in an airplane may jam the GPS navigation system. The FCC, the National Telecommunications and Information Administration (NTIA), and the Federal Aviation Administration (FAA) have estab-

lished a procedure with AMSC and other interested parties to address this problem; a memorandum of understanding that was concluded in November 1994 provides the means to prevent interference and allows AMSC to proceed with its system's deployment.[19]

Interference could be more serious between portable electronic devices and digital flight equipment, including navigation systems. These systems work with digital bit streams, which can be thought of as strings of ones and zeros. Interference might occur by inducing spurious currents and thus introducing new data to the normal data stream. Such data would probably be rejected by error-correcting routines in current avionics, resulting in an interruption rather than a deviation of normal aircraft control systems, but it is difficult to know with certainty that this would always occur. Even devices that are not designed as radio transmitters emit electromagnetic radiation. This has led to concern that uncontrolled use of any electronic device might cause interference. One recently publicized case involved a pilot who believed that a CD player in use in the first-class compartment interfered with the normal operation of the aircraft during landing.[20]

Because analog avionics systems are not dependent on data streams, they are not susceptible to such interference. Thus, where a digital cellular telephone may affect new Airbuses or Boeing planes, it is unlikely to affect an older Boeing 727. On the other hand, newer aircraft use fiber optic cabling for control systems and more fault-toler-

---

[16] Jerry Hannifin, "Hazards Aloft," *Time,* Feb. 22, 1993, p. 61.

[17] For example, verifiable cases of interference might resemble the following: when the flight crew notices something unusual occurring to the airplane, together with a passenger's use of an electronic device, they ask the passenger to turn the device off, and note whether the problem has disappeared. They then ask the passenger to turn the device back on to see if the interference occurs again. If it does, then this is an event to be explained. However, efforts to duplicate such effects on the ground have all been unsuccessful. John Sheehan, Pfaneuf Associates, CTIA consultant, chair of RTCA Special Committee 177, personal communication.

[18] The RTCA, an advisory body to the FAA on electronic matters, is meeting to set standards for electronic device emissions in aircraft in the wake of concern about consumer electronic devices. It expects to issue its report on nonradio device interference in the spring of 1995.

[19] Memorandum of Understanding between the FCC, NTIA, and the FAA, "Addressing Out-of-Band Emission Requirements for the Mobile-Satellite Service," effective Nov. 19, 1994.

[20] Jerry Hannifin, "Hazards Aloft," *Time,* Feb. 22, 1993, p. 61.

ant architectures, making them less susceptible to radio interference.

It is extremely difficult and costly to model these internal interference problems. Because there are so many variables—the type of emitting device; its power, frequency, and modulation schemes; the effectiveness of its filters; its place in the aircraft, the location of sensitive instruments, the location of wire or airframe with respect to the emitting device, and the activity the aircraft is performing (e.g. landing or cruising at altitude)—determining all the conditions for trouble-free operation of portable devices is nearly impossible.

UNANTICIPATED INTERACTIONS AMONG LARGE, COMPLEX SYSTEMS

A general issue in electromagnetic radiation is the unintended effects of radio waves. These involve compatibility problems that can, for the most part, be solved either by shielding devices, keeping radio waves away from people and sensitive equipment, or changing the modulation scheme

emitting devices use. However, with widespread deployment of small radio devices with complex operating characteristics, it is possible that at some point there will be interference leading to a system failure. Because of the large number of devices, the variety of ways they are used, and the complexity of the possible interactions, it is unlikely that every combination can be tested and potential problems anticipated.

New technologies will continue to be introduced that cannot be tested in all real-world situations. A recent example: the operator's manual for European-model BMW automobiles advises owners not to use a digital (GSM) cellular telephone while driving the car, because it may interfere with the car's electrical system and lead to premature deployment of the airbags. While this particular problem is no doubt fixable, it is one indication of the kinds of surprises that may crop up from time to time as wireless telecommunications technologies play a larger role in a complex technological world.

Appendix A: Radio Communication Basics[1] A

DEFINITIONS OF RADIOCOMMUNICATION TERMS

Amplitude: A measure of the value of a radio wave, measured in volts (see figure A-1).

Analog: In analog radio communication, the message or information to be transmitted is impressed onto (modulates) a radio carrier wave, causing some property of the carrier—the amplitude, frequency, or phase—to vary in proportion to the information being sent. Amplitude modulation (AM) and frequency modulation (FM) are two common formats for analog transmission. In order to send analog signals, such as voice and video, over digital transmission media, such as fiber optics or digital radio, they must first be converted into a digital format. See modulation.

Bandwidth: The process of modulating (see below) a radio wave to transmit information produces a radio signal, but also generates additional frequencies called "sidebands" on either side of the carrier (see figure A-2). The total width of frequencies, including the sidebands, occupied by a radio signal is its bandwidth. In practical terms, however, the bandwidth of a signal refers to the amount of spectrum needed to transmit a signal without excessive loss or distortion. It is measured in hertz. In figure A-2, the bandwidth of the signal is 4 kHz. The bandwidth of a radio signal is determined by the amount of information in the signal being sent. More complex signals contain more information, and hence require wider bandwidths. An A radio broadcasting signal, for example, takes 10 kHz, while an FM stereo signal requires 200 kHz, and a color television signal takes up 6 MHz. The bandwidth required by a television channel is 600 times greater than that of an AM radio channel.

Carrier: A radio wave that is used to transmit information. Information to be sent is impressed onto the carrier, which then carries the signal to its destination. At the receiver the carrier is filtered out, allowing the original message to be recovered.

---

[1] Material in this appendix is derived from Harry Mileaf (ed.), *Electronics One*, revised 2d ed. (Rochelle Park, NJ: Hayden Book Co., Inc., 1976); U.S. Congress, Office of Technology Assessment, *The Big Picture: HDTV & High-Resolution Systems*, OTA-BP-CIT-64 (Washington, DC: U.S. Government Printing Office, 1990); William Stallings, *Data and Computer Communications* (New York, NY: MacMillan Publishing Co., 1985).

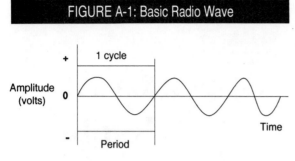

FIGURE A-1: Basic Radio Wave

Each cycle of a pure radio wave is identical to every other cycle.

SOURCE: Office of Technology Assessment, based on Harry Mileaf (ed.), *Electronics One*, revised 2d ed. (Rochelle Park, NJ: Hayden Book Co., 1976) p. 1-10.

Digital: Digital transmission formats can be used to transmit images and voice as well as data. For continuously varying signals such as voice or images, an analog/digital converter changes the analog signal into discrete numbers (represented in binary form by 0's and 1's). These binary digits, or bits, can then be sent as a series of "on"/"off" pulses or can be modulated onto a carrier wave by varying the phase, frequency, or amplitude according to whether the signal is a "1" or a "0." Data is sent in a similar fashion although it does not have to be converted into digital form first. (See figure A-3.)

Frequency: The number of cycles a radio wave completes in one second (see figure A-4). Frequency is measured in hertz (1 cycle per second equals one hertz). Radio frequencies are described as multiples of hertz:

kHz, kilohertz: thousand cycles per second;
MHz, megahertz: million cycles per second;
GHz, gigahertz: billion cycles per second.

The frequency of a radio wave is the inverse/reciprocal of its period. For example, if a wave had a period of 0.1 seconds, its frequency would be 10 hertz.

Modulation: The process of encoding information onto a radio wave by varying one of its basic characteristics—amplitude, frequency, or phase—in relation to an input signal such as speech, data, music, or television. The input signal, which contains the information to be transmitted, is called the modulating or baseband signal. The radio wave that carries the information is called the carrier wave. The radio wave that results from the combination of these two waves is called a modulated carrier. Two of the most common types of modulation are amplitude modulation (AM) and frequency modulation (FM) (see figure A-5).

Period: The length of time it takes a radio wave to complete one full cycle (see figure A-1). The inverse of the period is a radio wave's frequency.

Phase: A measure of the shift in position of a radio wave in relation to time (see figure A-6). Phase is often measured in degrees.

Spread Spectrum: Spread spectrum refers to various coding schemes used to modulate data information onto radio waves for transmission. Spread spectrum was originally used by the military to hide its communications in background "noise." Direct sequence spread spectrum systems encode each bit of information with a special code is known only to the transmitter and receiver. The transmitter sends these encoded bits over a

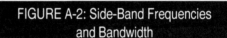
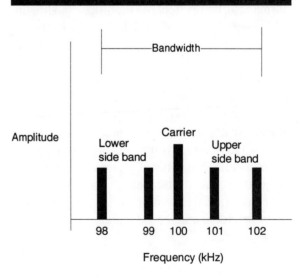

FIGURE A-2: Side-Band Frequencies and Bandwidth

NOTE: This figure represents a 100-kHz carrier wave modulated by 1- and 2-kHz frequencies.

SOURCE: Harry Mileaf (ed.), *Electronics One*, revised 2d ed. (Rochelle Park, NJ: Hayden Book Co., 1976) p. 1-31.

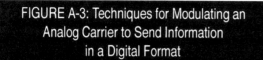

FIGURE A-3: Techniques for Modulating an Analog Carrier to Send Information in a Digital Format

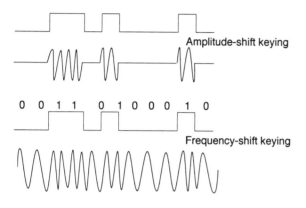

Amplitude-shift keying

0 0 1 1   0 1 0 0 0 1 0

Frequency-shift keying

SOURCE. U.S. Congress, Office of Technology Assessment, *The Big Picture: HDTV & High-Resolution Systems*, OTA-BP-CIT-64 (Washington, DC: U.S. Government Printing Office, June 1990), figure 3-3, p. 41.

wide range of frequencies assigned for the system. The receiver looks for the special coded bits and reassembles them in the proper order. In frequency-hopping spread spectrum systems, a wide range of frequencies is also used, but the system "hops" from frequency to frequency, transmitting bits of information on each frequency. Only the receiver knows the hopping pattern and how long the transmitter will stay on each frequency (as little as 100 milliseconds). This allows it to track the data across frequencies and reassemble the original signal.

Spectrum: Each radio signal is actually made up of a number of different radio waves at different frequencies. The spectrum of a radio signal refers to the range of frequencies it contains. In figure A-2, the spectrum of the signal extends from 98 to 102 kHz. The width of the spectrum is called the bandwidth of the signal. More broadly, the radio frequency spectrum consists of all the radio frequencies used for radio communications.

Wavelength: The distance between successive peaks of a continuous radio wave.

SPECTRUM BASICS[2]

■ Radio Waves

Radio waves are the basic unit of wireless communication.[3] By varying the characteristics of a radio wave—frequency, amplitude, or phase—these waves can be made to communicate information of many types, including audio, video, and data. Radio waves that carry information are called radio signals, and the process of encoding intelligence onto a radio wave so that it can be transmitted over the air is called modulation.[4] In the process of modulation, the information or message to be transmitted—a human voice, recorded music, or a television signal—is impressed onto (modulates) a "carrier" radio wave that is

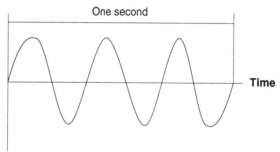

FIGURE A-4: Frequency of a Continuous Wave

One second

Time

Frequency = 3 cycles per second

SOURCE: Harry Mileaf (ed.), *Electronics One*, revised 2d ed. (Rochelle Park, NJ: Hayden Book Co., 1976) p. 1-10.

---

[2] Much of the material in this section comes from Richard Gould, "Allocation of the Radio Frequency Spectrum," contractor report prepared for the Office of Technology Assessment, Aug. 10, 1990.

[3] Although the term "radio" is most commonly associated with commercial radio broadcasting services (AM and FM radio), the term also properly encompasses the entire range of wireless communications technologies and services, including television, microwave, radar, shortwave radio, mobile, and satellite communications.

[4] Two of the most familiar modulation techniques are amplitude modulation (AM) and frequency modulation (FM).

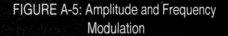

FIGURE A-5: Amplitude and Frequency Modulation

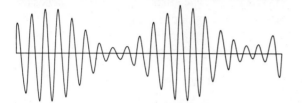

Amplitude-modulated wave

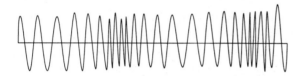

Frequency-modulated wave

SOURCE: U.S. Congress, Office of Technology Assessment, *The Big Picture: HDTV & High-Resolution Systems*, OTA-BP-CIT-64 (Washington, DC: U.S. Government Printing Office, June 1990), figure 3-1, p. 41.

then transmitted over the air. When a radio signal is received, the information is converted back into its original form (demodulated) by a receiver and output as sound, images, or data.

Radio waves are distinguished from each other by their frequency or their wavelength. Frequency represents the number of cycles a radio wave completes in one second, and is the most common description of a radiocommunication signal. The international unit of frequency measurement is the hertz (Hz), which represents one cycle per second.[5] Radio signals can also be identified by their wavelength. Signals with long wavelengths have lower frequencies, while those at higher frequencies have shorter wavelengths. Commercial AM radio signals, for example, consist of very long waves (approximately 100 to 300 meters), that

may complete a million cycles per second (1 megahertz (MHz)). Microwave signals, on the other hand, are very short (as little as 0.3 centimeters) and may complete hundreds of billions of cycles per second (100 gigahertz (GHz)). The relative nature of radio wavelengths is the origin of terms such as "short wave," which was given to radio frequencies around 2.8 MHz in the 1920s because the wavelengths in that frequency range were shorter than the wavelengths that had previously been used.

The radio spectrum is divided into "bands" that correspond to various groups of radio frequencies. These bands are identified by their frequencies or wavelengths (as above), or by descriptive terms that have been adopted over time. Several types of descriptive names have been attached to various portions of the spectrum (see figure A-7). One method denotes relative position in the spectrum: very low frequency (VLF), high frequency (HF), very high frequency (VHF), superhigh frequency (SHF), etc. Another method derives from usage developed in World War II to keep secret the actual

FIGURE A-6: Phase of a Continuous Wave

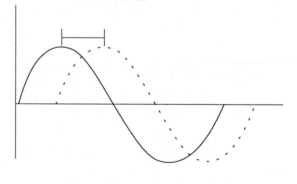

Phases = Difference between same points on different waves

SOURCE: Harry Mileaf (ed.), *Electronics One*, revised 2d ed. (Rochelle Park, NJ: Hayden Book Co., 1976) p. 1-10.

[5] Multiples of the hertz are indicated by prefixes (see box 2-A): "kilo" for one thousand, "mega" for one million, and "giga" for one billion. Thus, a million hertz—a million cycles per second—is expressed as one megahertz (abbreviated "MHz").

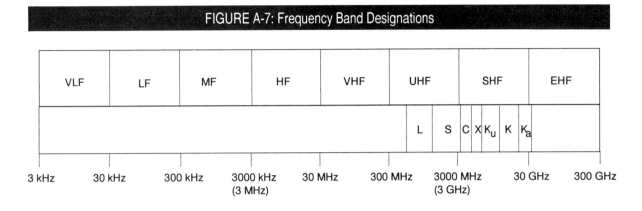

FIGURE A-7: Frequency Band Designations

SOURCE: Office of Technology Assessment, 1991, based on Richard G. Gould, "Allocation of the Radio Frequency Spectrum," OTA contractor report, Aug. 10, 1990.

frequencies employed by radar and other electronic devices: L-band, S-band, and K-band.[6] The International Telecommunication Union (ITU) classifies frequencies according to band numbers—Band 1, Band 2, etc. Frequency bands are also known by the services that use them—the FM radio broadcast band, for example, occupies the range (band) of frequencies from 88 to 108 MHz.

■ Transmission Characteristics

Several factors affect the transmission of radio signals, and, at different frequencies, some factors will affect radio waves more than others. Attenuation refers to the weakening of a radio signal as it passes through the atmosphere. All radio signals are attenuated as they pass through rain or any kind of water in the air (clouds, snow, sleet), but radio signals at higher frequencies will be attenuated more than those at lower frequencies. For instance, the attenuation of a radio signal passing through a rainstorm will be 10 times as great if the frequency of the signal is doubled from 5 GHz to 10 GHz. This makes radiocommunication, especially over long distances, extremely difficult in the upper (above 10 GHz) frequencies.

Radio waves are also bent and/or reflected as they pass through the atmosphere. Because of changes in the density of the atmosphere with height, radio signals bend as they pass from one atmospheric layer to the next. This bending is called refraction. In addition to refraction, if atmospheric conditions are right, radio waves are also reflected by the ionosphere, the top layer of the Earth's atmosphere. Ionospheric reflection enables some radio signals to travel thousands of miles, and accounts for the long-distance communication that is possible in the frequency range between about 3 and 30 MHz (the HF band—see below).

Although refraction and reflection are conceptually distinct, and refraction can occur without reflection, it is possible to think of reflection as an extreme case of refraction in the ionosphere.[7] The amount of refraction, or bending, experienced by a radio signal is related to its frequency. Lower frequencies bend (are refracted)

---

[6] These letter designations are not precise measures of frequency because the band limits are defined differently by different segments of the electronics and telecommunications industries.

[7] All radio waves are bent as they pass from a region of the atmosphere having a certain number of free electrons to a region with a different number of electrons. During the day, energy from the Sun splits the molecules of the gasses far above the surface of the Earth (in the troposphere and the ionosphere), producing many free electrons and creating layers of ionized particles. A radio wave from Earth entering one of these layers will be refracted, and if there are enough free electrons, the bending will be so great that the signal will be reflected back to Earth.

easily and are readily reflected back to Earth. Higher frequency signals experience less refraction than those at lower frequencies, and at progressively higher frequencies, there will be less and less bending. At a certain frequency, atmospheric conditions will be such that there is so little refraction that the signal will not be reflected back to Earth. The point at which this occurs is called the maximum usable frequency (MUF), and is generally in the range of 10 to 15 MHz, although it can be as high as 30 or 40 MHz or as low as 6 MHz, depending on time of day, season, and atmospheric conditions. Below the MUF, radio signals can be used for long-distance communication by reflecting the signal off the ionosphere. Above the MUF, the signal travels straight through the atmosphere and into space.

At higher frequencies, above the MUF, radio signals travel in almost straight lines from the transmitter to receiver, a transmission characteristic referred to as "line-of-sight."[8] Line-of-sight conditions affect radiocommunication above the MUF, but especially affect frequencies above one GHz. The distance a line-of-sight signal can travel is usually limited to the horizon or a little beyond. However, because the Earth is curved, the transmission distance will also be limited depending on the height of the transmitting antenna—the higher the antenna, the farther the signal can travel. For example, if the transmitting antenna is mounted on top of a mountain or a tall tower, the line-of-sight distance will be greater. Satellites, in simple terms, extend line-of-sight to the maximum distance (see figure A-8). Line-of-sight transmission requires that there be no obstacles between the transmitter and receiver—anything standing between the transmitter and receiver, e.g., a building or mountain will block the signal.

Atmospheric conditions have substantial impacts on line-of-sight radiocommunications. Differences in atmospheric temperature or the amount of water vapor in the air, for example, can cause radio signals to travel far beyond the "normal" line-of-sight distance. This condition is called ducting or superrefraction. At such times, signals travel for many miles beyond the horizon as though the Earth were flat. This condition is much more common over large bodies of water than over land. Atmospheric conditions can also bend the signal away from the Earth, shortening the practical transmission distance. The occurrence of these rare conditions complicates radio system design and spectrum management. For line-of-sight systems, too large a radius cannot be assumed for the service area because of the possibility that "subrefraction" or "negative" refraction may keep the signal from reaching the periphery of the service area. On the other hand, the same frequency cannot be used again many miles beyond the horizon because of the possibility that superrefraction may carry an interfering signal far beyond its accustomed limits. One of the basic functions of international spectrum management is to prevent or reduce such interference.

CHARACTERISTICS OF RADIO FREQUENCY BANDS

The physical properties of radio waves, combined with the various transmission characteristics discussed above, determine how far and where radio signals can travel, and make different radio frequencies better suited to certain kinds of communications services. The following is a brief description of the various radio bands, some of their uses, and the factors affecting transmission of radio signals in them.

■ Very Low, Low, and Medium Frequencies: 3 to 3000 kHz

In this portion of the spectrum, encompassing the bands denoted as VLF, low frequency (LF), and medium frequency (MF), radio signals are transmitted in the form of "groundwaves" that travel

8 It is important to note that refraction does not cease to affect radio waves above the MUF. Even at frequencies in the VHF and UHF bands, radio waves bend slightly as they move through the atmosphere.

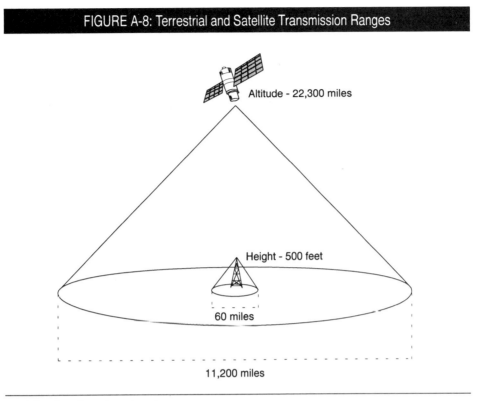

FIGURE A-8: Terrestrial and Satellite Transmission Ranges

Altitude - 22,300 miles

Height - 500 feet

60 miles

11,200 miles

NOTE: This figure is not drawn to scale.

SOURCE: Office of Technology Assessment, 1991, based on Richard G. Gould, "Allocation of the Radio Frequency Spectrum," OTA contractor report, Aug. 10, 1990.

along the surface of the Earth, following its curvature. Groundwaves lose much of their energy to the Earth as they travel along its surface, and high power is required for long-distance communication throughout this portion of the spectrum. Groundwaves travel farther over water than over land.

At the lower end of this region, transmissions are used for low data-rate communications with submarines and for navigation. The maritime mobile service, for example, has allocations in this band for communication with ships at sea. Conventional AM radio broadcasting stations also operate in a part of this band, at MF, typically between 540 and 1605 kHz. Attenuation during daylight hours limits the range of these AM stations, but at night, when attenuation is lower, AM radio signals can travel very long distances, sometimes even hundreds of miles. To prevent interference at these times to distant radio stations using the same frequency, some stations may be required to reduce the power of transmissions in the direction of those distant stations.

■ High Frequencies: 3 to 30 MHz

In this frequency range, denoted as HF, propagation of a "skywave" supplements the groundwave. While the groundwave dies out at about 100 miles, the skywave can be bent back to Earth from layers of ionized particles in the atmosphere (the ionosphere). When the signal returns to Earth, it may be reflected again, back toward the ionized layers to be returned to Earth a second time. The signal can make several "bounces" as it travels around the Earth. It is this reflection that makes long-distance communication possible. However, there are occasional—and largely unpredictable—disturbances of the ionosphere, including sunspots, that interfere with HF communications.

Overall, the reliability of HF communications is low, and the quality is often poor.

The HF "shortwave" bands are used primarily by amateur radio operators, governmental agencies for international broadcasting (Voice of America, Radio Moscow), citizens' band radio users, religious broadcasters, and for international aviation and maritime communications. Overseas telephone links using HF radio have, for the most part, been replaced by satellites, and Inmarsat satellites have taken over a major portion of the maritime communications previously provided by HF systems. Likewise, future aeronautical mobile-satellite service (AMSS) systems may also supplement or replace the HF channels now used by airplanes when they are out of range of the VHF stations they communicate with when over or near land.

While little use is made of HF radio systems for domestic communications in industrialized countries like the United States, developing countries still find HF cost-effective for some of their domestic radiocommunication needs. This has led to a conflict over allocating the HF band internationally: the developed world wants to use the band for international broadcasting and long-distance mobile communication, while the developing countries want to retain it for their domestic point-to-point systems.

■ Very High, Ultrahigh, and Superhigh Frequencies: 30 MHz to 30 GHz

The groundwave, which permits communication beyond the horizon at lower frequencies (VLF, LF, MF), dies out after a short distance in this frequency range. Moreover, the skywave—which is reflected from the ionospheric layers at lower frequencies—tends to pass through the atmosphere at these higher frequencies. Communication in this band is thus limited to little more than line-of-sight distances. For short transmitting antennas,

the maximum distance a radio signal can travel may be no more than 25 miles, but this distance can be increased by raising the height of the antenna.

This limitation can also be an advantage: the same frequencies can be reused by stations beyond the normal transmission range. Unfortunately, the distances that these line-of-sight signals can sometimes travel can be quite large, especially if the path is over water. At times, atmospheric conditions may establish a "duct" over a large body of water (see above). As it travels down the length of the duct a signal will be reflected back and forth between the water and the top of the duct, which can be hundreds of feet above the Earth's surface. These trapped signals can travel hundreds of miles. To minimize interference from a ducted signal, stations on the same frequency must be spaced far apart. This requirement limits the frequency reuse that can be achieved.

This part of the spectrum is used by many important communication and entertainment services, including television broadcast signals, FM radio, and land mobile communications. These frequencies are also used by the radiolocation service for long-range radars (1350 MHz to about 2900 MHz), aircraft landing radar (around 9000 MHz), and for point-to-point radio relay systems (various bands between 2000 and 8000 MHz). In recent years, communication satellites have made increasing use of frequencies in this band.[9]

The portion of this band between approximately 1 and 10 GHz is particularly valuable. It is bounded by increasing cosmic and other background noise at its lower end, and by precipitation attenuation at its upper end, but in between, communications can be carried out very well. Today, because of its favorable transmission characteristics, the 1 to 3 GHz band is especially sought after for mobile communications, including personal communication services (PCS), and for new

[9] Satellites operating in the C-band, e.g., use frequencies around 4 and 6 GHz, and are heavily used for transmitting television programming to cable television operators. Ku-band satellites, which generally operate at frequencies around 12 and 14 GHz, are increasingly being used for private communication networks and the delivery of entertainment programming.

broadcasting technologies such as digital audio broadcasting (DAB).

■ Above 10 GHz

At 10 GHz and above, radio transmissions become increasingly difficult. Greater attenuation of the radio signal takes place because of rain, snow, fog, clouds, and other forms of water in the signal's path. Nevertheless, crowding in the bands below 10 GHz is forcing development of the region above 10 GHz. One desirable feature of the frequencies above 10 GHz, beside the fact that they are relatively unused, is the extremely wide bandwidths that are available. The 3 to 30 MHz, HF band, for example, is 27 MHz wide. That is enough bandwidth for about 9,000 voice channels (at 3 kHz each). However, the frequency range 3 to 30 GHz is 27,000 MHz wide. That bandwidth could accommodate about 9 million voice channels.

Appendix B: Federal Government Roles B

In its attempts to address rapidly changing technology, expanding user needs, and an outmoded regulatory structure, the federal government—including both the legislative and executive branches—will play three key roles in the development of radio-based systems and the National Information Infrastructure (NII): user, catalyst, and policymaker/regulator.

GOVERNMENT AS USER

The federal government is a major user of all kinds of wireless communication systems for defense, public safety, emergency preparedness, and space communications. In some cases, these systems are built and operated by the government. Increasingly, however, government communication needs are being met by private sector service providers.

The federal government has already taken steps to define its special requirements and to see how they may or may not be met by evolving NII systems and services. In this sense, the federal government may actually be *ahead* of many in the private sector in positioning itself for the coming explosion in the availability of wireless services and systems. The lessons learned by government policymakers may be important to Congress and others interested in ensuring that wireless technologies benefit not only government users, but all users.

Most federal government activity regarding wireless technologies has been coordinated through the Federal Wireless Users Forum (FWUF). Established in 1992, the FWUF is composed of government agencies with interests in using wireless communications as part of their missions. FWUF sponsors workshops on wireless services that bring together industry representatives and federal users in order to enhance technical understanding and define the emerging needs of the agencies, both civilian and defense, for wireless systems and services.

Another group, the Federal Law Enforcement Wireless Users Group (FLEWUG), is composed of more than 60 representatives from federal agencies with law enforcement responsibilities, and is open to state, county, and local agencies as well. The FLEWUG, which was formalized under the auspices of the National Performance Review (NPR), hopes to establish a National Law Enforcement and Public Safety Network, which has been advocated by Vice President Gore. The momentum to form this network came from the lack of compatibility between different law enforcement radio systems. The group conceivably could build its own system with pooled resources, but more likely would develop specifications for a common procurement for equipment and/or services.

In January 1995, the Government Information Technology Services (GITS) committee of the Information Infrastructure Task Force (IITF) approved the creation of a Joint Federal Wireless Review Office (JFWRO), although its charter and mandate are not formally set. One of the roles of this new office will be to consolidate federal wireless programs by eliminating duplication and incompatible systems and promote more spectrum-efficient systems across government agencies. The use of commercial systems, where possible, is one solution the office is likely to pursue. The JFWRO's activities could reach into areas traditionally within the jurisdiction of the National Telecommunications and Information Administration (NTIA) and the Interdepartment Radio Advisory Committee (IRAC, see below). As a result, opposition to the creation of the office has developed. The division of authority and the interactions between NTIA, IRAC, and JFWRO are currently uncertain.

GOVERNMENT AS CATALYST

Although the primary burden of developing and deploying NII technologies and services falls on the private sector, the federal government has initiated grant programs that provide financial assistance for demonstration, planning, and even operation of telecommunications systems. NTIA has two programs that may be used to fund the development of wireless systems. First, the Public Telecommunications Facilities Program awards grants to noncommercial organizations primarily for the expansion or upgrading of public broadcasting (radio and television) facilities, and for the establishment of distance-learning projects. For FY 1994, funding was just over $21 million.[1]

The Telecommunications and Information Infrastructure Assistance Program (TIIAP) provides funds, on a matching basis, for planning and demonstration of technologies and applications that support the broader goals of the NII. In FY 1994, the first year of the program, grants totaled over $24 million. When combined with the matching funds, the program is expected to generate almost $68 million toward the development of the NII in schools, health care institutions, libraries and museums, social service organizations, and state and local governments.[2]

Based on a brief review of TIIAP grants awarded, wireless appears to play a small role.[3] Of the 57 demonstration grants, most plan to use traditional wireline technologies to provide access to computer networks, including the Internet. One project uses broadcast and one uses microwave for backbone transmissions, etc., but none indicates a direct use of wireless technologies for access purposes.

GOVERNMENT AS POLICYMAKER AND REGULATOR

The third role the government plays is that of policy- and rule-maker. Responsibility is shared among Congress, the executive branch, and the Federal Communications Commission (FCC), an independent regulatory agency. In international issues dealing with spectrum and satellite services, the Department of State also plays a role in policy development. Congress periodically passes legislation outlining both broad policy directions and directing specific actions on the part of the FCC and the executive branch. Responsibility for day-to-day regulation and management of the country's radio spectrum is divided between the FCC, which regulates private sector and

---

[1] The 1994 grants went to support 61 public television, 50 public radio, and 29 distance-learning grants in 42 states, American Samoa, the Northern Marianas, and the District of Columbia. U.S. Department of Commerce, National Telecommunications and Information Administration, "NTIA Announces FY1994 PTFP Grant Awards," news release, Sept. 19, 1994.

[2] U.S. Department of Commerce, National Telecommunications and Information Administration, "Public Institutions Receive Millions to Deploy Information Superhighway," news release, Oct. 12, 1994.

[3] This does not necessarily reflect a bias in the selection process. More likely it reflects applicants favoring traditional (wireline) solutions. The source of this favoritism may lie in costs, which could be higher for wireless; lack of knowledge about wireless alternatives; or a need that cannot be met by wireless applications. With over 1,000 applications received, no comprehensive data are available that reliably indicate which specific technologies are to be used. In addition, for the planning grants, the result of the proposed planning activity is to select appropriate technologies—in other words, no technology was necessarily selected in each application.

state/local government use of the spectrum, and NTIA, which oversees federal government spectrum use.[4]

■ Congressional Action

In the past several years, Congress has taken a more aggressive role in telecommunications policymaking—recognizing the increased importance and visibility that telecommunications has achieved as a contributor to U.S. business, international competitiveness, and quality of life. Addressing radio communications specifically, Congress passed the Omnibus Budget Reconciliation Act of 1993, which contained three major wireless policy initiatives.[5] The act:

1. directed the Secretary of Commerce to transfer at least 200 MHz of spectrum from federal government uses to the private sector,[6]
2. laid out the principles for regulating commercial mobile radio services (CMRS),
3. authorized the FCC to use competitive bidding (auctions) as a method for assigning portions of the radio frequency spectrum.

As a result of the act, NTIA transferred 50 MHz of spectrum to the FCC for reallocation to private use, and identified another 185 MHz of spectrum to be transferred; the FCC laid the foundation of CMRS regulation (although some issues are still being considered); and auctions have been held in narrowband (data/messaging) personal communications services (PCS), voice PCS, and interactive video data services (IVDS).

Congress is currently debating several bills that would substantially change how various parts of the nation's telecommunications infrastructure are regulated.[7] Generally, this legislation focuses on opening up the various segments of the telecommunications industry to more competition—in the belief that in-

creased competition will bring new services and low costs. The treatment of wireless communications in these bills is limited. Specific provisions relating to broadcasters' use of the spectrum are defined, but only a few paragraphs relating to CMRS providers are included—mainly to clarify the new legislation's relation to the Omnibus Budget Reconciliation Act of 1993.

■ Federal Communications Commission Proceedings

The FCC regulates all private sector and state/local government use of the radiofrequency spectrum. It *allocates* specific blocks of spectrum for use by different radio services, and it *assigns* to individual licensees the right to use specific frequencies or channels. The FCC has proceedings in progress that will affect almost every type of radio-based communication. Those proceedings will not be detailed here; the following represents only a summary of the wireless issues the FCC is currently considering. More discussion on the most important of these proceedings can be found in the specific sections of the report that deal with those issues:

- High-Definition Television (HDTV) proceeding
- Various PCS and CMRS proceedings
- Low-Earth orbit (LEO) satellite licensing
- Satellite digital audio broadcasting (DAB) licensing
- Enhanced-911
- zoning (petitions and comments have been filed; not yet a formal proceeding)
- public safety spectrum needs
- spectrum "refarming."

Until recently, the FCC has been unable or unwilling to tackle long-term spectrum planning issues. Critics have long accused the Commission of doing little

---

[4] The Communications Act of 1934 established the Federal Communications Commission and divided responsibility for spectrum management between it and the President. 47 U.S.C., sections 151,152, 305 (1989). In 1978, Executive Order 12,046 transferred Presidential authority for spectrum management to the Secretary of Commerce and established the National Telecommunications and Information Administration (NTIA). Finally, in 1992, Congress passed the National Telecommunications and Information Administration Organization Act, formally delegating federal government spectrum assignment authority to the head of NTIA. Public Law 102-538, Oct. 27, 1992.

[5] Public Law No. 103-66.

[6] The 200 MHz was specified to be taken from government allocations below 5 GHz, and at least 100 MHz will come from below 3 GHz—some of the most sought-after frequencies due to transmission characteristics (see app. A). Public Law 103-66, Aug. 10, 1993, Title VI.

[7] S. 652, H.R. 1555, and H.R. 1528.

more than reacting to technology developments, and observers have commented that the FCC will not act until someone forces it to by filing a petition for change. In the past several years, the FCC has become more willing to plan more aggressively. It initially took an aggressive approach to developing standards and an implementation schedule for HDTV, although that schedule has slipped. The FCC's Office of Plans and Policies has written studies on the future of fiber optics and the broadcasting industry. And in early 1992, the FCC proposed the creation of a "spectrum reserve" in order to promote the development of new radiocommunication technologies and services. Much of that spectrum is now devoted to future PCS systems. Most recently, the FCC has opened a proceeding to examine how to use radio frequencies above 40 GHz.[8]

■ Executive Branch Efforts

NTIA is responsible for developing and promoting executive branch telecommunications policy. It serves as the President's principal adviser on telecommunications policies and is also responsible for managing the federal government's use of the radio frequency spectrum.[9] In this role, it works closely with the FCC to develop policies and procedures that are consistent and that allow many portions of the spectrum to be shared by both government and non-government wireless users. To help it carry out its responsibilities for spectrum management, NTIA draws on the expertise of the members of the IRAC, which is made up of 20 federal agencies that use wireless communications, and the Spectrum Planning and Policy Advisory Committee, which consists of private sector and federal government members who advise NTIA on radiocommunication issues.[10]

The executive branch has taken steps to revitalize spectrum planning. NTIA established the Strategic Spectrum Planning Program to develop a long-range spectrum plan that will include both federal and non-federal users. In 1992, as part of this initiative, NTIA requested comments and information on "Current and Future Requirements for the Use of Radio Frequencies in the United States." In this proceeding, NTIA clearly notes the importance of improved planning of the spectrum resource:

> . . . planning helps ensure that adequate spectrum will continue to be available for public safety needs, other non-commercial uses such as amateur radio and scientific research, and local, state, and federal government uses. Moreover, improved planning is essential for the U.S. government to represent effectively the interests of all U.S. spectrum users in international spectrum negotiations.[11]

As a result of its inquiry, NTIA released a report, *U.S. National Spectrum Requirements: Projections and Trends*, and efforts to identify radio frequencies to meet the needs identified in the report has begun.[12] NTIA and IRAC have also completed analyses as part of their mandate to transfer 200 MHz (235 MHz was actually transferred) of federal government spectrum to the FCC for private sector use.[13]

To develop policy specifically for wireless communications, the executive branch has established several committees to address specific issue areas.

- **Federal Wireless Policy Committee.** Established in 1993, the Federal Wireless Policy Committee (FWPC) serves as a focal point for wireless policy development, both for the federal government and in relation to FCC activities. FWPC draws its mem-

---

[8] Federal Communications Commission, *Amendment of Parts 2 and 15 of the Commission's Rules to Permit Use of Radio Frequencies Above 40 GHz for New Radio Applications*, Notice of Proposed Rulemaking, ET Docket 94-124, released Nov. 8, 1994.

[9] The potential conflicts with this dual role are discussed in U.S. Congress, Office of Technology Assessment, *The 1992 World Administrative Radio Conference: Issues for U.S. International Spectrum Policy—Background Paper*, OTA-BP-TCT-76 (Washington, DC: U.S. Government Printing Office, November 1991).

[10] For a more complete discussion of the IRAC and SPAC, see ibid.

[11] U.S. Department of Commerce, National Telecommunications and Information Administration, *Current and Future Requirements for the Use of Radio Frequencies in the United States*, Notice of Inquiry, Docket No. 920532-2132, released June 1, 1992.

[12] U.S. Department of Commerce, National Telecommunications and Information Administration, *U.S. National Spectrum Requirements: Projections and Trends*, Special Publication 94-31 (Washington, DC: U.S. Government Printing Office, March 1995).

[13] U.S. Department of Commerce, National Telecommunications and Information Administration, *Spectrum Reallocation Final Report*, Special Publication 95-32 (Washington, DC: U.S. Government Printing Office, February 1995).

bers from a wide range of federal agencies that have operational, procurement, or policy interests in evolving wireless communications systems. Its mandate is to further the deployment of a digital, ubiquitous, interoperable, transparent, and secure (DUITS) wireless communications network for the federal government. It has, thus, become the focal point of efforts to procure mobile services for the federal government. A single procurement is envisioned—to be completed by the end of 1995— that will give federal users access to a wide range of cellular, specialized mobile radio, PCS, satellite, and emerging wireless communication services.

▪ In September 1994, FWPC produced a statement of *Current and Future Functional Requirements for Federal Wireless Services in the United States*. This document was based on information gathered from the various meetings and workshops sponsored by FWUF and information provided by FWPC members. It describes both specific and generic needs that the federal agencies think could be met with commercial or special wireless systems, and is intended as a guide for future procurement efforts.

▪ **Untethered Networking Working Group.** The National Science and Technology Council, in coordination with the Technology Policy Working Group of the IITF, set up an Untethered Networking Working Group to examine the impact of wireless technologies (satellite and terrestrial) on the evolution of the NII and the Global Information Infrastructure. It is unclear if this group ever met or what products it produced.

Appendix C: Acronyms and Glossary of Terms C

Allocation: The designation of a band of frequencies for a specific radio service or services. Allocations are made internationally at World Radio Conferences and are incorporated into the international Table of Frequency Allocations. Domestic allocations are made by the Federal Communications Commission (FCC) and the National Telecommunications and Information Administration (NTIA).

AMPS: Advanced Mobile Phone Service. AMPS is the existing U.S. analog cellular telephone standard.

Analog: In analog radio communications, information is transmitted by continuously varying the phase, amplitude, or frequency of a radio carrier wave.

Assignment: The granting by a government of the right to use a specific frequency (or group of frequencies) to a specific user or station. Each television station, for example, is assigned a small group of frequencies that correspond to a specific channel or number on the television dial.

Attenuation: The loss of power of electromagnetic signals between transmission and reception points. Attenuation is exacerbated by physical barriers, such as rain, buildings, and trees.

Bandwidth: The total range of frequencies required to transmit a radio signal without undue distortion. The required bandwidth of a radio signal is deter-

mined by the amount of information in the signal being sent. More complex signals contain more information, and hence require wider bandwidths. The bandwidth required by a television channel is 600 times greater than that of an AM radio channel.

BETRS: Basic Exchange Telecommunications Radio Service. BETRS is used as a wireless substitute to copper loops for providing basic telephone service.

Bps: Bits per second. The rate at which digital data are transmitted over a communications path. Speeds are usually designated kbps (thousands of bits per second), Mbps (millions of bits per second), and Gbps (billions of bits per second).

BSS: Broadcasting-Satellite Service. An ITU-defined service that refers to the delivery of information or programming directly from satellites to user receivers. The BSS includes new systems planned to deliver high-definition television services (BSS-HDTV) and audio services (BSS-Sound).

CAI: common air interface. Refers to the standard (there are many) that allows a mobile unit such as a cellular phone to communicate with a base station.

Cellular Telephony: A mobile radio service in which a geographic area is divided into smaller areas known as "cells." A transmitter in each cell provides radio coverage to the users in the cell. Calls are handed off from one transmitter to the next as the user moves between cells.

CDMA: Code division multiple access. CDMA is a radio communication format that uses digital technology and spread spectrum transmission to send information. Each radio signal is assigned its own unique code and is then spread over a range of frequencies for transmission.

CDPD: Cellular Digital Packet Data. Announced in 1992 by McCaw Cellular, IBM, and a group of eight other major cellular companies, CDPD uses the idle time in the analog cellular telephone system to transmit packetized data at rates up to 19.2 kbps.

Cloning: Cloning is the practice of reprogramming a phone with a MIN/ESN pair from another phone.

CMRS: Commercial Mobile Radio Service. A new regulatory classification for mobile telephone service created by the FCC in response to the Omnibus Budget Reconciliation Act of 1993. Under the new rules, cellular, SMR/ESMR, and PCS will be brought under the same regulatory umbrella.

Common Carrier: A company that is recognized by an appropriate regulatory agency as providing communications service to the general public on a nondiscriminatory basis. Common carriers cannot exercise any control over content of the messages they carry.

DAB: Digital Audio Broadcasting. DAB refers to the transmission of audio broadcasts in digital form as opposed to today's (AM and FM) analog form. DAB promises compact disc quality sound over the air. DAB systems may use terrestrial, satellite, or hybrid transmission.

DBS: Direct Broadcast Satellite. Medium- to high-power satellites that are designed to transmit programming directly to small satellite receiver dishes at users' homes.

Digital: In digital communication, information is sent by modulating the carrier frequency in such a way that there are discrete changes in the phase, frequency, or amplitude.

Downlink: In satellite communications, the signal that travels from the satellite down to the receivers on earth is called the downlink. The direction the downlink signal travels is also called space-to-Earth. *See* Uplink.

Encryption: The process of electronically altering or "scrambling" a signal, usually for security purposes.

EPA: Environmental Protection Agency.

ESMR: Enhanced Specialized Mobile Radio. The next generation of SMRs, ESMR systems take advantage of digital technology combined with a cellular system architecture to provide greater capacity than existing SMR systems. *See* Specialized Mobile Radio.

ESN: Electronic serial number. A number encoded in each cellular phone that uniquely identifies each cellular telephone manufactured.

FCC: Federal Communications Commission. Established by the Communications Act of 1934, the FCC is an independent federal agency that regulates all electronic interstate communications, including telephony, cable television, and broadcasting. The FCC is also responsible for assigning the radio frequencies used by all non-federal users of the spectrum.

FDA: Food and Drug Administration.

FDMA: Frequency Division Multiple Access. FDMA allows multiple users to share a band of radio frequencies by dividing the spectrum into separate channels. Analog cellular systems, for example, use separate frequencies for each call in each cell.

GPS: Global Positioning System. GPS is a network of satellites that provides precise location determination to receivers.

GSM: Global System for Mobile Communications, formerly Groupe Special Mobile. GSM is a second-generation digital system adopted as a European standard in the mid-1980s and introduced in 1992. Now deployed across Europe, GSM is intended to replace existing analog cellular telephone services. GSM allows systems in different countries to interoperate, permitting consumers to use their cellular phones anywhere in Europe.

HDTV: High-definition television. Refers to future generations of television that will have higher picture resolution, a wider aspect ratio, and digital quality sound.

HIPERLAN: High-Performance Radio Local Area Network. HIPERLAN is a European standard for a short-range (50 meters) high-performance radio local area network. The current specification is for operation in the 5.1 - 5.3 GHz band. Another band from 17.1 - 17.3 has been designated for HIPER-LAN use, but detailed specifications have not yet been developed.

Hz: Hertz. Cycles per second. Radio frequencies are described in multiples of Hertz:

kHz, kilohertz: thousand cycles per second;

MHz, megahertz: million cycles per second;

GHz, gigahertz: billion cycles per second.

IEEE: Institute of Electrical and Electronics Engineers. The IEEE is the professional society for electrical engineers. It produces standards for a range of communications technologies.

Internet: Refers to a large collection of interconnected computer networks that use a common transmission protocol, allowing users to communicate across networks.

IRAC: Interdepartment Radio Advisory Committee. Established in 1922 and now located in the Department of Commerce, the IRAC consists of 20 representatives from the various government agencies involved in or using the radio frequencies.

ISDN: Integrated Services Digital Network. A new digital service offered by local phone companies that allows users to send digital data over copper wires.

ITS: Intelligent Transportation System (formerly referred to as IVHS—see below).

ITU: International Telecommunication Union. The ITU is a specialized agency of the United Nations responsible for international regulation of telecommunications services of all kinds, including telegraph, telephone, and radio.

IVHS: Intelligent Vehicle Highway System. IVHS uses information technology and sensors to improve the management of traffic flow.

LAN: Local area network. Computers are connected so that they can talk to each other and share a central file server and printer. A LAN is confined to a limited area, usually a single office, or building, or campus.

LATA: Local Access and Transport Area. Refers to the local exchange areas developed in connection with the divestiture of AT&T, within which the Bell Operating Companies (BOCs) may provide service. Pursuant to the Modified Final Judgment (MFJ), BOCs are not permitted to transport calls across LATA boundaries but rather must connect these calls to interexchange carriers.

LEC: Local Exchange Carrier. The LEC is the local telephone company. There are over a thousand LECs, ranging in size from the very small independent telephone companies that serve rural areas to the much larger Bell Operating Companies (BOCs).

LEO: Low-Earth orbiting satellite. LEO satellites are smaller and cheaper to design, build, and launch than traditional geosynchronous satellites. Networks of these small satellites are being planned that will provide data ("little" LEOS) and voice ("big" LEOS) services to portable receivers all over the world.

LMDS: Local Multipoint Distribution Service. An experimental service using low-power transmitters, configured in a cellular-like arrangement, to transmit video to receivers in homes and businesses.

Microwave: Radio frequency spectrum signals between 890 MHz and 20 GHz. Point-to-point microwave transmission is commonly used as a substitute for copper or fiber cable.

MIN: Mobile identification number. A number encoded in each cellular telephone that represents the telephone number.

Modulation: The process of encoding information onto a radio wave by varying one of its basic characteristics—amplitude, frequency, or phase—in relation to an input signal such as speech, music, or video. Two of the most common types of modulation are amplitude modulation (AM) and frequency modulation (FM).

MMDS: Multi-Channel, Multi-Point Distribution Service. Also known as "wireless cable," MMDS uses high-power transmitters to broadcast up to 33 channels of subscription video programming to receiving equipment in homes and businesses. By using digital technology, MMDS operators may be able to transmit a much larger number of channels.

MSS: Mobile-Satellite Service. MSS is an ITU-defined service in which satellites are used to deliver communications services (voice or data, one- or two-way) to mobile users such as cars, trucks, ships, and planes. It is a generic term that encompasses several types of mobile services delivered by satellite, including Maritime MSS (MMSS),

Aeronautical MSS (AMSS), and Land MSS (LMSS).

NTIA: National Telecommunications and Information Administration. NTIA is the President's adviser on communications and is responsible for administering all federal government use of the radio frequency spectrum, including military communications. NTIA is located in the Department of Commerce.

NTSC: National Television Systems Committee. A committee composed of industry representatives that established the NTSC standard for black-and-white television in 1940, and color television in the early 1950s.

Part 15: Part 15 refers to a section of the Rules enacted and administered by the Federal Communications Commission. Part 15 rules govern unlicensed radio communications in certain frequency bands. Examples of Part 15 communications devices include: cordless telephones, spread spectrum ISM band devices, low-power wireless microphones, and baby monitors.

PBS: Public Broadcasting Service.

PCS: Personal Communications Service. A radio service broadly defined by the FCC to be a "family" of communications services providing mobile and incidental fixed services for voice and data applications.

PSTN: Public switched telephone network. The publicly accessible dial-up telephone network.

Roaming: Roaming is the practice of using a cellular phone in cellular networks outside the user's home system.

SMR: Specialized Mobile Radio. The FCC established the SMR service in 1974 to provide dispatch service to trucking, taxi and similar industries, government entities, and to indivi- duals on a for-profit basis. SMR systems can also connect to the PSTN. The FCC's ongoing CMRS proceeding will bring SMR under the same regulatory umbrella as cellular and new PCS. *See* Commercial Mobile Radio Service and Enhanced Specialized Mobile Radio.

Spectrum: The spectrum consists of all the radio frequencies that are used for radio communication.

Spread Spectrum: Spread spectrum modulation uses a wide band of frequencies to send radio signals. Instead of transmitting a signal on one channel, spread spectrum systems process the signal and spread it across a wider range of frequencies.

TDMA: Time division multiple access. Refers to a form of multiple access where a single communications channel is shared by segmenting it by time. Each user is assigned a specific time slot.

Tumbling: Tumbling is the practice of programming a phone with ESN/MIN pairs until a valid combination is found.

Uplink: In satellite communications, the signal that travels from the Earth transmitting station up to the satellite. The direction the uplink signal travels is also known as Earth-to-space. *See* Downlink.

VBI: Vertical blanking interval. After a television image has been displayed, it takes a certain amount of time for the electron gun to be moved into position to scan the next image. No picture information is sent during this time, allowing data for other types of information services to be sent.

VSAT: Very small aperture terminal. Refers to small (less than 6 feet in diameter) satellite receive dishes that can send and receive voice, data and video communications. VSATs are usually deployed in networks, allowing tens or even hundreds of sites to be connected in one network.

Wireless Local Loop: Wireless systems can be used instead of copper loops to provide basic telephone service to households.

Appendix D: Reviewers and Contributors

D

John Abel
National Association of Broadcasters

Michael Alpert
Alpert & Associates

Michael Altschul
Cellular Telecommunications Industry Association

Ray Barnett
U.S. Secret Service

Rich Barth
Motorola

Robert Bonometti

Charles Bostian
Center for Wireless Telecommunications
Virginia Polytechnic Institute
    and State University

Larry Bowman
Defense Information Systems Agency

Sandra Braman
Institute of Communications Research
University of Illinois

Robert Briskman
CD Radio, Inc.

Donald C. Brittingham
Bell Atlantic

Dale Brown
MCI

Charles Cape
Office of the Secretary
U.S. Department of Commerce

Russ Coffin
Northern Telecom

David Cohen
National Telecommunications and
    Information Administration
U.S. Department of Commerce

Richard Cotton
NBC

George Curtis

W. Russell Daggatt
Teledesic Corporation

Richard Dean
National Security Agency

Robert Dilworth
Metricom, Inc.

Bert Dumpé
Ergotec, Inc.

Rick Ellinger
MFS Datanet, Inc.

Rob Euler
Ardis

Timothy Fain
Office of Management and Budget

Alex Felker
Time Warner Telecommunications

Douglas Fields
United Parcel Service

Russell Fox
Gardner, Carton, & Douglas

Howard Frank
Advanced Research Projects Agency

Rob Frieden
School of Communications
Pennsylvania State University

Jerry Fritz
Albritton Communications

David Furth
Wireless Telecommunications Bureau
  Federal Communications Commission

Steve Garcia

Joseph Gattuso
National Telecommunications and
  Information Administration
U.S. Department of Commerce

David Goodman
Wireless Information Network Labs
Rutgers University

Gary Green
Metricom, Inc.

Robert Gurss
Wilkes, Artis, Hedrick & Lane

Kerry Hanson
Texas Instruments

Larry Harris
MCI

Dale Hatfield
Hatfield Associates, Inc.

Kathy Hawk

Mark Jamison
Sprint

Douglas Johnson
EIA

Fred Karnas
National Coalition for the Homeless

Randy Katz

Kevin Kelley
Qualcomm, Inc.

Michael Keyes
Burson-Mirsteller

Brian Kidney
AirTouch Communications

Brian Kovalsky
Mobile Solutions, Inc.

Lawrence Krevor
Nextel Communications, Inc.

John LaRoche
Infrared Data Association

Lon Levin
American Mobile Satellite Corporation

Theodore Litovitz
Catholic University of America

Jennifer Manner
Akin, Gump, Strauss, Hauer & Feld, LLP

John Marinho
AT&T Network Wireless Systems

Andrew Marino
School of Medicine
Louisiana State University

Sandi Martin

Cynthia Nila
National Telecommunications and
    Information Administration
U.S. Department of Commerce

Jack Oslund
COMSAT

Michael Papillo
Houston Associates, Inc.

Richard Parlow
National Telecommunications and
    Information Administration
U.S. Department of Commerce

Jon Peha
Carnegie Mellon University

Robert Pepper
Office of Plans and Policy
Federal Communications Commission

Pamela Portin
US WEST NewVector Group, Inc.

Robert Rasor
Financial Crimes Division
U.S. Secret Service

Michael Rau
EZ Communication

Paul Robinson
Tansin A. Darcos & Co.

Bob Roche
Cellular Telecommunications Industry Association

Walda Roseman
CompassRose International, Inc.

Gregory Rosston
Office of Plans and Policy
Federal Communications Commission

Marco Rubin
Booz–Allen & Hamilton, Inc.

Michael Ruger
Baker and Hostetler

Eric Schimmel
Telecommunications Industry Association

Jennifer Schmidt
East Coast Migrant Health Project, Inc.

Richard Sclove
The Loka Institute

Joseph Sedlak
Volunteers In Technical Assistance

Lawrence Seidman
GM Hughes Electronics Corp.

Katherine Sgroi
United Parcel Service

Thomas Stanley
Federal Communications Commission

Charles Steinfield
Michigan State University

Amy Stephan
Personal Communications Industry
 Association

Raymond Strassburger
Northern Telecom

David Strom
Consultant

Harry Thibedeau
Satellite Broadcasting and Communications
 Association

Maria Tilves–Aguilera
Northern Telecom

LaRene Tondro
Society of Satellite Professionals
 International

Linda Townsend Solheim
CompassRose International, Inc.

Philip Verveer
Willkie Farr & Gallagher

Thomas Wanley
Personal Communications Industry
 Association

Mark Weiser
Xerox PARC

Douglas Weiss
Corporation for Public Broadcasting

Rolf Wigand
School of Information Studies
Syracuse University

Kurt Wimmer
Covington & Burling

OTA REVIEWERS

Steven Bonorris

David Butler

Michael Callaham

Michael DeWinter

Wendell Fletcher

Kathleen Fulton

Betsy Gunn

Joan Winston

Fred Wood

Appendix E: Workshop Participants | E

Russ Coffin
Northern Telecom

Richard Dean
National Security Agency

Mark Epstein
Qualcomm, Inc.

Mike D. Franklin
Bell Atlantic Mobile

William Garner
American Mobile Satellite Corporation

John (Nick) Gorham
Motorola, Inc.

David Johnson
School of Computer Science
Carnegie Mellon University

Randy Katz
ARPA

Brian Kovalsky
Mobile Solutions, Inc.

Arvind Krishna
IBM

John A. Marinho
AT&T Network Wireless Systems

R. Michael Schmalz
ARDIS

Mobility and Wireless Telecommunication Technologies Workshop - October 31, 1994

Philip Aspden
Bellcore

Allen Batteau
Wizdom Systems, Inc.

Andrew Blau
Benton Foundation

Sandra Braman
Institute of Communications Research
University of Illinois

John Carey
Greystone Communications

James Katz
Bellcore

Sara Kiesler
Department of Social and Decision Sciences
Carnegie Mellon University

Patricia L. Mokhtarian
Department of Civil and Environmental Engineering
University of California, Davis

George Morgan
Pamplin College of Business
Virginia Polytechnic Institute

John Robinson
Department of Sociology
University of Maryland, College Park

Index

Superintendent of Documents **Publications** Order Form

Order Processing Code:
*7758

P3
Telephone orders (202) 512-1800
(The best time to call is between 8-9 a.m. EST.)
To fax your orders (202) 512-2250
Charge your order. It's Easy!

☐ **YES**, please send me the following:

_____ copies of *Wireless Technologies and the National Information Infrastructure (300 pages),*
S/N 052-003-01421-1 at $19.00 each.

The total cost of my order is $_____. International customers please add 25%. Prices include regular domestic postage and handling and are subject to change.

_____
(Company or Personal Name)   (Please type or print)

_____
(Additional address/attention line)

_____
(Street address)

_____
(City, State, ZIP Code)

_____
(Daytime phone including area code)

_____
(Purchase Order No.)

Please Choose Method of Payment:

☐ Check Payable to the Superintendent of Documents

☐ GPO Deposit Account ☐☐☐☐☐☐☐ – ☐

☐ VISA or MasterCard Account

☐☐☐☐☐☐☐☐☐☐☐☐☐☐☐☐☐☐☐☐☐

☐☐☐☐ (Credit card expiration date)   *Thank you for your order!*

_____
(Authorizing Signature)   (8/95)

                                                        YES NO
May we make your name/address available to other mailers? ☐ ☐

Mail To: New Orders, Superintendent of Documents, P.O. Box 371954, Pittsburgh, PA 15250-7954

THIS FORM MAY BE PHOTOCOPIED

Superintendent of Documents **Publications** Order Form

Order Processing Code:
*7758

P3
Telephone orders (202) 512-1800
(The best time to call is between 8-9 a.m. EST.)
To fax your orders (202) 512-2250
Charge your order. It's Easy!

☐ **YES**, please send me the following:

_____ copies of *Wireless Technologies and the National Information Infrastructure (300 pages),*
S/N 052-003-01421-1 at $19.00 each.

The total cost of my order is $_____. International customers please add 25%. Prices include regular domestic postage and handling and are subject to change.

_____
(Company or Personal Name)   (Please type or print)

_____
(Additional address/attention line)

_____
(Street address)

_____
(City, State, ZIP Code)

_____
(Daytime phone including area code)

_____
(Purchase Order No.)

Please Choose Method of Payment:

☐ Check Payable to the Superintendent of Documents

☐ GPO Deposit Account ☐☐☐☐☐☐☐ – ☐

☐ VISA or MasterCard Account

☐☐☐☐☐☐☐☐☐☐☐☐☐☐☐☐☐☐☐☐☐

☐☐☐☐ (Credit card expiration date)   *Thank you for your order!*

_____
(Authorizing Signature)   (8/95)

                                                        YES NO
May we make your name/address available to other mailers? ☐ ☐

Mail To: New Orders, Superintendent of Documents, P.O. Box 371954, Pittsburgh, PA 15250-7954

THIS FORM MAY BE PHOTOCOPIED